MOLECULAR HUMAN CYTOGENETICS

ACADEMIC PRESS RAPID MANUSCRIPT REPRODUCTION

ICN–UCLA Symposia on Molecular and Cellular Biology
Vol. VII, 1977

MOLECULAR HUMAN CYTOGENETICS

edited by

ROBERT S. SPARKES
Department of Medicine
University of California, Los Angeles
Los Angeles, California

DAVID E. COMINGS
Medical Genetics
City of Hope National Medical Center
Duarte, California

C. FRED FOX
Department of Bacteriology
and Molecular Biology Institute
University of California, Los Angeles
Los Angeles, California

ACADEMIC PRESS INC. New York San Francisco London 1977
A Subsidiary of Harcourt Brace Jovanovich, Publishers

ACADEMIC PRESS, INC.
111 Fifth Avenue, New York, New York 10003

United Kingdom Edition published by
ACADEMIC PRESS, INC. (LONDON) LTD.
24/28 Oval Road, London NW1

Library of Congress Cataloging Publication Data

Main entry under title:

Molecular human cytogenetics.

(ICN-UCLA symposia on molecular and cellular biology; v. 7)
Proceedings of a symposium held in Keystone, Colo., Mar. 6–11, 1977.
1. Human genetics–Congresses. 2. Cytogenetics –Congresses. I. Sparkes, Robert S. II. Comings, David E. III. Fox, C. Fred IV. Series.
QH431.H8343 573.2′21 77-14429
ISBN 0-12-656350-0

PRINTED IN THE UNITED STATES OF AMERICA

Contents

III. CHROMOSOME MAPPING

VII. CLINICAL CYTOGENETICS

Preface

Recent years have seen very rapid progress in cytogenetics. Many of the advances have been at the molecular level and this accounts for the focus of the present conference. Knowledge and advances in all areas of genetics have implications for human genetics; it thus seemed desirable to bring human cytogeneticists together with scientists active in basic genetics research on other species. The desire to create a mutual awareness of the current status and problems faced by human cytogeneticists and geneticists working in nonhuman systems was gratifyingly achieved by the participants in the conference. Most of the plenary papers and some of the poster sessions presented at the meeting are included in this volume.

We wish to acknowledge the much appreciated yearly financial support from ICN Pharmaceuticals, Inc., which has sustained the series, and a special grant from the National Foundation March of Dimes, which greatly enriched this conference. Furthermore, we thank the staff of the conference office for their general support and Robert Sinow, who graciously accepted the task of indexing the present volume. Finally, we wish to indicate our appreciation to the speakers for their excellent presentations and to all participants for their essential contributions to the many stimulating discussions that were held at the conference.

MOLECULAR HUMAN CYTOGENETICS

INTERNAL AND HIGHER-ORDER STRUCTURE OF CHROMATIN NU BODIES

Donald E. Olins

University of Tennessee-Oak Ridge Graduate School of Biomedical Sciences, and the Biology Division, Oak Ridge National Laboratory, Oak Ridge, TN 37830

ABSTRACT. Based upon current biophysical data (including recent laser-Raman studies) of isolated nu bodies and inner histones, we have proposed that the chromatin subunit consists of a DNA-rich outer domain surrounding a protein core composed of α-helical-rich histone globular regions, close-packed with dihedral point-group symmetry. Analysis of the effects of urea on isolated nu bodies suggest that these two domains respond differently: the DNA-rich shell exhibits noncooperative destabilization; the protein core undergoes cooperative destabilization. This differential response of the two regions of a nu body to a simple chemical perturbant (i.e., urea) may furnish a model for the conformational differences in nu bodies postulated for "active" chromatin.

Nu bodies are believed to organize into 20-30 nm higher-order fibers in condensed regions of chromatin. However, the integrity of subunits in these thick fibers has recently been seriously challenged. Evidence from our laboratory, presented here, confirms that the 20-30 nm chromatin fibers consists of a close-packing of nu bodies. The chromatin subunits, therefore, retain their integrity within the higher-order fibers.

INTRODUCTION

The nucleohistone component of eukaryotic chromatin is organized as a string of close-packed subunits, called nu bodies or nucleosomes. Each nu body consists of about 140 nucleotide pairs (np) of DNA enveloping a protein core which is made up of two molecules each of H4, H3, H2A, H2B (the "inner histones"). Adjacent nu bodies are connected by a nuclease-sensitive stretch of DNA (30-70 np), to which are bound the lysine-rich histones, H1 or H5 (1,5,6,11,17). Large quantities of monomeric nu bodies (ν_1) have recently been prepared, subfractionated and characterized by a variety of biophysical techniques (17,18). These well-characterized preparations furnish the material for current studies on the conformational states of nu bodies.

Internal Structure of Chromatin Nu Bodies. In collaboration with B. Prescott and G. J. Thomas, Jr. (21,24), we have completed a laser-Raman study of chicken DNA, inner histones, ν_1 and chromatin. Laser-Raman spectroscopy is a valuable tool for determining the secondary structures of nucleic acids and proteins in aqueous solutions (8,23). By laser light scattering one obtains a vibrational spectrum characteristic of the structure or environment of macromolecular subgroups. Figure 1 represents a comparison of the Raman spectrum of ν_1 in 0.2 mM EDTA, a spectrum of ν_1 in 2.0 M NaCl, and the sum of spectra of the constituent DNA and inner histones. From these, and other data, we have made the following conclusions:

a) The DNA of ν_1 and chromatin is in the B-genus conformation (i.e., including the B and C forms of double-stranded DNA). There appears to be negligible amounts of A conformation.

b) The secondary structure of the inner histones remains largely unchanged whether associated with DNA in 0.2 mM EDTA, in 2.0 M NaCl, or devoid of DNA in 2.0 M NaCl. Estimates of secondary structure (12) of the inner histones in high salt are 51 $\pm$ 5% α-helix, 36 $\pm$ 4% "random chain" and 13 $\pm$ 9% β-sheet. This conclusion is based upon an analysis of the positions and relative intensities of Amide I and III Raman lines in H_2O and in D_2O solutions.

c) The cysteine residues of chicken H3 histone are in a similar state within ν_1 or in the isolated inner histones. The spectral evidence is not supportive of S-S bridges, but would be consistent with a hydrogen-bonded interaction.

d) Intense Raman lines characteristic of tyrosine, argue that these residues [i.e., 9 tyrosines/heterotypic tetramer (26)] are hydrogen-bonded within the inner histones (in 2.0 M NaCl) and in ν_1.

The extensive NMR studies of isolated inner histones have permitted identification of the regions involved in histone-histone interactions (1,13). These structured apolar regions are each about 70-85 amino acid residues in length, or approximately 60% of the histone sequences. The remaining basic portions of the histones, believed to be the sites of interaction of the histones with DNA, are not appreciably structured in high salt conditions, and are, therefore, not likely sites for extensive α-helicity. If 50% α-helix is confined to 60% of the histone, these structured regions would approach 80% α-helical content, which is comparable to the helix content of myoglobin, hemoglobin and myohemerythrin.

Extending our previous model of the protein-core of a nu body (18), we suggest that this core consists of closely packed α-helical-rich globular domains of about equal size (i.e., 25-30 Å diameter globular regions per histone). We further suggest that the basic segments of the histones possess little α-helix or β-sheet structure whether dissociated or associated with the DNA. Consideration of the

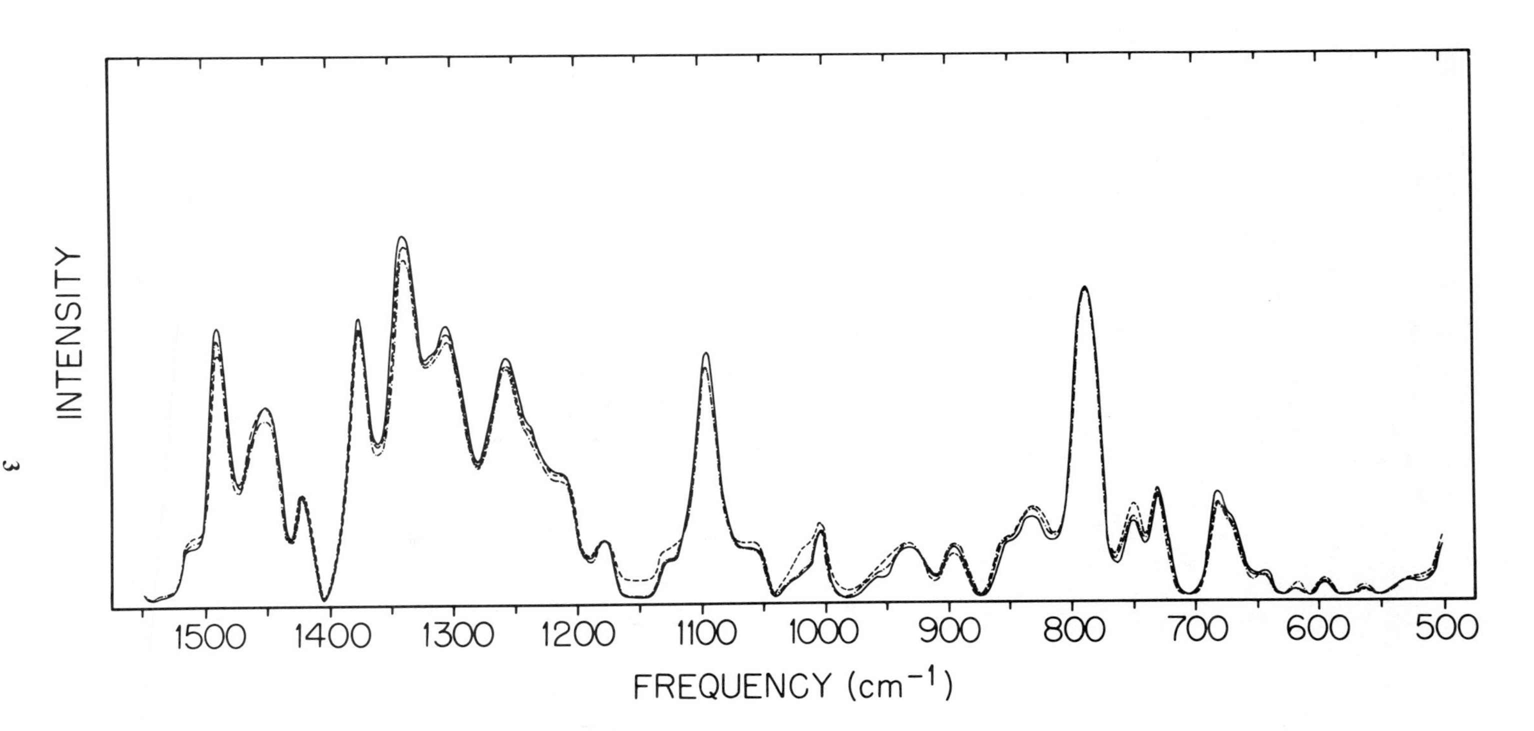

Fig. 1. Laser-Raman spectra of ν_1 in 0.2 mM EDTA (-·-), in 2.0 M NaCl (—) and the sum of spectra of DNA and inner histones (----). Conditions: excitation wavelength 488.0 nm; spectral slit width 10 cm^{-1}; radiatn power 300 mW; temperature 32°C. Note the weak line at 1240 cm^{-1} (Amide III, indicating little β structure), and lines at 1094 and 832 cm^{-1} (indicating that the DNA is in the B-genus conformation).

stoichiometry of the inner histone has led to models (18,27) that postulate a single true dyad axis per ν_1. Eight globular units of comparable size could close-pack with dihedral point-group symmetry (e.g., cubic or square anti-prism). Such models of the protein-core of ν_1 were employed in interpreting the various high resolution electron microscopic images (18). Our current view is that the protein-core of the nu body exhibits histone-histone associations into symmetric structures, reminiscent of hemoglobin or hemerythrin, with the basic portions extending outward into the DNA-rich shell. Although the path of the DNA in the nu body is unknown, it is attractive to assume that its folding is determined by the complementary DNA binding sites, and that this path preserves an overall twofold symmetry for the chromatin subunit.

Conformational States of the Nu Body. Recent biochemical studies of "actively transcribing" chromatin have indicated the possibility that the nu bodies persist, but in altered conformational states (9,25). Since so little is currently known of the conformational states of nu bodies, we have initiated experiments on the structural consequences of simple chemical perturbants. In this way we hope to identify and define conformational states, and subsequently search for biological molecules that effect related structural transitions. We have started these studies with urea (i.e., 0-10 M urea, 0.2 mM EDTA, pH 7), conditions which do not dissociate histones and DNA. These studies have involved the mutual efforts of P. N. Bryan, R. E. Harrington, W. E. Hill, A. L. Olins and myself (19,20).

Increased urea concentration exhibits a diversity of physical effects on ν_1. These changes fall into two categories: non-cooperative (i.e., gradual effects, occurring over most or all of the urea concentration range); cooperative (i.e., abrupt structural transitions occurring over a narrow range of urea molarity). Non-cooperative effects were observed in studies of the thermal stability and hydrodynamics of ν_1, and in the effects of urea on the contribution of DNA to the circular dichroic (CD) spectra. Cooperative transitions due to urea appeared in the protein 2° structure contribution to CD, the reactivity of H3 thiol groups to N-ethylmaleimide (NEM) and the ultrastructure.

In studies on the thermal stability of ν_1, monitored at 260 nm, about 2/3 of the total hyperchromicity exhibits a strikingly non-cooperative destabilization by urea (Figure 2). The melting temperature (Tm) of this fraction of the nu body DNA reveals a linear decrease with urea molarity. By comparison with the melting of naked chicken DNA in different NaCl solutions, it is clear that $\Delta Tm/\Delta M$ urea of ν_1 in 0.2 mM EDTA is most comparable to DNA in 2.0 M NaCl. Thus, the DNA of

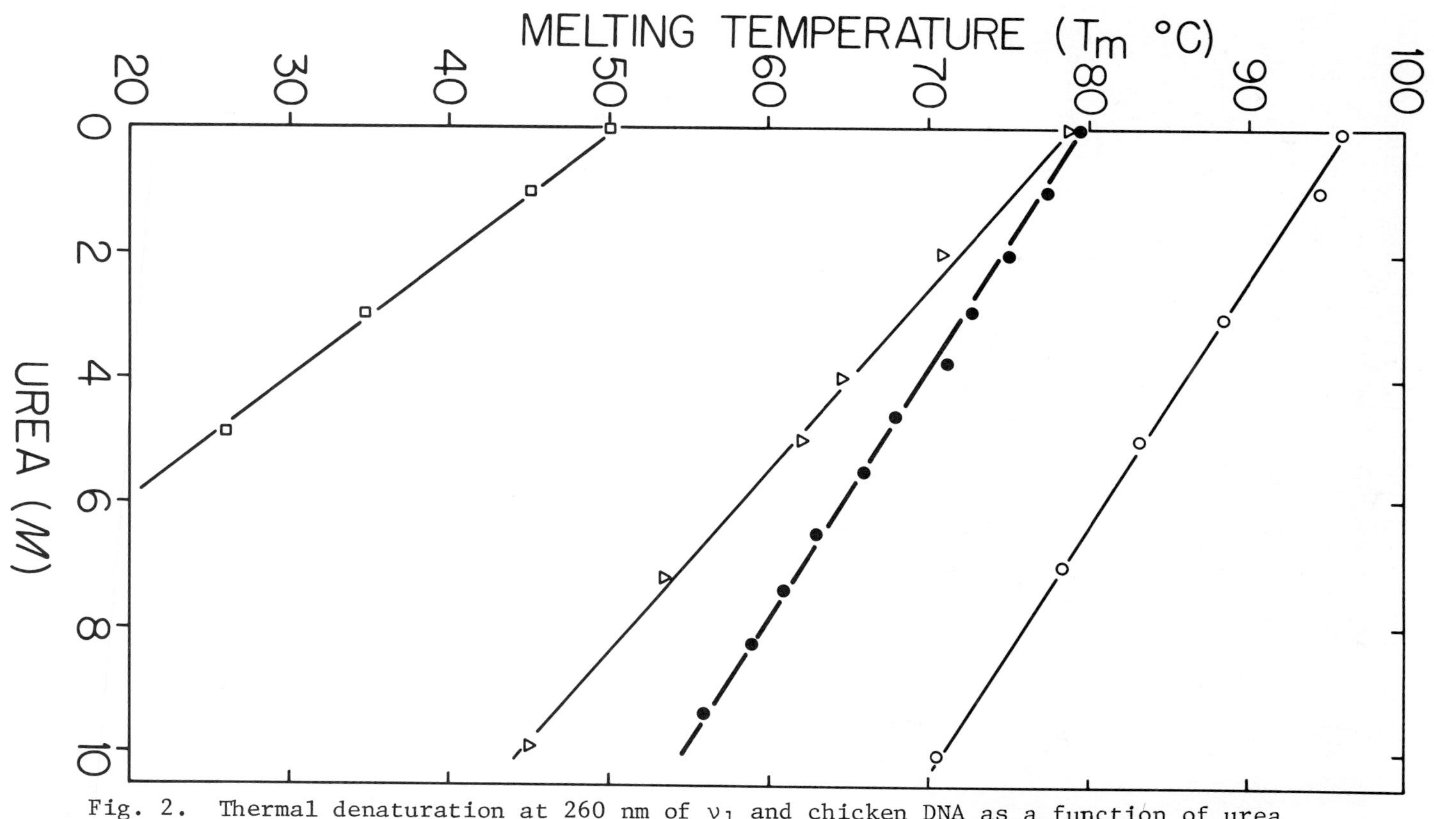

Fig. 2. Thermal denaturation at 260 nm of ν_1 and chicken DNA as a function of urea molarity. At least 2/3 of the hyperchromicity of the DNA within ν_1 exhibits a linear destabilization with urea molarity (●). The ν_1 are in 0.2 mM EDTA; chicken DNA in 2(O), 0.065 (△) and 0.0015 (□) M NaCl, all containing 0.2 mM EDTA.

of nu bodies is considerably stabilized by association with histones (e.g., in 0.2 mM EDTA: DNA has a Tm of 50°C; ν_1 has a Tm of 80°C), yet remains as sensitive as naked DNA to the destabilizating effects of urea.

Hydrodynamic studies (s, η and apparent partial specific volume) of ν_1 as a function of urea indicated gradual changes in particle size and shape. From these data, and employing the theory of Scheraga and Mandelkern (22), we conclude that ν_1 particles undergo some increase in asymmetry with increasing urea concentration up to 6-8 M. Over this range, the particles are best described as going from a spherical shape, in no urea, to a structure resembling a prolate ellipsoid of revolution.

Circular dichroic spectra of ν_1 as a function of urea molarity exhibited both cooperative and non-cooperative effects (Figure 3). The $[\theta]_p$ at 280 nm, reflecting DNA conformation, increased linearly with urea molarity, up to 8 M. The $[\theta]_p$ at 222 nm, reflecting the sizeable contribution of histone α-helix, reveals a cooperative loss of 2° structure with increased urea molarity. Parallel studies on the isolated inner histone heterotypic tetramer (26), indicated a similar cooperative destabilization by urea (although the molarity of 50% denaturation was a sensitive function of NaCl concentration). These data, combined with the conclusion of laser-Raman spectroscopy, argue that the inner histones retain similar 2° structure and sensitivity to urea denaturability, whether removed from DNA in high NaCl, or associated in ν_1.

The reactivity of the cysteine residues of H3 to the highly specific reagent N-ethylmaleimide (NEM) is markedly affected by increased urea molarity. The number of moles NEM/ν_1 increases from 0.1-0.2 in 0-5 M urea to about 2.2 in 8-10 M urea. The release of the thiol group is highly cooperative, and the final level of reaction is consistent with the known average cysteine content of chicken erythrocyte histones.

Electron microscopy of ν_1 similarly exhibits dramatic ultrastructural changes with increased urea molarity, exhibiting little structural damage until 4 M urea (Figure 4). The destruction of normal morphology appears to follow a pattern involving particle swelling and unravelling, ultimately leading to short rodlets in high urea solutions.

Our current conception of the diverse effects of increasing urea (and of high NaCl) on ν_1 are schematically represented in Figure 5. In the "native" state, $\sim$ 140 np of DNA are shown wrapped around the outside of the α-helical-rich globular domains of $(H4, H3, H2A, H2B)_2$ close-packed with dihedral-point-group symmetry (18). A postulated true dyad

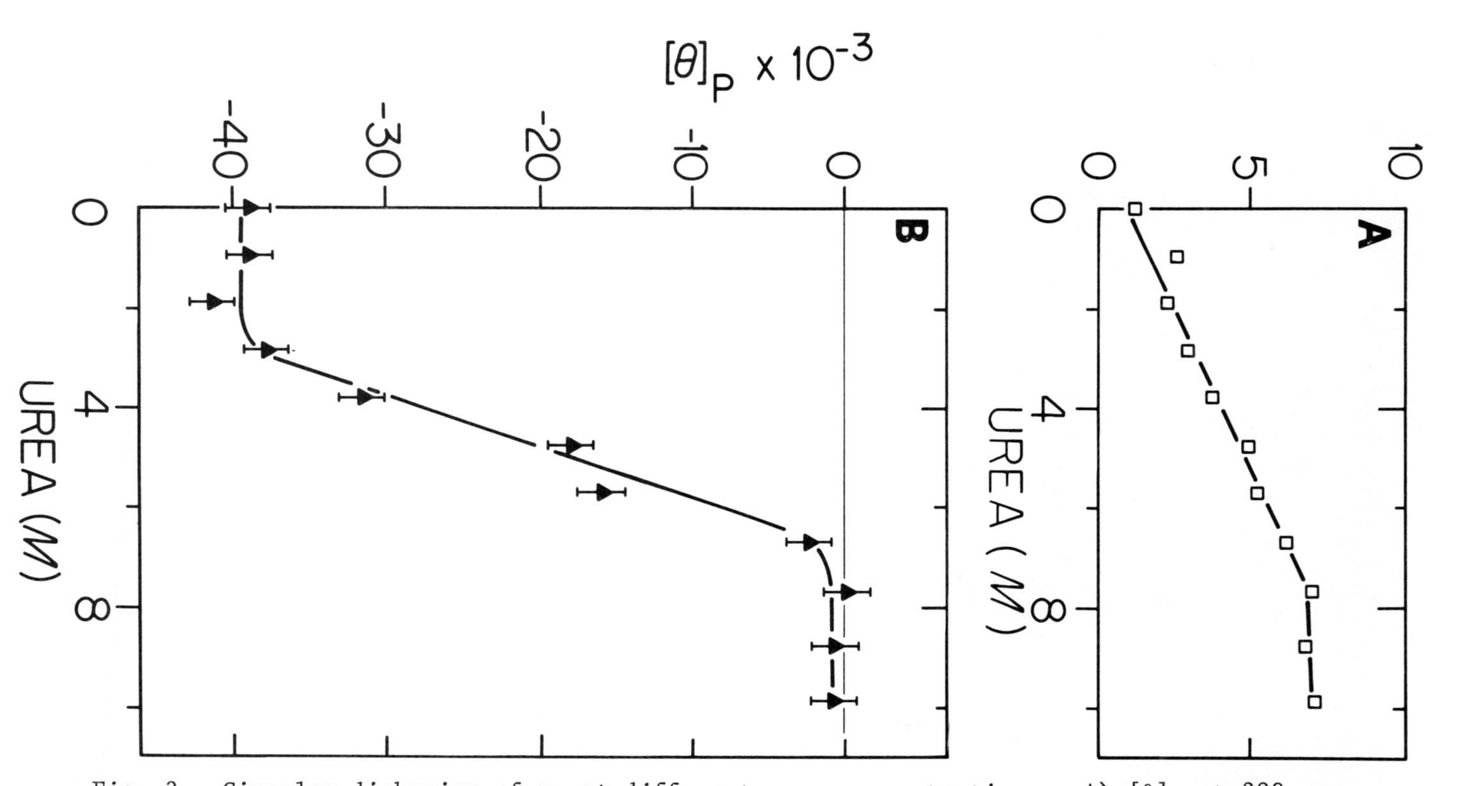

Fig. 3. Circular dichroism of ν_1 at different urea concentrations. A) $[\theta]_P$ at 280 nm. B) $[\theta]_P$ at 222 nm. Conditions: 1 cm cuvettes; $A_{260} \leq 1.0$; Jasco circular dichrograph SS-10; room temperature.

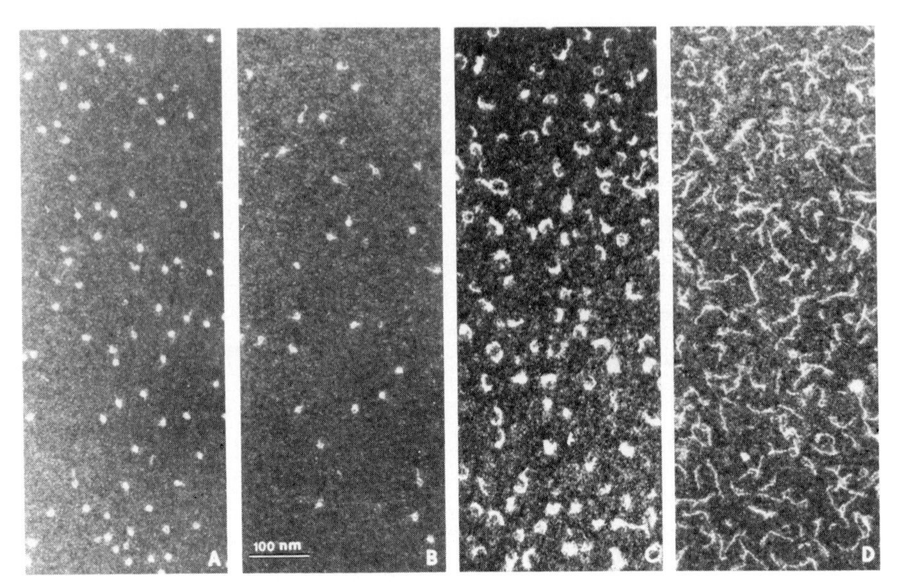

Fig. 4. Dark-field electron microscopy of ν_1 exposed to varying urea concentrations. A drop of ν_1 ($A_{260} = 1.0$) in 0.2 mM EDTA or urea buffer was placed on a freshly glowed carbon film for 30 seconds. The grid was rinsed in Kodak Photoflo (pH 7.0) and dried. Contrast was achieved by negative staining with 5 mM uranyl acetate for 30 seconds. Samples: A) 0.2 mM EDTA; B) 2 M urea; C) 6 M urea; D) 8 M urea.

axis (18,27) relates the heterotypic tetramers, and relates the portions of DNA associated with each heterotypic tetramer. The path of the DNA is arbitrary in this scheme, but preserves the equivalent single superhelical turn (10) and the postulated dyad axis. The H3 thiol residues are buried and H-bonded to some other histone residues (24). In "low urea" concentrations (3-4 M): the ν_1 have swollen and increased particle asymmetry; DNA has become more B-like; the α-helical content remains essentially unchanged; and the H3 thiols remain buried. Two possible "low urea" states are diagrammed--one with the histone globular regions not "swollen" (i.e., maintaining native 2°, 3° and 4° structure); the other, indicating a "swelling" of the globular regions (i.e., loss of 3° and 4° structure) with no loss of 2° structure. In "high urea" solutions (8-10M) : particle swelling and unravelling has been completed; by circular dichroism the DNA looks like its "normal" B configuration (except that its melting temperature is still considerably greater than naked DNA in

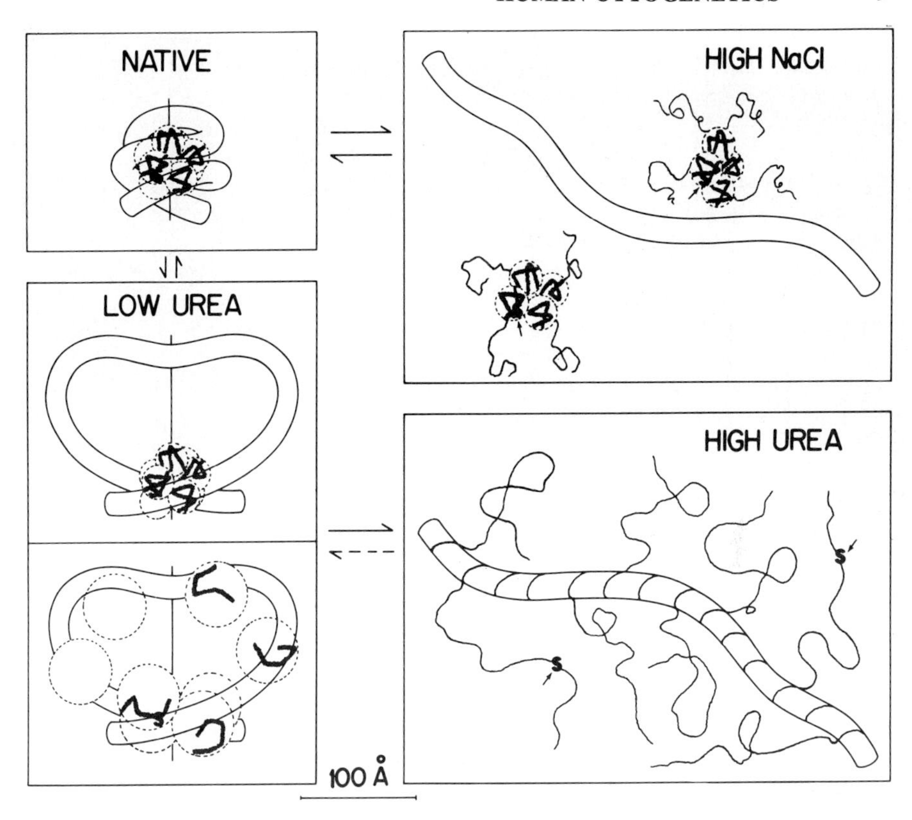

Fig. 5. Scheme of the effects of urea and of high NaCl on ν_1. The postulated true dyad axis is represented as a vertical line penetrating the Native and Low urea structures. The equivalent single superhelical turn of DNA per ν_1 is conceived as being stabilized at the ends of the DNA fragment, presumably by interaction with H3 and H4. Additional folding of the DNA, while still maintaining the single equivalent superhelical turn, can be accomplished by twisting the DNA clockwise or counter-clockwise around the particle dyad axis and binding to the histone core, in the Native structure. α-helical segments are represented in the globular domains of the histone molecules. Two different possible Low Urea states are represented with the DNA coil untwisting and expanding into a more asymmetric and open structure. In High Urea, the histone globular regions have become random coils, while the basic regions are attached to the extended DNA rod. The H3 thiol groups (S) are exposed and reactive (⟶). In High NaCl (pH 7.0) ν_1 dissociates into the DNA fragment and two heterotypic tetramers.

the same low ionic strength buffer); the α-helix is disrupted; and the H3 thiol groups are exposed and reactive. Also shown in Figure 5 is the probable consequence of treatment of ν_1 with "high salt" (e.g., 2 M NaCl): the particles dissociate into heterotypic tetramers and DNA; 2° and 3° structure of the histone globular regions are retained, and the H3 thiol groups are exposed, although still H-bonded.

The present suggestion that the different domains of a ν body (i.e., the DNA-rich shell and the protein-rich core) exhibit differential response to a simple chemical perturbant (urea), offers a basis for viewing the apparently altered conformation of transcriptional chromatin (9,25). Modulations of the DNA-rich shell (e.g., expansion and destabilization of the DNA) might occur with only minor alterations in the conformation of the protein-rich core, permitting rapid renaturation of the transcriptionally-inactive chromatin configuration.

Higher-Order Structure of Chromatin Nu Bodies. In our earliest electron microscopic studies of nu bodies (16), we noted that chromatin fibers contained regions of close-packing of subunits. We suggested that "further packaging of the DNA might represent a folded or helical close-packing of the spherical ν bodies under the influence of metal cations and non-covalent interactions" (16). These ideas were developed in a theoretical study from our laboratory (2), which demonstrated that various close-packed lattices of spherical particles could form the basis of the low-angle X-ray reflections observed with chromatin or nuclei. The acceptable arrangements included some two-, four-, and six-contact helicies of nu bodies, which had to meet the additional constraints of generating a 20-30 nm fiber with a DNA packing ratio of about 30/1. Subsequent studies (3) on neutron diffraction from oriented fibers of chromatin appeared to demonstrate a "helical cross" pattern. These authors interpreted the data as being consistent with a helical arrangement of nu bodies. At about the same time, an electron microscopic study (7) of isolated chromatin fragments was interpreted as a helix (or solenoid) of a "continuous density rod" (or nucleofilament). These authors argued that such a model would be "consistent with the lack of contrast seen along the nucleofilaments in (their) electron micrographs." These micrographs (i.e., their Figure 3), however, appear to us to contain regions that show clear contrast along a string of nu bodies. The solenoidal model of a continuous nucleofilament, nonetheless, became intellectually satisfying to H. G. Davies, writing in Nature (4). He commented, "Those electron microscopists who, for one reason or another, did

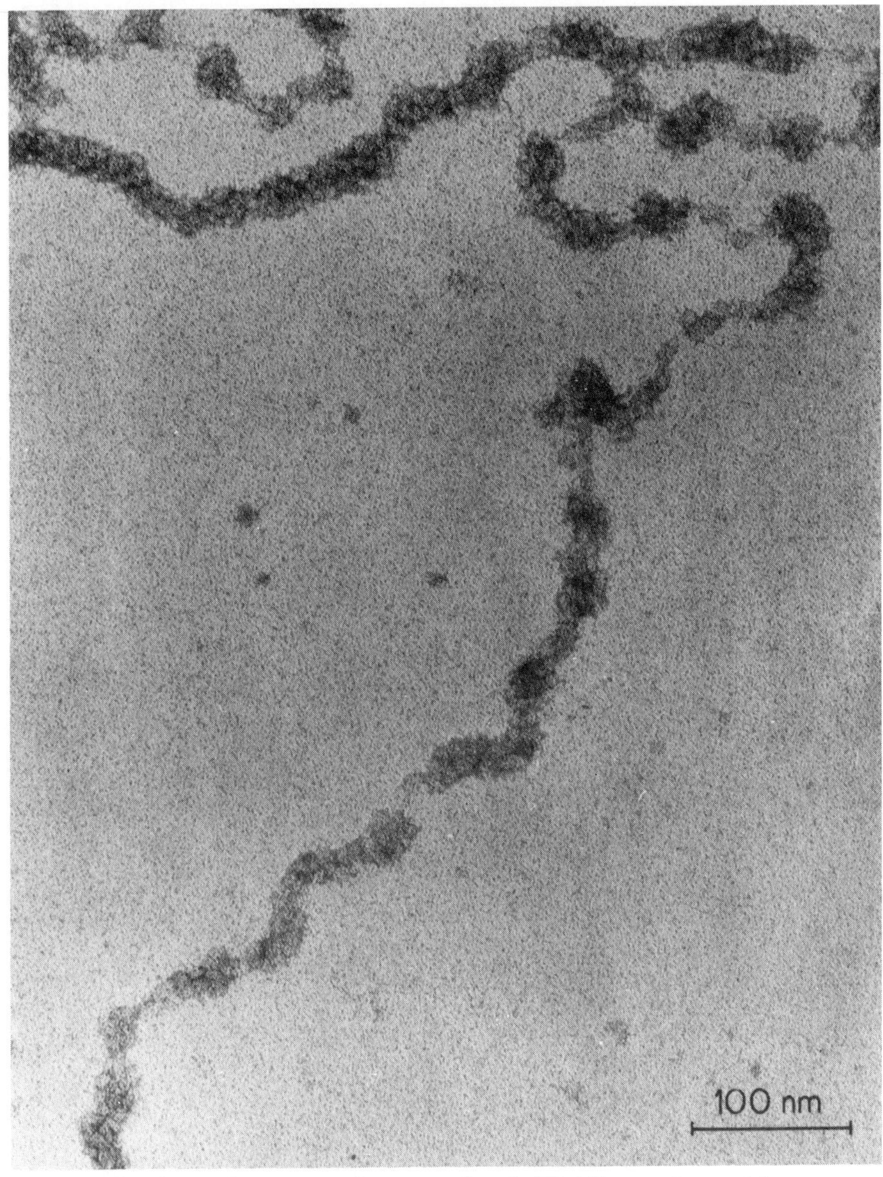

Fig. 6. Electron micrograph of 20-30 nm chromatin fibers spilling out of a chicken erythrocyte nucleus. Isolated nuclei were washed in 0.2 M KCl and diluted 1:150 in 0.02 M KCl, 0.5 mM $MgCl_2$. The swollen nuclei were centrifuged onto freshly glowed carbon-coated grids, rinsed with Photoflo, dried and negatively stained with 5 mM uranyl acetate.

not see the thin fibers as beaded, but continous, a result confirmed by Finch and Klug, will be unable to resist a smile at this latest turn of events." We believe that this joy was premature and insufficiently justified.

Over the past few years Dr. A. L. Olins has striven to visualize a higher-order arrangement of nu bodies from freshly spread nuclei, rather than from chromatin fragments (14,15). An example of thick (20-30 nm) chromatin fibers spilling from a chicken erythrocyte nucleus is presented in Figure 6. Nu bodies can be readily visualized in these fibers ("noodles") although no obvious helical parameters can be easily extracted from the micrograph. We believe that this, and other micrographs (see Figure 8) demonstrate convincingly the clear stain contrast completely around the nu body. The evidence is not consistent with the solenoidal model of a continuous density nucleofilament.

Employing our conditions of spreading nuclei (i.e., 20 mM KCl $\pm$ 0.5 mM $MgCl_2$), most of the chromatin fibers exhibit short regions of thick fibers separated by less-close-packed arrays of nu bodies. We frequently observe strings of clusters of nu bodies (Figure 7). These clusters generally exhibit diameters of 20-30 nm. Their origin is unclear. They could reflect weak points in a continuous thick fiber and/or inherent point-group symmetry in the arrangement of nu bodies, possibly determined by the histones H1 or H5. Figure 8 presents a scheme of the possibilities for regular arrangements of nu bodies in the 20-30 nm chromatin fibers.

We believe that the fundamental chromatin subunit, the nu body, maintains its integrity in the higher-orders of chromatin. Close-packing, per se, does not destroy its individuality. The "special conditions" (4) of microscopy employed to visualize the nu body-pulled apart or packed together--are special because they have succeeded where others have failed.

ACKNOWLEDGEMENTS

The studies of our laboratory represent a team effort, for which I am its present spokesman. The names of my collaborators are presented within the text. They have helped to make these experiments both productive and pleasurable. E. B. Wright and M. Hsu-Hsi have facilitated these investigations by contributing their wealth of talent. Support for these studies was derived in part from research grants to DEO (NIH) and ALO (NSF) and in part from ERDA under contract with Union Carbide Corporation.

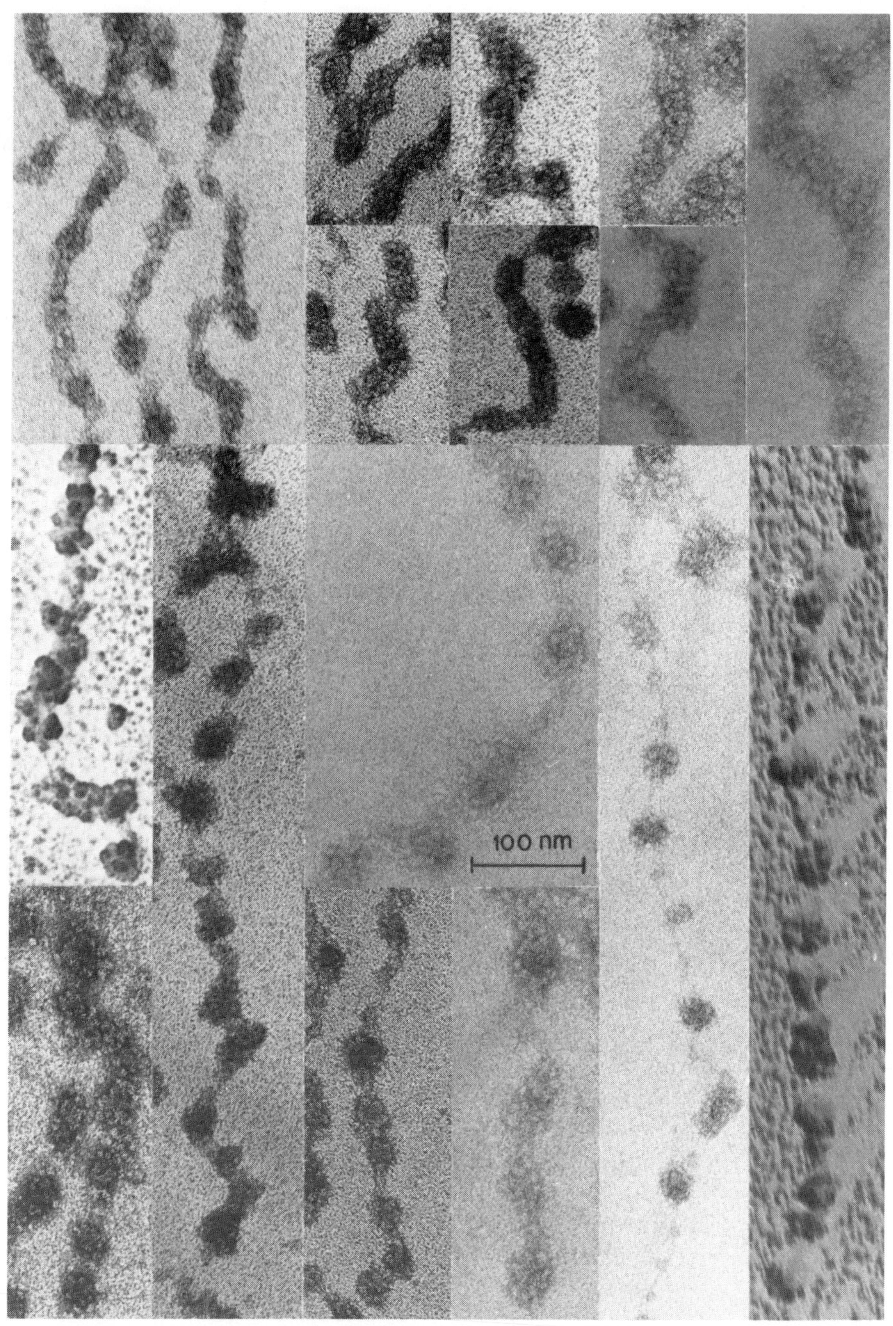

Fig. 7. Chromatin fibers spilling out of chiken erythrocyte nuclei. The upper third of the figure illustrates continuous fibers; the bottom two-thirds of the figure illustrates discontinuous (clustered) fibers.

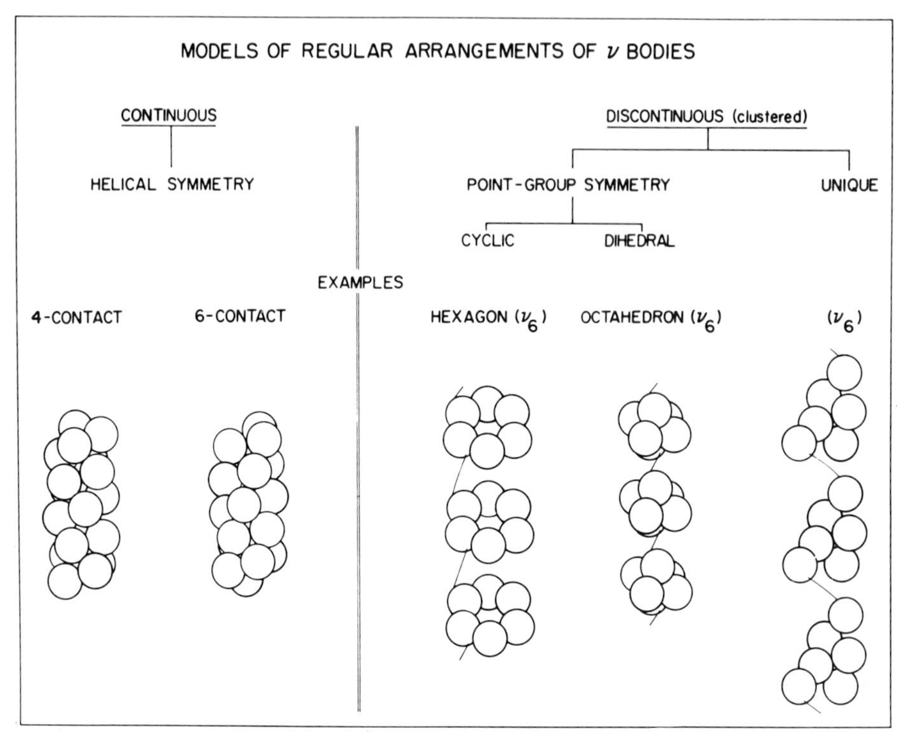

Fig. 8. Scheme of possible regular arrangements of nu bodies in 20–30 nm chromatin fibers. It is, of course, necessary to note the possibility that the actual arrangement of nu bodies *in vivo* in the thick fibers is highly irregular. The examples shown are entirely arbitrary, and merely chosen to illustrate the logical possibilities of ordered fibers.

REFERENCES

1. Bradbury, E. M. (1976) *Trends Biochem. Sci.* 1, 7.
2. Carlson, R. D. and Olins, D. E. (1976) *Nucleic Acids Res.* 3, 89.
3. Carpenter, B. G., Baldwin, J. P., Bradbury, E. M. and Ibel, K. (1976) *Nucleic Acids Res.* 3, 1739.
4. Davies, H. G. (1976) *Nature (Lond.)* 262, 533.
5. Elgin, S. C. R. and Weintraub, H. (1975) *Ann. Rev. Biochem.* 44, 725.
6. Felsenfeld, G. (1975) *Nature (Lond.)* 257, 177.
7. Finch, J. T. and Klug, A. (1976) *Proc. Natl. Acad. Sci. U.S.A.* 73, 1897.
8. Frushour, B. G., and Koenig, J. L. (1975) in *Advances in Infrared and Raman Spectroscopy*, R. J. H. Clark and R. E. Hester, eds. London: Heyden & Sons.
9. Garel, A. and Axel, R. (1976) *Proc. Natl. Acad. Sci. U.S.A.* 73, 3966.
10. Germond, J. E., Hvit, B., Oudet, P., Gross-Bellard, M. and Chambon, P. (1975) *Proc. Natl. Acad. Sci. U.S.A.* 72, 1843.
11. Isenberg, I. and Van Holde, K. (1975) *Acct. Chem. Res.* 8, 327.
12. Lippert, J. L., Tyminski, D. and Desmeules, P. J. (1976) *J. Am. Chem. Soc.* 98, 7075.
13. Moss, T., Cary, P. D., Crane-Robinson, C. and Bradbury, E. M. (1976) *Biochemistry* 15, 2261.
14. Olins, A. L. (1977) *Biophys. J.* 17, 115a.
15. Olins, A. L. (1977) submitted for publication.
16. Olins, A. L. and Olins, D. E. (1974) *Science* 183, 330.
17. Olins, A. L., Carlson, R. D., Wright, E. B. and Olins, D. E. (1976) *Nucleic Acid Res.* 3, 3271.
18. Olins, A. L., Breillatt, J. P., Carlson, R. D., Senior, M. B.,Wright, E. B. and Olins, D. E. (1977) in *The Molecular Biology of the Mammalian Genetic Apparatus (Part A)*, P. O. P. T'so, ed. Amsterdam: Elsevier/North Holland, in press.
19. Olins, D. E., Bryan, P. N., Harrington, R. E., Hill, W. E. and Olins, A. L. (1977) submitted for publication.
20. Olins, D. E., Bryan, P. N., Olins, A. L., Harrington, R. E. and Hill, W. E. (1977) *Biophys. J.* 17, 114a.
21. Prescott, B., Thomas, Jr., G. J., and Olins, D. E. (1977) *Biophys. J.* 17, 114a.
22. Scheraga, H. A. and Mandelkern, L. (1953) *J. Am. Chem. Soc.* 75, 179.
23. Thomas, Jr., G. J. (1975) in *Vibrational Spectra and Structure (Volume 3)*, J. R. Durig, ed. New York: Dekker.

24. Thomas, Jr., G. J., Prescott, B. and Olins, D. E. (1977) submitted for publication.
25. Weintraub, H. and Grondine, M. (1976) *Science* 193, 848.
26. Weintraub, H., Palter, K. and Van Lente, F. (1975) *Cell* 6, 85.
27. Weintraub, H., Worcel, A. and Alberts, B. (1976) *Cell* 9, 409.

THE APPROACH TO A FUNCTIONAL DEFINITION OF HISTONES: A CURRENT STOCK-TAKING

Irvin Isenberg and Steven Spiker

Department of Biochemistry and Biophysics
Oregon State University
Corvallis, Oregon 97331

ABSTRACT. A summary is given of the present-day status of the identification of histones and proteins that are histone-like. At the present time, the situation has clarified considerably with respect to H3, H4, H2a and H2b, but serious difficulties still remain for H1 and the HMG proteins.

INTRODUCTION

A protein is best defined in terms of its function. But if its function is not known, other and less satisfactory criteria must be used.

Histones have been known since 1884 (25), but prior to the discovery of the subunit structure of chromatin, histones were, of necessity, defined in an imprecise manner: They were most often known simply as the basic proteins associated with DNA. The individual histones were defined operationally by various physical and chemical criteria, particularly those used to prepare the histones. However, difficulties arose when comparisons were made between the histones of different species. Some of the histones of plants and lower eukaryotes had new properties and serious ambiguities developed.

The discovery of the subunit structure of chromatin (24, 38,39,41,46,59,61), and the elucidation of at least some of the properties of the nucleosomes, has changed this. The DNA of a nucleosome is at or near the surface (1,36,44,59), surrounding a core made up of the inner histones: H2a, H2b, H3 and H4. The interactions between the inner histones maintain the structure of the nucleosome. These histones may therefore now be defined by criteria which are at least partially functional: These are the histones that are in the core of the nucleosome.

WHICH HISTONES ARE HOMOLOGOUS?

The classic work of DeLange et al. (10) demonstrated a remarkable evolutionary conservation of H4 in going from animals to plants. There can be little doubt as to which histones are H4. Similar remarks may be made about H3 (45).

In contrast, H2a and H2b do vary considerably in amino acid composition (53). Furthermore, a knowledge of the sequence of an H2a or H2b in a plant is still lacking. In addition, plant H2a and H2b have higher molecular weights than animal H2a and H2b (49). All of these differences have, in the past, blocked a definitive identification of the plant histones (6,33,40,43,52,50,51,54). Which protein was H2a and which was H2b was by no means clear and, in fact, even whether plants had an H2a or an H2b was questioned. Tentative attempts at an identification were made, based on gel patterns (6,50,52,55) but the difficulties prevented workers from posing a number of significant questions: Are the interactions between histones universal throughout all of the eukaryotic kingdoms? Are some interactions essential and others not? Are there invariant regions in each histone that are involved in histone-histone interactions?

Recently, Spiker et al. (54) attempted an identification of plant histones by means of a set of criteria: staining properties of gels, electrophoretic mobility, and solubility characteristics. It is important to note here that these criteria do indeed identify the calf inner histones.

We have recently examined the cross complexing pattern of pea histones (53) and have found that, if the identification methods of Spiker et al. (54) are used, then the cross-complexing pattern of pea histones is the same as the cross-complexing pattern of calf thymus histones (9,19,58).

The operational criteria for identifying calf and pea histones is therefore fully consistent with the idea that the pattern of complexing of the inner histones is conserved in going from animals to plants. Either both are right, or both are wrong.

In fact, there is every reason to believe that both are right. The amino acid analyses of what we may now identify as pea H2a and H2b are much closer to their calf counterparts than a reverse identification would have them (53). (We should note, however, that H2a and H2b are not nearly as evolutionarily invariant as H3 and H4 are. They, in fact, show appreciable variation (53)).

Plant chromatin has a sub-unit structure (30,34,35) and, indeed, nucleosomes have now been found in a wide variety of species (15,20,29,30,32,34,35,37,56). Just how widespread the pattern of histone interactions is, and whether we now have, or can obtain, a universal set of operational identifying criteria, valid for all species, are subjects for further study.

CAN HISTONE H1 AND THE HMG PROTEINS BE DEFINED AT THE PRESENT TIME?

H1 is known simply as the very lysine-rich histone. This simple chemical identification has apparently created no difficulty thus far, but one must recognize that we have, as yet, no functional or structural determinant of H1. There is, of course, abundant evidence that H1 is involved in the condensation of chromatin (2,11,12,28,31), and there are reports that the phosphorylation of H1 may trigger mitosis (3-5,16,17, 26,27), but these are not yet sufficient to serve as identifiers of H1. Furthermore, there are remarkable sequence homologies (60) between HMG17, a protein classed as a _non_-histone chromosomal protein, and a histone, histone T, found in trout (18). Should HMG17 be classed as a histone? Should histone T be classed as a non-histone? Do they, perhaps, have similar functions in divergent species, or have they only evolved from the same protein?

The object of classification is to obtain clues as to the function of these proteins, and not to exercise semantic skill in taxonomy. On the one hand it might therefore be best simply to avoid too fine a classification scheme at the present time. On the other hand, the hints that we have make it reasonable to put H1 and some, or all, of the HMG proteins in one and the same category. Johns and his co-workers (13,14, 21), in fact, class the HMG proteins as "histone-like", based on solubility characteristics. One may go even further.

H1 has subfractions (7,8,22,23,47) (Fig. 1). The concept of having H1 and HMG1 classified similarly has been considerably strengthened by the finding that HMG1 and HMG2 have specific interactions with the various subfractions of calf thymus H1 (48). Some subfractions bind very strongly to HMG1; others bind only weakly. Furthermore, the subfractions that bind strongly to HMG1 do not bind to any appreciable extent to HMG2. Other subfractions bind weakly to both.

The specificity of the interactions shows, of course, that H1 subfractions can indeed show dramatic differences in binding. Even should it turn out that the H1-HMG interactions have no biological significance as such, the fact that the subfractions can bind so differently will remain as a significant physical property of H1, perhaps expressing itself _in vivo_ by variations of binding to other proteins. However, and more germane to the present paper, the interaction pattern suggests that H1 might in fact associate with HMG proteins _in vivo_, although there is, at present, no evidence to support such a speculation, only a very weak clue that has already been pointed out (48).

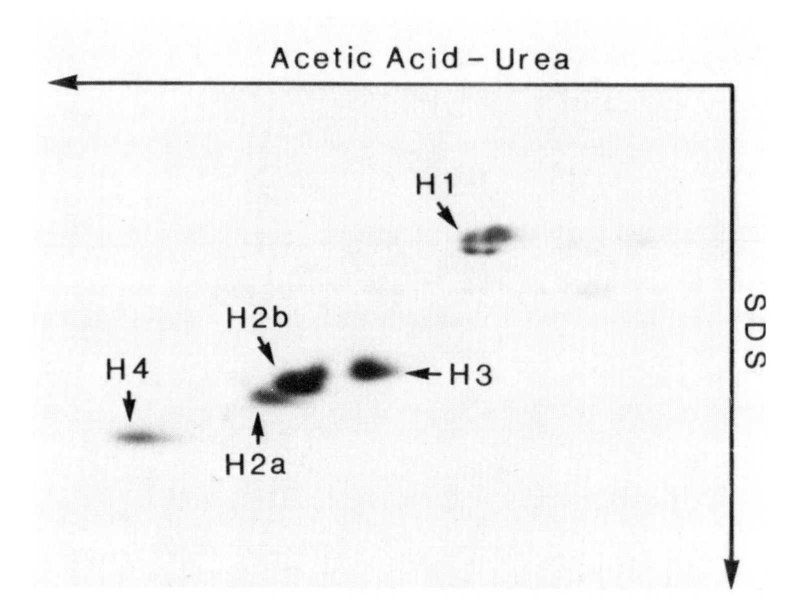

Fig. 1. Two-dimensional electrophoresis of whole histone from calf thymus. Four of the H1 subfractions are clearly resolved. The first dimension (right to left) was an acetic acid-urea gel according to Panyim and Chalkley (42). The second dimension (top to bottom) was an SDS gel according to Thomas and Kornberg (57).

We see, however, that there are two groups of proteins, H1 and HMG, both associated with chromatin, interacting *in vitro* specifically, and sometimes very strongly. This suggests that these may associate with one another, and serve common functions. Whether they do, or do not, will emerge from future studies. However, at present, there is as much reason, and perhaps more, to regard both H1 and the HMG proteins as members of the same family, as there is to think of one as "histone" and the other as "non-histone." In any case, we have only hints as to the function of H1, and not even hints for the HMG proteins. Operational criteria for their definitions must be viewed with caution; they are but partial, temporary, expedients in an attempt to understand their function.

ACKNOWLEDGEMENTS

This work was supported by Grant CA 10872 awarded by the National Cancer Institute, Department of Health, Education and Welfare.

REFERENCES

1. Baldwin, J.P., Boseley, P.G., Bradbury, E.M. and Ibel, K. (1975) *Nature* 253,245.
2. Billett, M.A. and Barry, J.M. (1974) *Eur. J. Biochem.* 49,477.
3. Bradbury, E.M., Inglis, R.J. and Matthews, H.R. (1974) *Nature* 247,257.
4. Bradbury, E.M., Inglis, R.J. and Matthews, H.R. (1974) *Nature* 249,553.
5. Bradbury, E.M., Inglis, R.J., Matthews, H.R. and Sarner, N. (1973) *Eur. J. Biochem.* 33,131.
6. Brandt, W.F. and Von Holt, C. (1975) *FEBS Lett.* 51,84.
7. Bustin, M. and Cole, R.D. (1968) *J. Biol. Chem.* 243,4500.
8. Bustin, M. and Cole, R.D. (1969) *J. Biol. Chem.* 244,5286.
9. D'Anna, J.A., Jr. and Isenberg, I. (1974) *Biochem.* 13,4992.
10. DeLange, R.J., Fambrough, D.M., Smith, E.L. and Bonner, J. (1969) *J. Biol. Chem.* 244,5669.
11. Elgin, S.C.R. and Weintraub, H. (1975) *Annu. Rev. Biochem.* 44,725.
12. Finch, J.T. and Klug, A. (1976) *Proc. Natl. Acad. Sci. USA* 73,1897.
13. Goodwin, G.H. and Johns, E.W. (1973) *Eur. J. Biochem.* 40,215.
14. Goodwin, G.H., Sanders, C. and Johns, E.W. (1973) *Eur. J. Biochem.* 38,14.
15. Gorovsky, M.A. and Keevert, J.B. (1975) *Proc. Natl. Acad. Sci. USA* 72,3536.
16. Gurley, L.R., Walters, R.A. and Tobey, R.A. (1974) *J. Cell Biol.* 60,536.
17. Gurley, L.R., Walters, R.A. and Tobey, R.A. (1975) *J. Biol. Chem.* 250,3936.
18. Huntley, G.H. and Dixon, G.H. (1972) *J. Biol. Chem.* 247,4916.
19. Isenberg, I. (1977) *in Search and Discovery - A Volume Dedicated to Albert Szent-Gyorgyi.* B. Kaminer, Ed., New York, N.Y., Academic Press.
20. Jerzmanowski, A., Staron, K., Tyniec, B., Bernhardt-Smigielska, J. and Toczko, K. (1976) *FEBS Lett.* 62,251.
21. Johns, E.W., Goodwin, G.H., Walker, J.M. and Sanders, C. (1975) *Ciba Found. Symp.* 28,95.

22. Kinkade, J.M., Jr. and Cole, R.D. (1966) *J. Biol. Chem.* 241,5790.
23. Kinkade, J.M., Jr. and Cole, R.D. (1966) *J. Biol. Chem.* 241,5798.
24. Kornberg, R.D. (1974) *Science* 184,868.
25. Kossel, A. (1884) *Hoppe-Seylers Z. Physiol. Chem.* 8,511.
26. Lake, R.S. (1973) *Nature New Biol.* 242,145.
27. Lake, R.S. and Salzman, N.P. (1972) *Biochem.* 11,4817.
28. Littau, V.C., Burdick, C.J., Allfrey, V.G. and Mirsky, A.E. (1965) *Proc. Natl. Acad. Sci. USA* 54,1204.
29. Lohr, D. and Van Holde, K.E. (1975) *Science* 188,165.
30. McGhee, J.D. and Engle, J.D. (1975) *Nature* 254,449.
31. Mirsky, A.E., Burdick, C.J., Davidson, E.H., and Littau, V.C. (1968) *Proc. Natl. Acad. Sci. USA* 61,592.
32. Morris, N.R. (1976) *Cell* 8,357.
33. Nadeau, P., Pallotta, D. and Lafontaine, J. (1974) *Arch. Biochem. Biophys.* 161,171.
34. Nagl, W. (1976) *Experientia* 32,703.
35. Nicolaieff, A., Philipps, G., Gigot, C. and Hirth, L. (1976) *J. Microscopie Biol. Cell.* 26,1.
36. Noll, M. (1974) *Nucleic Acids Res.* 1,1573.
37. Noll, M. (1976) *Cell* 8,349.
38. Olins, A.L. and Olins, D.E. (1973) *J. Cell Biol.* 59,252a.
39. Olins, D.E. and Olins, A.L. (1974) *Science* 183,330.
40. Oliver, D., Sommer, K., Panyim, S., Spiker, S. and Chalkley, R. (1972) *Biochem. J.* 129,349.
41. Oudet, P., Gross-Bellard, M. and Chambon, P. (1975) *Cell* 4,181.
42. Panyim, S. and Chalkley, R. (1969) *Arch. Biochem. Biophys.* 130,337.
43. Panyim, S., Chalkley, R., Spiker, S. and Oliver, D. (1970) *Biochim. Biophys. Acta* 214,216.
44. Pardon, J.F., Worcester, D.L., Wooley, J.C., Tatchell, K., Van Holde, K.E. and Richards, B.M. (1975) *Nucleic Acids Res.* 2,2163.
45. Patthy, L., Smith, E.L. and Johnson, J. (1973) *J. Biol. Chem.* 248,6834.
46. Sahasrabuddhe, C.G. and Van Holde, K.E. (1974) *J. Biol. Chem.* 249,152.
47. Smerdon, M.J. and Isenberg, I. (1976) *Biochem.* 15,4233.
48. Smerdon, M.J. and Isenberg, I. (1976) *Biochem.* 15,4242.
49. Sommer, K.R. and Chalkley, R. (1974) *Biochem.* 13,1022.
50. Spiker, S. (1975) *Biochim. Biophys. Acta* 400,461.
51. Spiker, S. (1976) *Nature* 259,418.
52. Spiker, S. and Chalkely, R. (1971) *Plant Physiol.* 47,342.
53. Spiker, S. and Isenberg, I. (1977) *Biochem.* (In Press).
54. Spiker, S., Key, J.L. and Wakim, B. (1976) *Arch. Biochem. Biophys.* 176,510.
55. Spiker, S. and Krishnaswamy, L. (1973) *Planta* 110,71.

56. Thomas, J.O. and Furber, V. (1976) *FEBS Lett.* 66,274.
57. Thomas, J.O. and Kornberg, R.D. (1975) *Proc. Natl. Acad. Sci. USA* 72,2626.
58. Van Holde, K.E. and Isenberg, I. (1975) *Acc. Chem. Res.* 8,327.
59. Van Holde, K.E., Sahasrabuddhe, C.G. and Shaw, B.R. (1974) *Nucleic Acids Res.* 1,1579.
60. Walker, J.M., Hastings, J.R.B. and Johns, E.W. (1976) *Biochem. Biophys. Res. Commun.* 73,72.
61. Woodcock, C.L.F. (1973) *J. Cell Biol.* 59,368a.

HISTONE ANTIBODIES AND CHROMATIN STRUCTURE

Michael Bustin

Laboratory of Nutrition and Endocrinology, Developmental Biochemistry Section, National Institute of Arthritis, Metabolism and Digestive Diseases, National Institutes of Health, Bethesda, Maryland 20014

ABSTRACT

Antibodies elicited by purified histone fractions specifically recognize each of the five main histone classes and can distinguish between various H1 subfractions present in a single tissue. These antibodies bind specifically to the nucleus of a cell, to isolated nuclei, purified chromatin, isolated metaphase and polytene chromosomes, and to purified nucleosomes. Thus, they can serve as probes for histone molecules in their "native" chromatin-bound state.

Antisera elicited in rabbits against histones derived from calf thymus cross react strongly with histones derived from a variety of tissues. It is possible therefore to use one type of antiserum as a general reagent to study the organization of histones in chromatin and chromosomes derived from various sources.

Indirect immunoflorescence studies reveal that each of the histone fractions is located along the entire length of each chromosome present in a tissue. When polytene chromosomes are stained with antihistone sera the resolution of individual bands observed by fluorescence microscopy is of the same order as that obtained by orcein staining or by phase contrast microscopy. The amount of exposed antigenic sites in chromosome-bound histones is proportional to the amount of DNA present in a band. Each band contains each type of histone. Puffing of specific bands results in changes which can be detected using antihistone sera.

The detailed organization and composition of histones in chromatin subunits can be directly visualized by immuno-electron microscopy. Upon specific binding of antibodies to chromatin subunits, a significant increase in the diameter of the nucleosome is observed. Each nucleosome contains histone H2B.

Nucleosomes which interact with antihistone antibodies can be visualized in transcribed regions of Drosophila embryos.

Antihistone antibodies purified by affinity chromatography interact with isolated nucleosomes to yield a particle of an increased mass which is separable from unreacted 11S nucleosomes by centrifugation on sucrose gradients. Using this technique it was conclusively shown that each nucleosome contains equal amounts of each of the histones H2A, H2B, H3 and H4.

INTRODUCTION

Chromatin is a complex macromolecular structure in which several types of proteins are complexed with DNA in a stringent manner. The chromatin fiber is composed of subunits (nucleosomes), each containing about 200 base pairs of DNA complexed with 8 molecules of histones. This "beaded" fiber (reviewed in 1) is coiled in a superstructure whose details are still undetermined. The coiled, beaded fiber is complexed with nonhistone proteins to yield a dynamic structure in which the expression of the genes is regulated. This structure can respond to external stimuli and changes its morphological appearance during the life cycle of a cell. Obviously it is difficult to study the organization of a particular component in such a dynamic macromolecular complex.

Immunological reactions occur under mild conditions, require relatively low amounts of material and are specific. Thus, they are particularly suitable to study the organization of a specific chromatin-bound component complexed in its native state, thereby providing information which is not easily obtainable by other techniques.

The application of serological techniques to the study of chromosomal components requires: 1) the antibodies have to be specific for the component, 2) these antibodies, even though elicited against a component purified from chromatin, should be able to react with the immunogen when it is complexed in its native state and 3) methods to detect and quantitate the reaction have to be available. Histone antibodies satisfy these three requirements. Each purified histone fraction elicits antibodies with a high degree of specificity for the immunogen (2,3). Antibodies elicited against purified histone fractions interact with chromosomes and chromatin-bound histones. This interaction can be studied by several techniques (4).

RESULTS AND DISCUSSION

ANTISERA TO CALF THYMUS HISTONES AS GENERAL REAGENTS

Histones extracted from calf thymus are well characterized proteins and have been frequently used as standards against which histones derived from other sources have been compared (5). With a few exceptions, antihistone sera were elicited against histone fractions also derived from calf thymus. The question arises whether antisera elicited against calf thymus histones can be used to study histones from other sources. The amino acid sequence of histones is highly conserved during evolution, therefore, it could be expected that antisera elicited against calf thymus histones would be suitable for studying histone organization in a variety of experimental systems. Indeed as Stollar and Ward (2) reported, the antisera to calf thymus histones reacted strongly with histones derived from species as diverse as human and lobster. We have studied the immunological cross reaction of Drosophila and calf thymus histones (6). With the exception of H1, the homologous histone fractions cross react very strongly. The data are summarized in Table 1. In all cases the immunological cross reaction was measured by the complement fixation technique which is very sensitive to structural differences (7). Since histones extracted from a variety of sources cross react very strongly, as measured by the sensitive complement fixation test, it is suggested that rabbit anticalf thymus histone H4, H3, H2B and H2A can be used as universal reagents to study the _in situ_ organization of histones in a variety of experimental systems. The use of standard immunological reagents is not only convenient but will also facilitate comparisons of results obtained in different laboratories. Furthurmore, since histones are poor immunogens, even slight contamination with tissue specific nonhistone proteins [which display a remarkable immunological specificity 8,9] may result in antisera preparations which contain antibodies directed against contaminants.

TABLE 1

IMMUNOLOGICAL COMPARISON OF HISTONES EXTRACTED FROM:

Serum	Human		Chicken		Lobster		Drosophila	
Anti-calf thymus	I.D.[b]	%[c]	I.D.	%	I.D.	%	I.D.	%
H1							120	27 (30)
H2A	0	0	0	0	7.9	2	30	6 (5)
H2B	2.5	1.2	2.5	1.2	0	0	0	0 (9)
H3							20	4 (2)
H4	0	0	14	3	0	0	0	0 (0)

a/ The immunological similarity between histones was assessed using the microcomplement fixation technique (10,11). A linear relation between immunological distance (I.D.) and amino acid differences of homologous proteins has been reported (11). The immunological distance between calf histones and histones obtained from human, chicken and lobster is calculated from the data presented by Stollar and Ward (Tables III-V), Ref. 2).

b/ I.D., Immunological distance (11).

c/ % amino acid differences between calf histones and histones extracted from the various tissues calculated from the immunological distance. It is assumed that the relation between I.D. and % amino acid differences for lysozyme (11) is applicable to histones. The values in parentheses in the Drosophila column are differences calculated from published amino acid composition. For H1 reference (27) for other histones (28).

The use of cross reacting sera will ensure that an immunological reaction is indeed due to histone antigenic determinants.

SPECIFICITY OF H1

Histone H1 differs from the other 4 histones in its molecular weight, multiplicity and species specificity. In the rat thymus there are 5 H1 subfractions (Fig. 1) which are immunologically distinct (10). The immunological distance (11) between the various subfraction was measured using antisera elicited against each purified H1 subfraction. From the immunological distance it could be calculated that neighboring fractions differ in amino acid composition by 6-18% (Fig. 1).

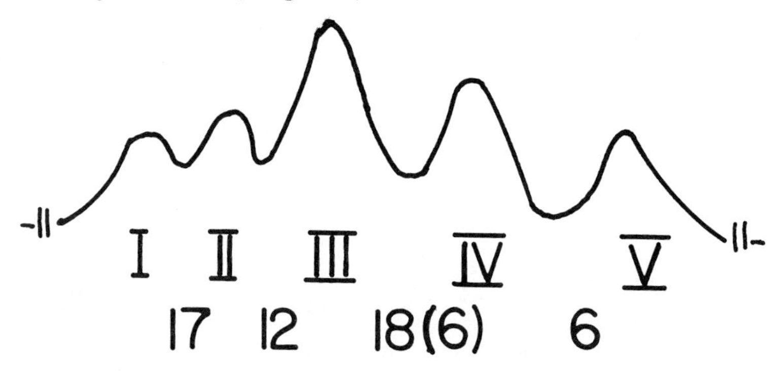

Fig. 1. Immunological specificity of rat H1 subfractions. A typical elution profile of rat thymus H1 fractionated on Amberlite IRC-50 is presented. The Roman numerals indicate the order of elution of the peaks. The Arabic numerals between two peaks indicate the percent amino acid difference between two adjacent peaks, based on immunological distance (10). The value in parenthesis was obtained by amino acid analysis (13).

These values are higher than would be predicted from amino acid analysis of H1 subfractions derived from several sources (12, 13). It is possible, however, that some sequence substitutions are not reflected in the amino acid composition. Indeed, sequence substitutions frequently involve interchanges of amino acids (quoted in Ref. 5).

In conclusion, purified histone fractions derived from calf thymus elicit specific antibodies.

With the exception of anti-H1, these antibodies cross react very strongly with homologous histone fractions derived from a variety of organisms. The antisera to calf thymus histone fractions, elicited in rabbits, can serve as standard reagents to study histones in a variety of experimental systems. Anti-H1, while still showing cross reaction among various species, reflects the species specificity of this protein fraction. Sequence differences between H1 subfractions within one tissue can be detected by the microcomplement fixation technique.

SPECIFIC BINDING OF ANTIHISTONE SERA TO CHROMATIN

Chromatin-bound histones are poor immunogens since antisera elicited by chromatin does not contain detectable amounts of antihistone antibodies (3,8). Antihistone sera are elicited by histone:RNA complexes. The histones used, were purified by procedures which could lead to denaturation. It is possible, therefore, that these antibodies will not bind to antigenic sites in chromatin-bound histones. Indeed, in the microcomplement fixation test most of the antihistone sera reacted poorly with chromatin and required large amounts of chromatin for reaction (3). The reaction between chromatin-bound histones and antihistone is best studied by an immunoadsorbtion assay (Fig. 2).

Quantitative measurements on the ability of chromatin-bound histones to interact with antihistone sera revealed that only a small portion of the determinants present in the immunogen are also exposed in chromatin. The relative number of histone antigenic determinants exposed in chromatin can be defined as the "equivalent antigenicity" (3). Compared to isolated histones in solution, the "equivalent antigenicity" of chromatin-bound histones was 9.6%, 3.2%, 0.9% and 0.9% for histones H1, H2B, H3 and H4 respectively. The ability to measure the exposure of the histone determinants in chromatin offers an additional analytical tool for detecting differences between chromatins obtained from various sources.

Table 2 summarizes the results obtained with chromatins derived from several organs of the rat as well as with chromatin obtained from calf thymus (14). Slight differences in the reactivity of the various chromatins were observed.

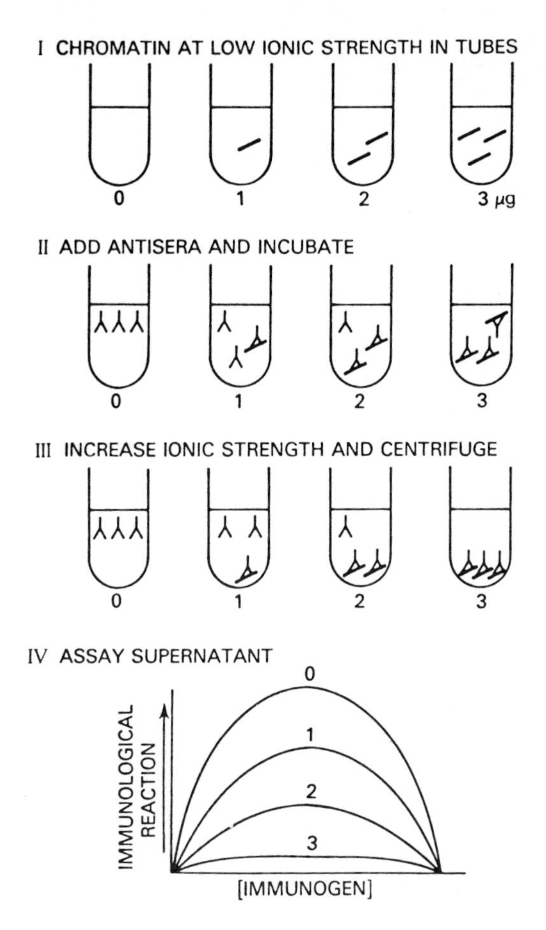

Fig. 2. Measurement of antibody binding to chromatin by immunoadsorbtion. Step I: Various amounts of chromatin are placed in a constant volume at low ionic strength (i.e., 5 mM Tris-HCl, pH 8). Step II: A constant amount of antiserum is added to each tube and the mixture is incubated. Step III: The mixtures are made 0.15 M in NaCl and the chromatin-antibody complex sedimented by centrifugation. Step IV: Supernatant is assayed for the remaining unadsorbed antibodies. The degree of immunological reaction is inversely proportional to the amount of chromatin used as immunoadsorbant.

Table 2 - Relative Degree of Exposure of Histone Determinants in Chromatin from Different Sources

Histone	Source of Chromatin
H1	Rat brain = Rat liver > Rat thymus = Calf thymus
H2A	Rat brain > Rat liver = Rat thymus < Calf thymus
H2B	Rat brain < Rat liver > Rat thymus < Calf thymus
H3	Rat brain > Rat liver < Rat thymus < Calf thymus
H4	Rat liver > Rat brain = Rat thymus < Calf thymus

An alternative, potentially more exact, way to measure the availability of histone determinants in chromatin is to use an ^{125}I-labeled antibody (15). Specific anti-H1 antibodies are purified by affinity chromatography on H1-Sepharose columns. The antibodies, which are of the IgG class, can be labeled with ^{125}I by either the chloramine-T or peroxidase method. The ^{125}I labeled antibodies specifically bind to chromatin. ^{125}I-IgG obtained from non-immunized rabbits do not bind to chromatin.

The amount of ^{125}I antibody bound to chromatin is directly proportional to the amount of antibody added (15). As shown in Fig. 3, 85% of the antibody can be adsorbed by increasing the amount of chromatin used as immunoadsorbant. Chromatin extracted from rat thymus was more efficient in binding antibodies than chromatin extracted from rat brain.

The binding of a specific antibody to its chromatin-bound histone determinant is dependent on numerous parameters. The slight differences in immunoadsorbtion may reflect differential shielding of histone determinants by nonhistone proteins or differences in histone modification.

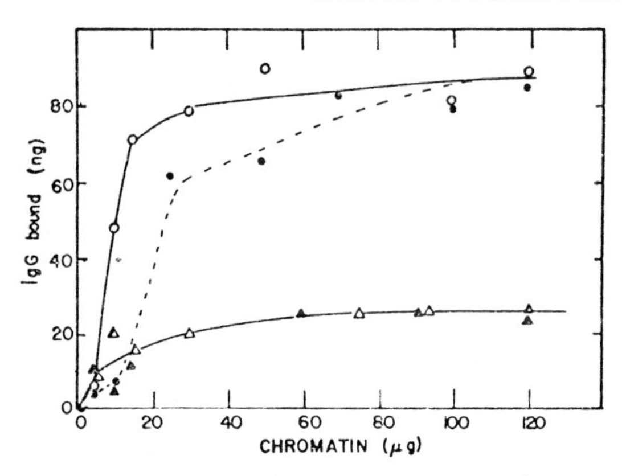

Fig. 3. Measuring histone determinants in chromatin by radioimmunoadsorbance. Variable amounts of rat thymus (open symbols) or rat brain (closed symbols) chromatin were added to tubes containing 1.0 g ^{125}I-anti-H1 antibodies (●,O) or ^{125}I-normal rabbit IgG (▲,△). After incubation at low ionic strength the mixtures were made 0.15 M in NaCl and centrifuged. The radioactivity present in the washed precipitate is a measure of the amount of antibody bound to chromatin (reprinted from 15 with permission).

LOCALIZATION OF HISTONES IN CHROMOSOMES BY IMMUNOFLUORESCENCE

Because antihistone sera specifically bind to chromatin, it seemed feasible to attempt to visualize the organization of histones in chromosomes by the indirect immunofluorescence technique. Metaphase chromosomes are usually fixed and spread in alcoholic solutions containing acetic acid, under conditions which lead to extraction of histones from chromosomes or to rearrangement of histone molecules along the DNA fiber. Therefore, chromosomes have been isolated at neutral pH and the proteins fixed by cross-linking with glutaraldehyde. Under these conditions we found (17) that each of the metaphase chromosomes from a variety of tissues, contains each of the histone fractions evenly distributed along the entire length of the chromosome. This finding is consistent with the view that all histones are necessary for maintaining the structure of the chromosome. Brief treatments with acetic acid:methanol mixtures leads to extraction of histones from chromosomes in such a way that a spotty fluorescent pattern is obtained. The fluorescence pattern reveals differences both in the intensity and the distribution of the fluorescent spots among the various chromosomes. Most probably the fluorescence pattern arises from the differential extractability of histones due to the folding of the chromatin fiber and to the presence of nonhistone proteins. Since the fluorescence pattern varied among chromosomes, it is feasible that the immunofluorescence obtained with antihistone sera may be of use in cytological studies.

Immunofluorescence studies on polytene chromosomes allow insight into the organization of histones in the unit of genetic activity. The unit of genetic and transcriptional activity in polytene chromosomes appears to be the individual bands (chromomeres) resolvable by light microscopy. Each of the antihistone sera specifically binds to the histone determinants in the polytene chromosomes of _Chironomus thummi_ (18). The resolution of individual bands obtained by immunofluorescence is of the same order as that observed by phase contrast microscopy (see Fig. 4). Antisera to histones H2B, H3 and H4 produced bands which closely resemble the banding pattern observed by phase contrast microscopy or by aceto-orcein staining. These 3 antisera stain intensely the same bands suggesting that the gross organization of the histones is similar in the various bands. The dense bands stain more intensely than the interband regions suggesting that the intensity of fluorescence is correlated with the amount of histone in a band and that the accessability of histone determinants in the bands is of the same order than that in the interbands. The permanently puffed region in chromosome 4 of _Chironomus thummi_ stains very weakly. That the exposure of histone determinants in puffed regions may be different from the exposure of the determinants in the bands or interbands is supported by experiments done on heat shocked _Drosophila_ chromosomes. The bands neighboring the puffed region stain with anti-H3 and anti-H4, while the puffs do not. They do stain, however, with anti-RNA polymerase even after antihistone sera was added (19).

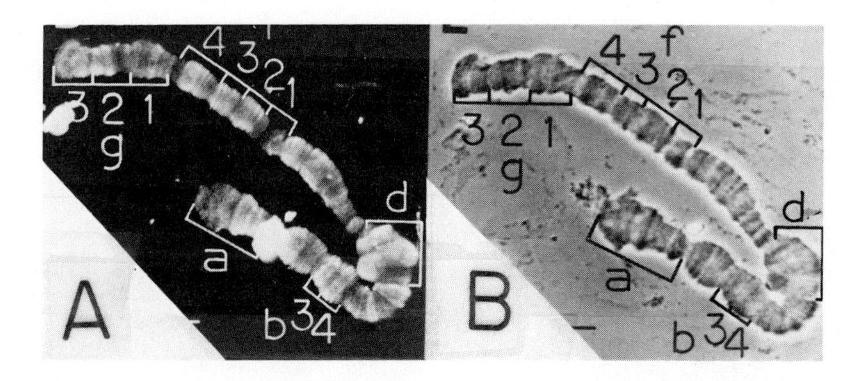

Fig. 4. Immunofluorescence of polytene chromosomes. Chromosome I from Chironomus thummi stained by the indirect immunofluorescence technique with anti-H4 sera (18). Various regions on the chromosome are identified according to the nomenclature of Keyl (29). A) fluorescent micrograph B) corresponding phase contrast micrograph.

ORGANIZATION OF HISTONES IN CHROMATIN VISUALIZED BY IMMUNO-ELECTRON MICROSCOPY

The immunofluorescence studies using the light microscope can yield information on the gross arrangement of histones in chromosomes. A more detailed visualization of the organization of histones in the chromatin fiber requires application of immuno-electron microscopic techniques.

Addition of antihistone sera to chromatin particles spread on electron microscope grids precoated with bovine serum albumin brings about a significant enlargement in the particles (20). The nucleosome, which is 100 Å in diameter, is coated with IgG molecules whose essential morphology is that of a Y with a maximum length of 110 Å. Using ferritin labeled goat antirabbit gamma globulin it was possible to show that the enlargement of the diameter of the nucleosome results from specific binding of antibodies to nucleosomes (Fig. 5).

At antibody saturation, the number of nucleosomes with a diameter of over 150 A is a measure of the number of nucleosomes containing a histone antigenic determinant which is exposed and available to interact with antihistone antibodies. With rat liver chromatin the percent of nucleosomes reacting with anti H2B, H3, H4 and H2A sera was about 90, 80, 55 and 40 respectively (21). Thus, the antihistone sera detect some type of heterogeneity in the rat liver nucleosomes which have been spread on electron microscopic grids. The nature of this heterogeneity is not presently known.

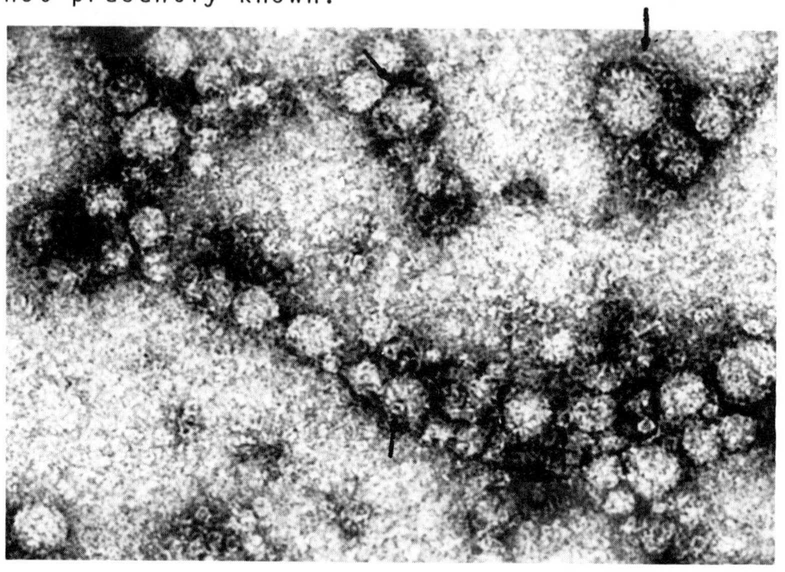

Fig. 5. Electron micrograph of rat liver chromatin reacted with anti-chromatin and with ferritin labelled goat anti-rabbit gamma globulin (20). The arrows point to ferritin molecules. The dark center of the ferritin molecule is 70 A in diameter. (Micrograph taken by Dr. R. Sperling)

The immuno-electron microscopic techniques have been applied to the study of organization of histones in transcriptionally active chromatin (22). McKnight and Miller (23) have reported that in *Drosophila* embryos, nonribosomal transcribed regions contain beaded chromatin which is morphologically similar to nontranscribed regions. The transcribed regions indeed contain nucleosomes which specifically interact with antibodies to histones (22). While these results conclusively show that the transcribed regions contain histones within nucleosomes, they do not yield information on possible structural rearrangements occurring during transcription.

HOMOGENEITY OF HISTONE COMPOSITION IN HELA NUCLEOSOMES

We have recently shown that each nucleosome isolated from HeLa cells contains histone H2B (24). In view of our finding that not all the nucleosomes isolated from rat liver and spread on electron microscopy grids react with antisera to H3, H4, and H2A and in view of other data (25, 26) implying heterogeneity within nucleosomes, we have used the immunosedimentation technique (24) with anti-H3 and anti-H4 sera. Anti-H3 and anti-H4 antibodies were purified by affinity chromatography on Sepharose columns to which the respective histone was covalently linked by the CNBr procedure. The purified antibodies were reacted with nucleosomes in which the DNA was radioactively labeled. The binding of a single IgG molecule to a nucleosome creates a heavy complex which can be separated from unreacted nucleosomes. As seen in Fig. 6 each type of antibody moved all of the nucleosomes from the original position.

The binding of antibodies to nucleosomes is specific since IgG molecules, devoid of antihistone activity do not cause a shift in the sedimentation of the nucleosomes. It is therefore concluded that each nucleosome isolated from HeLa cells contains a full complement of histones.

Thus, the heterogeneity in the interaction of antihistone sera with chromatin particles, isolated from rat liver and spread on electron microscope grids, is not due to heterogeneity in the histone composition of the nucleosomes. It is possible that the heterogeneity detected by immuno-electron microscopy reflects heterogeneity in the exposure of histone antigenic determinants due to complexing with nonhistone proteins or due to heterogeneity in the interaction between the histone core protein and the DNA. Alternately, the heterogeneity could be due to the various orientations of the nucleosome on the grid suggesting asymmetry in the organization of the histone within the nucleosome.

CONCLUSION

Antibodies elicited against purified histone fractions can serve as versatile probes for the study of the organization of histones in isolated chromatin, in metaphase and polytene chromosomes and

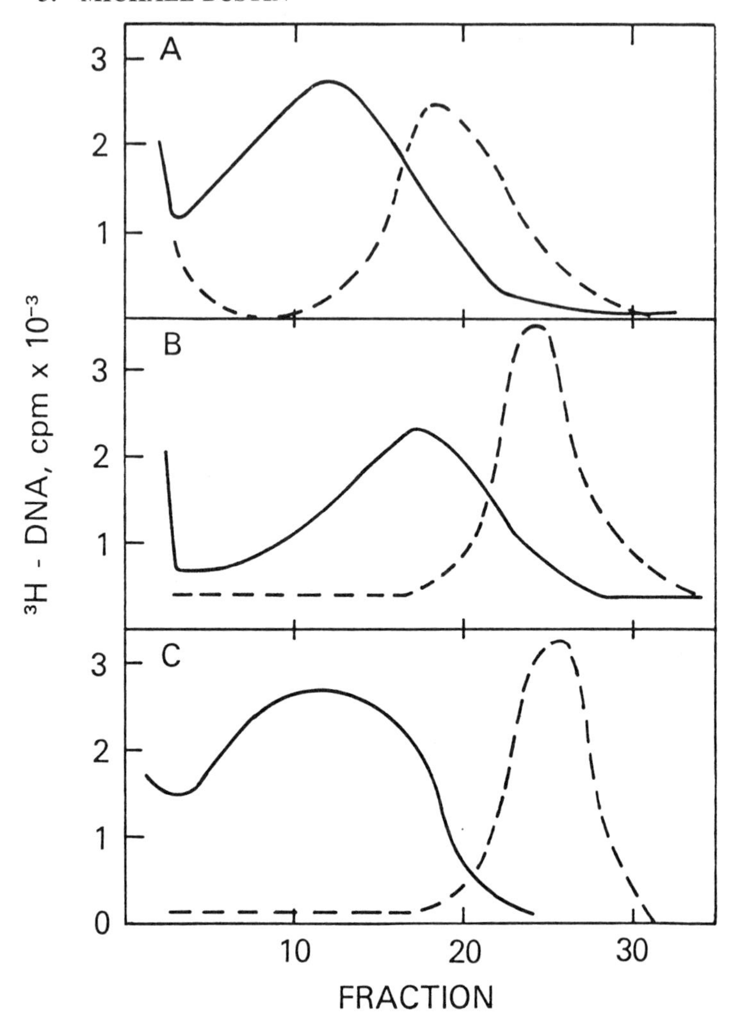

Fig. 6 Interaction of antisera to histones with ^{3}H-thymidine labeled nucleosomes studied by immunosedimentation. Antisera to A) histone H3, B) histone H4, C) histone H2B reacted with nucleosomes and subjected to sedimentation on isokinetic sucrose gradients (for methology see ref. 24).

--- Nucleosomes reacted with control IgG.

___ Nucleosomes reacted with antibody purified by affinity chromatography. Sedimentation from right to left.

in the nucleosomes which build the chromatin fiber. Serological techniques have been used to detect changes in chromosomes during specific puffing and to demonstrate the existence of nucleosomes in transcriptionally active genes. Table 3 lists some of the uses of antihistone sera.

Table 3 - Uses of Antihistone Sera

---

1. Comparative Studies of Histones.
2. Comparative Studies of Chromatin.
3. Organization of Histones in Chromosomes.
4. Organization of Histones in Nucleosomes.
5. Organization of Histones in Repressed, Transcribed, and Replicating Chromatin.

It is clear that the usefulness of serological techniques in the study of chromatin and chromosomes is not limited to antihistone antibodies. The methodology used with these sera can be applied to the study of any chromosomal component against which specific and well characterized antibodies can be elicited.

*Acknowledgement: I am grateful to Dr. R.T. Simpson for hospitality, helpful discussions and critical evaluation of this manuscript. Part of the research was supported by grants from the Israeli Commission for Basic Research and from the U.S.-Israel Binational Science Foundation.

REFERENCES

1. Felsenfeld, G. (1975) Nature 257, 177-178.
2. Stollar, B.D., and Ward, M. (1970) J. Biol. Chem. 245, 1261-1266.
3. Goldblatt, D. and Bustin, M (1975) Biochemistry 14, 1689-1695.
4. Bustin, M. (1976) FEBS Letters 70, 1-10.
5. Elgin, S.C.R., and Weintraub, H. (1975) Ann. Rev. Biochem. 44, 726-774.
6. Bustin, M., Reeder, R.H., and McKnight, S.L., unpublished.
7. Crumpton, M.J. (1974) In The Antigens, Vol. II, M. Sela, ed., Academic Press, p.1, 1974.
8. Chytil, F. and Spelsberg, T.C. (1971) Nature

New Biol. 233, 215-218.
9. Wakabayashi, K., Wang, S., and Hnilica, L.S. (1974) Biochemistry 13, 1027-1032.
10. Bustin, M. and Stollar, B.D. (1973) J. Biol. Chem. 248, 3506-3510.
11. Champion, A.B., Soderberg, K.L., Wilson, A.C. and Ambler, R.P. (1975) J. Mol. Evol. 5, 291-305.
12. Bustin, M., and Cole, R.D. (1968) J. Biol. Chem. 243, 4500-4505.
13. Kinkade, J.M., Jr. (1969) J. Biol. Chem. 244, 3375-3386.
14. Goldblatt, D. and Bustin, M. unpublished.
15. Bustin, M. and Kupfer, H. (1976) Biochem. Biophys. Res. Commun. 68, 718-723.
16. Zick, Y., Goldblatt, D. and Bustin, M. (1975) Biochem. Biophys. Res. Commun. 65, 637-643.
17. Bustin, M., Yamasaki, H., Goldblatt, D., Shani, M., Huberman, E. and Sachs, L. (1976) Exp. Cell Res. 97, 440-444.
18. Kurth, P.D., Moudranakis, E.N., and Bustin, M. unpublished.
19. Plagens, U. and Bustin, M., unpublished.
20. Bustin, M., Goldblatt, D. and Sperling, R.(1976) Cell 7, 297-304.
21. Goldblatt, D., Sperling, R. and Bustin M., unpublished.
22. McKnight, S.L., Bustin, M. and Miller, O.L. Jr. (1977) This symposium.
23. McKnight, S.L. and Miller, O.L. Jr. (1975) J. Cell Biol. 67, 276a.
24. Simpson, R.T. and Bustin, M. (1976) Biochemistry 15, 4305-4312.
25. Cohen, L.H., Newrock, K.M., and Zweidler, A. (1975) Science 190, 994-997.
26. Weintraub, H., and Groudine, M. (1976) Science 193, 848-856.
27. Oliver, D., and Chalkley, R. (1972) Exp. Cell Res. 73, 295-302.
28. Alfageme, C.R., Zweidler, A., Mahowald, A. and Cohen, L.H. (1974) J. Biol. Chem. 249, 3729-3736.
29. Keyl, H.G. (1957) Chromosoma 8, 738-756.

SCANNING TRANSMISSION ELECTRON MICROSCOPY STUDIES ON CHROMATIN ARCHITECTURE

John C. Wooley* and John P. Langmore+

Department of Biophysics and Theoretical Biology, The University of Chicago, Chicago, Illinois 60637

ABSTRACT. Dark field scanning transmission electron microscopy studies on unfixed, unstained chromatin fibers and nucleosomes (unit particles or monomer subunits) suggest that nucleosomes are probably discs, 110Å wide by 55Å high, with two tightly packed DNA loops or turns each about 70 - 80 base pairs long and at a radius between 35 and 45Å. We find nucleosomes generally to be closely spaced but separated by a thin 20 - 30Å wide fiber although other structural classes are also observed on the microscopic grid including large slightly-ordered nucleosomal aggregates.

INTRODUCTION

The electron microscopic observations by Olins and Olins (1) and Woodcock (2) of regularly spaced nucleoprotein beads on a thin DNA-like string, soon substantiated by nuclease-digestion studies (3,4) have catalyzed rapid progress in the understanding of chromatin architecture (see 5 for a review). Using the scanning transmission electron microscope (STEM), we have been attempting to characterize further the fine structure of the chromatin bead or particle and its assembly into the chromatin fiber. The properties of the STEM have allowed us to visualize unstained, unfixed chromatin and DNA and to determine the height of the particle and its molecular weight as well as the amount and apparent location of the DNA (6-9). In this paper, we summarize our qualitative and quantitative measurements on chromatin architecture and relate these to other studies in the field. We use the term nucleosome to refer to the circular nucleoprotein unit visualized by microscopy in keeping with current usage (10); this term is synonomous with the term unit particle used previously (6,7).

*Present address: The Biological Laboratories, Harvard University, Cambridge, MA 02138

+Present address: MRC Laboratory of Molecular Biology, Cambridge, United Kingdom CB22QH

The biochemical methods for preparation of the chromatin samples, the details for specimen preparation for microscopy, and the methods for analysis of the microscopic data have been described previously (7-9). The scanning transmission electron microscopy images were obtained using the microscope developed in the laboratory of Dr. A.V. Crewe (11). Conventional transmission electron microscopy images were obtained using a Hitachi HU11E or a Phillips 200 equipped with a dark field beam tilt.

RESULTS

Chromatin Fiber Architecture

To obtain the best indication possible of the composition of our chromatin samples we chose to record images of all structures found. In most cases, more than 80 to 90% of the material appeared as circular particles of the order of 100Å in diameter periodically and closely spaced along the chromatin fiber, i.e., separated by a short stretch of a thin 20-30Å wide DNA-like fiber. Nonetheless, a few other kinds of structures were found. Samples can be described and assigned qualitatively into a total of five structural classes. Representative examples are shown in Figure 1. The features of the different classes are shown in Table I. The table also gives the average center-to-center spacing between nucleosomes in the different classes. Class 3 is the most striking and is probably equivalent to the structure found for 80-90% of the chromatin by nuclease digestion. Nonetheless, the other classes may be indicative of other states of chromatin in the nucleus. In a few class 4 structures individual nucleosomes can be visualized; some of the structures assigned to this class might be indicative of a higher order packing. Local regions of class 5 structures sometimes exhibit two dimensional order and appear to be partially regular aggregates of nucleosomes. In some of these regions many nucleosomes are stacked in cylindrical structures perpendicular to the microscopic grid (8).

When samples of our chromatin are dried directly from ethanol or amyl acetate or are critical point dried, no nucleosomes are visualized and only knobby or continuous fibrous structures are seen (Figure 2). This result probably arises from a loss of tertiary structure upon exposure to organic solvents since the characteristic low-angle x-ray diffraction pattern is lost when chromatin samples are exposed to solutions containing more than about 40% ethanol (Wooley and Pardon, unpublished observation).

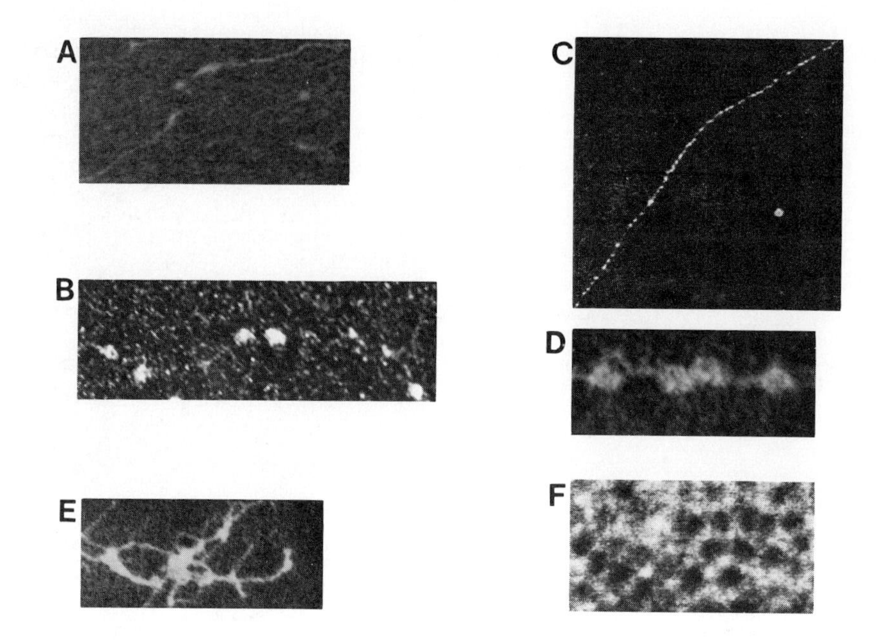

Fig. 1. Examples of dark field images of the structural classes found in our chromatin specimens: A) class 1, full width = 800Å; B) class 2, full width = 2700Å; C) low magnification image of class 3, full width = 1020nm; D) high magnification image of class 3, full width = 1025Å; E) class 4, with some strand aggregation present, full width = 7500Å; F) class 5, full width = 836Å.

Fig. 2. An example of an unstained, unfixed critical point dried chromatin specimen prepared according to Dupraw and Bahr (12). Full width = 1430nm.

TABLE 1

Chromatin Structural Class	Characteristic Features	Center-to-center Spacing of Nucleosomes
1	Uniform, thin (about 20Å wide) fibers, often very long	——
2	Unbranched, extended fibers (20-30Å wide containing circular particles (roughly 100Å wide) widely spaced	400-800Å
3	Closely and regularly spaced round particles (roughly 100A or wider) along a thin (20-30A wide) fiber	200Å
4	Thick, often variable (50-350Å) filaments	probably 110Å or less (55Å?)
5	Aggregated regions, 300Å to 10,000Å in size, with a looping substructure, often possessing class 3 regions near edges	110Å

THE STRUCTURE OF NUCLEOSOMES

We first wish to ascertain the widths of the nucleosomes we visualize in chromatin. (Figure 3 gives some representative images of nucleosomes.) However, in general the true width of a biological object cannot be readily determined by microscopy. Our unstained images, even though they do not present the ambiguities in interpretation associated with stained objects, are difficult to measure since the intensity of the image is not a step function but falls off slowly toward the outer edges of a nucleosome. The class 5 regions (see Figure 1) permit us to circumvent these difficulties since we can readily ascertain the center of a nucleosome in such a structure and thus measure the center-to-center spacings of adjacent nucleosomes. Measuring the separation of adjacent nucleosomes in a number of fields of class 5 structures indicates that the average width of a nucleosome is about 110Å (± 10Å).

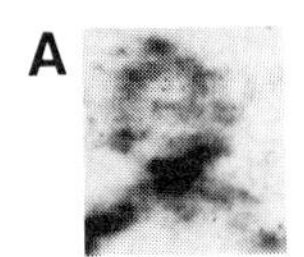

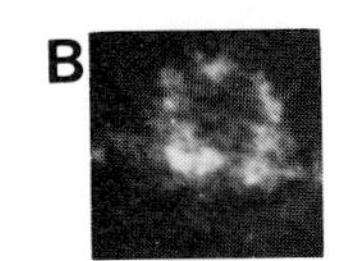

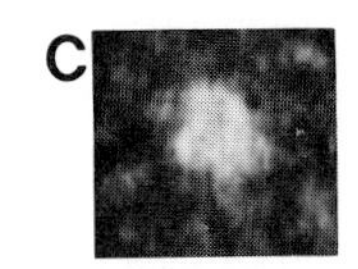

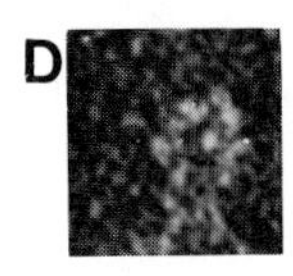

Fig. 3. Examples of images of nucleosomes:
(A) Reverse contrast print of dark field image of unstained, unfixed nucleosome, full width = 160Å;
(B) Dark field image of unstained, unfixed nucleosome exhibiting a looping image, full width = 250Å;
(C) Dark field image of unstained, unfixed isolated monomer subunit purified after nuclease digestion, full width = 320Å;
(D) Dark field image of uranyl actate stained, formaldehyde fixed (13) nucleosome, full width = 260Å.

Next, in order to estimate the height of nucleosomes in solution, we need to measure the height of the dehydrated nucleosomes. The height of a biological object can be calculated from measurement of the inelastic scattering properties (8). The scattered electron currents from an object are converted to a probability for inelastic scattering along lines in the image. Examples of such line scans are shown in Figure 4. The mass thickness at a particular point in a nucleosome is directly proportional to the height of the trace at that point. In contrast to the peak in the center expected for a spherical object, the trace is fairly constant or even decreases in the center. From the inelastic scattering properties (8), we calculated that the nucleosomes are 35 (± 10)Å high (6-7) as visualized on the electron microscope grid. To calculate the height of nucleosomes in solution, we first assume that the internal hydration of chromatin is about 0.4 g H_2O per gram chromatin. This is the value calculated from the DNA and amino acid composition of chromatin and predicted from indirect measurements of the hydration of nucleohistone (8). The internal hydration measured for a similar nucleoprotein object, the 30S ribosomal subunit, is also 0.4 g/g (14). A discussion of the probable hydration has been given previously (8). Secondly, we assume, based on experience with other biological objects prepared for microscopy by our techniques (8) that dehydration causes only a collapse in height and not a change in width. The partial specific volume of the nucleosome is 0.66 (Van Holde, personal communication; 15). The height of a hydrated, cylindrical nucleosome, based on the above assumptions, is then simply the anhydrous height times the ratio of the partial specific volume plus the internal hydration to the partial specific volume. Hence, the nucleosome is probably about 55Å high in solution (± 20Å).

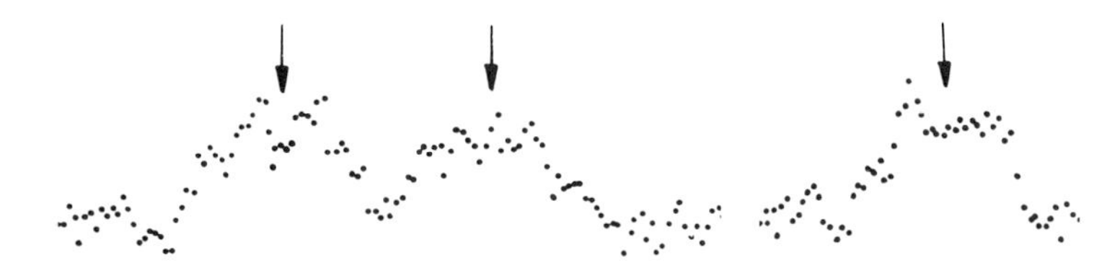

Fig. 4. Representative traces of line scans giving the probability for inelastic scattering across nucleosomes. The height of the trace is directly proportional to the mass thickness (7,8). The arrow indicates the center of the nucleosome.

The images of unstained nucleosomes frequently have a less dense center, leading to a looping appearance (see Figures 1 and 3). The "looping image" and lack of substantial mass in the center led us to suggest that the DNA must be near the periphery of the nucleosome (7). More direct evidence comes from an analysis of the radial mass distribution provided by line scans of nucleosomes like those shown in Figure 4 and by comparing those line scans to line scans of the internucleosomal fiber or of naked DNA fibers. A diagram explaining the experimental methodology is shown in Figure 5. Comparing the probability of electron scattering for DNA to that for a nucleosome, we find that

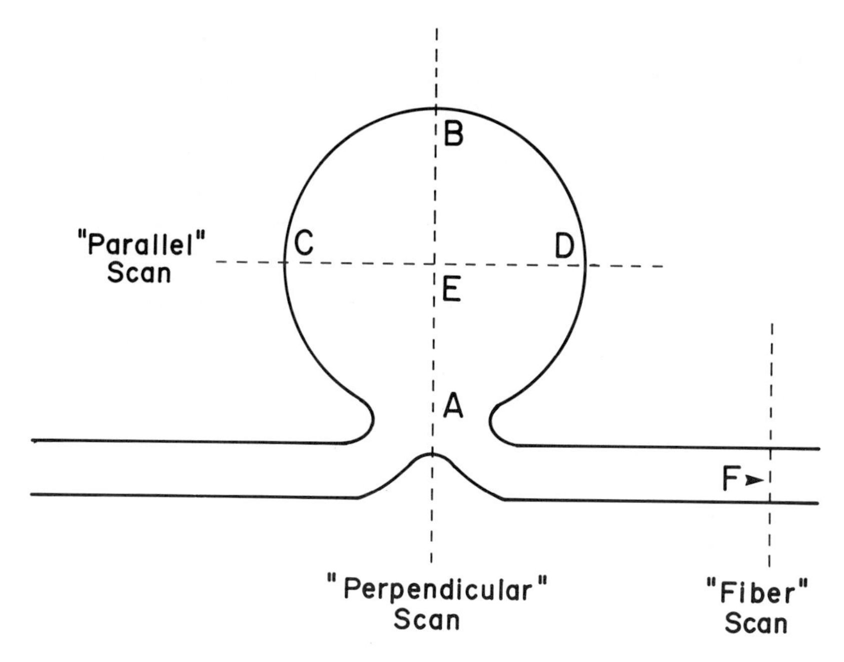

Fig. 5. Diagrammatic representation of the experimental measurements made in attempting to determine the radial distribution of the DNA and also the number of loops of annuli of DNA per nucleosome. The directions chosen for analyzing the probability for scattering are indicated.

the radius of the DNA must be between about 30Å and 50Å since there is too little material at radii less than 30 or greater than 50Å for even one DNA molecule to be present. Assuming that the peak in the electron scattering corresponds to the most probable location of the DNA, the radius of the DNA is between 35 and 45Å.

Many of the nucleosomes in class 3 chromatin fibers appear to possess a common region for entry and exit of the internucleosomal DNA (e.g., see Figure 3). Furthermore, A. Olins (16) has obtained convincing images in preparations of gently-lysed nuclei showing many examples of nucleosomes with the DNA entering and leaving the particle on one side. These observations suggest that there is an integral number of turns in a nucleosome. We wish to ascertain how many turns of DNA there are in a nucleosome. Nuclease digestion studies have found the monomer subunit to contain 140 base pairs of DNA (17,18). We have determined in the STEM the molecular weight of nucleosomes in H1-depleted chromatin and the weight of isolated monomer subunits (purified after nuclease digestion of chromatin) from their electron scattering properties and found the number average molecular weight of both nucleosomes and monomers to be very similar. These data in conjunction with the stoichiometry of our chromatin samples and the center-to-center spacing of nucleosomes along chromatin fibers indicates that the nucleosomes we visualize also contain about 140 base pairs of DNA. (The results will be described in detail elsewhere: 19). Given the limits calculated for the radial distribution of the DNA, one loop or turn of DNA is inadequate to account for the amount of DNA present and there would be too much DNA in three loops or turns. Thus, there are probably roughly two turns of DNA in a nucleosome.

The probability for electron scattering across nucleosomes in scans parallel to or perpendicular to the chromatin fiber axis provides a quantitative measure of the number of loops of DNA per nucleosome (see Figure 5 for a diagram of the experimental measurement). A comparison of such line scans (in the two orthogonal directions) indicates that whereas the ratio obtained from the parallel scan is essentially one, the ratio perpendicular to the fiber axis is about 1.5 (19). This result argues strongly that there must be about two tightly-packed loops of DNA in a nucleosome. Striking micrographs showing nucleosomes, some of which appear to be lying on edge (in puddles of negative stain) and are bipartite or striped in nature (suggestive of two loops of DNA), have been obtained by Varshavsky and Bakayev (20).

DISCUSSION

Our experimental results on the size, shape and internal composition of the nucleosome are summarized by the diagram in Figure 6a. The data suggest that the monomer subunit or nucleosome is a disc, about 110Å wide by 55Å high, with two loops, turns, or annuli, each roughly 70 base pairs of DNA (with associated segments of histones) surrounding the histone core. The apparent central depression or cleft is indicated. In air-dried specimens of chromatin, the nucleosomes almost always lie in one orientation on the electron microscopic grid as shown in Figure 6b. This figure also indicates the distinction between the nucleosomal DNA and the internucleosomal DNA, a distinction also found in studies by *in situ* nuclease digestion (17, 18).

The width we estimate for the nucleosome differs from other published electron microscopic estimates (a compilation

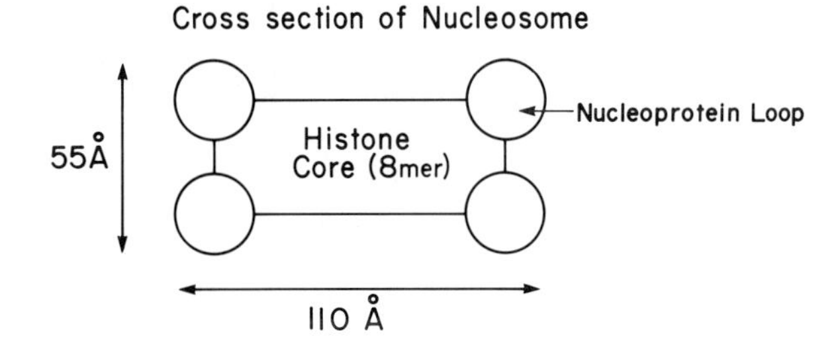

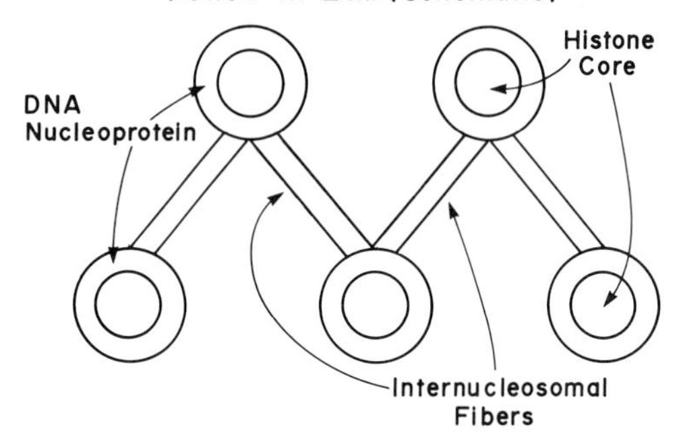

Fig. 6. Summary of our electron microscopic data. The schematic view of H1-containing chromatin represents those fiber sections with nucleosomes (containing about 160 base pairs based on their spacing) in which the DNA appears to enter and exit from about the same region.
For further discussion of our model see Note Added in Proof.

of the measured values is given in 16). The differences between our value and those obtained from studies on stained specimens probably reflect in part the nature of the interaction between the different stains and chromatin. Positive stains reflect the topology of charged groups (e.g. the DNA phosphates), whereas negative stains can penetrate biological structures (21). The situation is further complicated since stain-specimen interactions cannot be completely controlled and a compound can simultaneously both positively and negatively stain the same or adjacent regions of a specimen. For example, it is not clear if the intense central spot frequently observed in uranyl acetate stained nucleosomes (e.g., Figure 3d) arises from positive staining of histone carboxylic acid groups or by negative staining of the central cleft or depression suggested by the line scans across particles

(Figure 4). Thus, negative and positive staining techniques, although revealing useful information on the architecture, do not necessarily reveal the outer diameter of an object. An additional source of the differences in observed widths for nucleosomes might be the difference in spatial orientation of air-dried nucleosomes, largely lying as diagrammed in Figure 6b, and nucleosomes in puddles of negative stain, which presumably could lie in all orientations (compare 20).

We originally suggested that the thin internucleosomal fibers might provide binding sites for non-histone proteins without requiring extensive restructuring of histone-histone, histone-DNA, or nucleosome-nucleosome interactions (7). More rapid recognition of very important DNA control element sequences by sequence-specific regulatory proteins would be facilitated by a mechanism ensuring the presence of such sequences within the internucleosomal DNA. In the absence of a phase relationship between the chromatin repeat unit and DNA sequence, and assuming that important DNA recognition sequences are less than 60 base pairs in length, at least one copy of the DNA sequence for a given control element would be accessible if there are either serially repeated copies of that element or multiple copies periodically spaced out of phase with the 200 base pair chromatin repeat.

ACKNOWLEDGEMENT

We wish to acknowledge the encouragement and support of R.B. Uretz and A.V. Crewe. We thank C. Sahasrabuddhe, B. Shaw and K.E. Van Holde for providing the purified 140 base pair monomer subunits (core particles). The work was supported by The Fay Hunter Cancer Research Fund and a U.S. Public Health Service grant CA 2739 (to R.B.U.), and a U.S. Atomic Energy Commission, Biology Division, grant (to A.V.C.).

REFERENCES

1. *Olins, A.L. and Olins, D.E. (1974) Science 183, 330.*
2. *Woodcock, C.L.F. (1973) J. Cell Biol. 59, 368a.*
3. *Sahasrabaddhe, C. and Van Holde, K.E. (1974) J. Biol. Chem. 249, 152.*
4. *Noll, M. (1974) Nature 251, 249.*
5. *Elgin, S. and Weintraub, H. (1975) Ann. Rev. Biochem. 4.*
6. *Langmore, J.P. and Wooley, J.C. (1974) J. Cell Biol. 63, 135a.*

7. *Langmore, J.P. and Wooley, J.C. (1975) Proc. Natl. Acad. Sci. 72, 2691.*
8. *Langmore, J.P. (1975) Ph.D. Thesis, The University of Chicago.*
9. *Wooley, J.C. (1975) Ph.D. Thesis, The University of Chicago.*
10. *Oudet, P., Gross-Bellard, M. and Chambon, P. (1975) Cell 4, 281.*
11. *Wall, J., Langmore, J.P., Isaacson, M. and Crewe, A.V. (1974) Proc. Natl. Acad. Sci. 71, 1.*
12. *Dupraw, E.J. and Bahr, G.F. (1969) Acta Cytol. 13, 188.*
13. *Brutlag, D., Schlehuber, C. and Bonner, J. (1969) Biochemistry 8, 3214.*
14. *Van Holde, K.E. and Hill, W.E. (1974) in Ribosomes, eds. Nomura, M., Tissieres, A. and Lengyel, P., (Cold Spring Harbor Laboratory, Cold Spring Harbor, N.Y.) 53.*
15. *Olins, A.L., Carlson, R.O., Wright, E.B. and Olins, D.E. (1976) Nuc. Acids Res. 3, 327.*
16. *Olins, A.L., Breillatt, J.P., Carlson, R.D., Senior, M., Wright, E.B. and Olins, D.E. The Molecular Biology of the Mammalian Genetic Apparatus, Elsevier/North Holland, The Netherlands, in press.*
17. *Shaw, B.R., Herman, T.M., Kovacic, R.T., Beaudreau, G.S. and Van Holde, K.E. (1976) Proc. Natl. Acad. Sci. 73, 505.*
18. *Noll, Marcus (1976) Cell 8, 349.*
19. *Langmore, J.P. and Wooley, J.C., manuscript in preparation.*
20. *Varshavsky, A.J. and Bakayev, V.V. (1975) Mol. Biol. Reports 2, 247.*
21. *Klug, A. and Finch, J.T. (1965) J. Mol. Biol. 11, 403.*

NOTE ADDED IN PROOF

The term core particle or core nucleosome seems now more generally used for the 140 base pair chromatin unit [Lohr et al (1977) Proc. Natl. Acad. Sci. 74,79]. Recently, a disc model similar to the structure we suggest for the nucleosome has been found to provide a "best fit" to the wide angle neutron scattering profile observed from solutions of nucleosomes [Richards et al (1977) Cell Biol. Intl. Rpts. 1,107]; these studies further suggest some fine structure for the arrangement of histones within the nucleosome.

SOME PROGRESS IN OUR UNDERSTANDING OF CHROMATIN ORGANIZATION

James Bonner

Division of Biology, California Institute of Technology
Pasadena, California 91125

ABSTRACT. Methods for the separation of the template active expressed portion of the mammalian genome from the template inactive nonexpressed portion of the mammalian genome are described. The two portions have in common a composition which includes DNA histones and some nonhistone proteins, but differ in physical and structural properties. The expressed portion of the genome, which is a subset of the whole genomal sequences, contains DNA in an extended form resembling B-form DNA, melts at a temperature not greatly different from that of pure DNA and is, of course, transcribable by RNA polymerase. The nonexpressed portion of the genome is physically in typical nubody or nucleosomal configuration, the "beads on a string" conformation, which the expressed portion of the genome is not. We pose the question: What chemical transformation alters the nucleosomal configuration of the unexpressed portion of the genome to the non-nucleosomal configuration of the expressed portion of the genome? It is shown below that acetylation of histones results in loosening of the nucleosomal structure presumably by allowing the N-terminal basic ends of the histones to let go of the DNA to which they are bound. The physical alterations of the DNA follow as a consequence. It becomes extended, low melting, and is transcribable by RNA polymerase. Acetylation would appear to be a key process in the activation and expression of genes.

In this paper I review our present understanding and project our future understanding of not only the structure of chromatin but also the way in which nonexpressed genes are converted to expressed genes and vice versa. The study of chromatin is a subject which has not always been popular in the world of modern biology. When I started my studies in this field in 1960 in collaboration with Professor R.C. Huang (10) there was Alfred E. Mirsky and, from time to time, Vincent Allfrey. Alfred Mirsky had published papers about chromosomes since the early 1950's; the chemical properties of chromosomes, but little about their biology. In 1960 Professor Huang and I showed that isolated interphase chromosomes, namely, chromatin, possesses an RNA polymerase activity and the ability to transcribe RNA from chromosomal DNA. In 1962 we reported (9) that DNA complexed with histones in the way that the majority of the DNA of chromatin is complexed with histones is not available for transcription by RNA polymerase. This is true now as it was then. During the intervening years, the last 15 years, the number of papers published about chromatin per year has increased from 1 or 2 to, as of last count in 1976, about 450. Chromosomology is becoming a major, important topic of modern biology. We see this reflected in the interest of the young people who apply for admission to graduate school. About two-thirds of those that apply to our Division of Biology wish to work on the control of gene expression and to do so by the study of interphase chromosomes.

We know that any subject that starts out small and becomes popular and grows big, like the study of chromatin, will ultimately peak and interest in the subject will thereafter gradually diminish. If we plot number of papers published per year vs. time, the net result will be a symmetrical bell-shaped curve. I do not think we are at the high point of number of papers per year on chromosome structure and chromosome biology as yet, perhaps in a year or two. Let's say maybe about 1980 when there are, let us say, a thousand papers published per year. From then on the popularity of chromosomal research will diminish, dying away by the year 2000. In the interval knowledge of chromosome structure, the regulation of gene expression, etc. will have become an integral part of our knowledge of biology; it will be taught in high school or maybe junior high school. Knowledge of chromosomology will not diminish just because fewer papers are published. It will become a part of textbooks instead of papers in Cell, Biochemistry, and Journal of Molecular Biology.

Now to more recent and more specific matters concerning studies on chromatin. We all know that there are five principal histones (6), six if one includes histone 5, that of erythrocytes of birds, etc., but five, let us say, for mammals. Each of these five species of histones is considerably conserved in primary structure over evolution. Here we must pay tribute to Douglas Fambrough and Robert DeLange who determined the primary structure of histone 4 of peas and cows. Histone 4 is the most conserved, histone 3 the next most, histones 2b1 and 2b2 the next most, and histone 1 the least. Even so they are considerably conserved. We all know also that histones can be modified by phosphorylation and by acetylation. Thus, histone 4 can be phosphorylated or not phosphorylated. It can be acetylated 0,1,2 or 3 times: four modes of acetylation and two modes of phosphorylation.

Histone 4 can be eight different proteins and these can be resolved if one uses the techniques of Gordon Dixon, and into at least two if one uses the techniques of Panyim and Chalkley (16), namely, 1 acetylated vs. 0 acetylated. The simplicity of the histones is only skin-deep. However, we know a really vast amount about the biochemistry and chemistry of the histones during the last few years primarily due to the work, on the one hand, of DiAnna and Isenberg (4) on histone-histone interaction, and on the other hand, to Don and Ada Olins (15) with their electron microscopic studies of chromatin structure. We know that interphase chromatin in its simplest form is organized into a "beads on a string" conformation. Each bead is about 100 Å in diameter and each bead consists of eight histones--2 molecules each of histone 2b1, 2b2, histone 3 and histone 4. These four species of histone molecules strongly interact with one another to form a tetramer, as illustrated in Figure 1 based on the work of Irvin Isenberg. Histones 3 and 4 strongly interact. Histone 4 interacts with 2b2, histone 3 with 2b1, and histones 2b1 and 2b2 interact with one another. We believe that a second tetramer interacts asymmetrically with the first, the second tetramer lying approximately above the first. In the octameric subunit the hydrophobic portions of the histone molecules (and this is one of the most important lessons that we have learned from sequencing of histones is that they have a hydrophobic end and a basic amino acid end) interact with one another to form the basic subunit leaving the positively-charged DNA complexing N-terminal sequences waving out in open air ready to combine to DNA. In the presence of DNA they do so. We know from work starting with that of Marcus Noll (14) that the amount of DNA engaged in wrapping around an octamer is 160 base pairs, more or less, and also the amount of DNA between one octamer and the next is 200 base pairs, so there are 40 base pairs left in the linker between the so-called nubodies or nucleosomes. Most students of chromosomology believe that histone 1 is complexed on the linker, namely, the 40 base pairs between nucleosomes. Jack Griffith quite a number of years ago showed (private communication) that histone 1 is spaced about 200 base pairs apart center to center on DNA to which it has been covalently linked and from which all the histones have been removed.

We must remember that at the salt concentration of the nucleus about 0.25 M NaCl, KCl, etc. chromatin of the structure and composition outlined above is insoluble in a supercoiled configuration in which the nucleosomes are packed in some kind of helical configuration close to one another. The DNA in nucleosomes itself is already condensed with a packing ratio of 7:1. The packing ratio of DNA in the nucleus is probably of the order of 50:1 or more.

The DNA that is super packed and supercoiled in a nucleosomal configuration is not accessible to transcription for RNA polymerase. In fact even if RNA polymerase binds to it, the polymerase cannot elongate whatever RNA chain it initiates. We know too, and have known for many years, that only a small fraction of the DNA of interphase chromatin is available for transcription either by homologous RNA polymerase or by exogenous RNA polymerase. This fraction varies from 2% to about 15 or 20% in a few exceptional cases, and averages, I would think, about 10% DNA of interphase chromatin that is available for transcription. What is different about the 10% available transcription from the 90% that is not available for transcription? This is the question to which many have addressed themselves. I will speak first of all about our own method of physically separating the transcribable from the nontranscribable fraction of chromatin.

In order to isolate a transcribable fraction of chromatin it is necessary that we shear the chromatin into fragments such that the average fragment length is less than the transcriptional unit length. Statistical calculations performed by Norman Davidson show that on the average the recoverable fragment length of transcribable chromatin will be one-third of the length of the transcriptional unit in intact chromatin. We have used fragmentation of chromatin by DNase II (13,7), a nuclease which makes double stranded clips in DNA. Others have used sonication, mechanical shear and so forth. Our method has proven, I think, superior in every respect.

We shear interphase chromatin with DNase II for such a time and with such a concentration of enzyme that the shortest fragments produced are about 500 base pairs long. We then add 2 mM magnesium chloride and centrifuge the precipitate away from the supernatant. The precipitate is transcriptionally little active, that is, RNA polymerase finds it an unsuitable substrate for transcription. The supernatant material soluble in 2 mM magnesium chloride is almost as active in supporting transcription by RNA polymerase as is deproteinized DNA itself. These facts indicate that we have separated two importantly different forms of chromatin from one another. In addition, the same separation from transcriptionally active from transcriptionally inactive chromatin may be achieved by the use not of 2 mM Mg/Cl but by use of .15 M sodium chloride, a concentration less than but approximately equivalent to the concentration of ions in the nucleus. It is an interesting fact that the 10% of rat liver chromatin DNA that is liberated as magnesium or sodium chloride soluble material contains all of the RNA polymerase of liver chromatin. It also contains a 10% subset of whole rat DNA single copy sequences and an equally equivalent subset of the repetitive sequences of the rat genome. It would appear, therefore, that DNase II selectively attacks a portion of interphase chromatin which is more readily attacked than is the remainder. Indeed, by the time the magnesium soluble material has been reduced to an average segment length of 500 base pairs, the magnesium chloride insoluble material has an average length of about 5000 base pairs. Other work in our group has shown that the 10% of rat liver chromatin DNA which is releasable as 500 base pair long segments by DNase II is also very rapidly degraded by DNase I, which makes single stranded endonucleated nicks in DNA. DNase I attacks the magnesium chloride soluble fraction of chromatin also 10-100-fold times more rapidly than it does the magnesium chloride insoluble material.

So, we have fractionated chromatin into two fractions. One is a 10% subset of whole genomal rat DNA. The experiments of Gottesfeld _et al_. (7) show that the 10% subset, which is Mg^{++} soluble, constitutes a 10% subset of rat DNA single copy sequences. The DNase II excised sequences hybridize to a high degree, about 60% of total, with whole cell RNA and the level of hybridization achieved herewith is about 10-fold higher than that achieved for the hybridization of whole cell RNA to the magnesium insoluble chromosomal DNA. We have reason to believe, therefore, that our fractionation has separated 10% of chromosomal DNA sequences which are expressed and have left behind a population of sequences (90% of all) which are impoverished in expressed sequences.

A great deal of further work is being done on this fractionation procedure and this interpretation. For example, my colleague, Bruce Wallace, has gone on to show that if we study the chromatin of induced (by DMSO) Friend cells, the globin gene sequence is contained in the template active fraction, that is, the magnesium soluble portion of Friend cell chromatin is excised by DNase II (19). What is true for whole cell RNA hybridization is true for the hybridization of messenger RNA, actually the cDNA to the messenger, of a particular gene. Bruce Wallace has shown this to be true for one other group of genes also.

Now we come to the problem of what is the difference in chemical composition, physical conformation, etc. between expressed DNA, that is, DNA available for transcription, on the one hand, and nonexpressed DNA not available for transcription, on the other hand? There has been a great deal of confusion on this subject raised in recent years (the last 2 or 3) by virtue of the fact that most people who have studied the structure of chromatin have worked with whole chromatin, that is a mixture of 90% transcriptionally inactive material and 10% or 5% transcriptionally active. There has been a tendency, I think, to assume that what is true of the 90% is true of the 100% and that all DNA of chromatin is organized as nucleosomes. We, therefore, go back to the investigations of the nontranscribed and the transcribed fractions of chromatin as defined by DNase II cleavage and separation of the two fractions by 2 mM magnesium chloride precipitation of the template inactive material. The template inactive material contains histones and DNA in a ratio of about 1-1.2 to 1. The template active contains histones of DNA in a ratio which varies according to investigator from about 0.9:1 to about 1:1--slightly less than the ratio found for template inactive chromatin. This difference may be an artifact of degradation or analysis, as will be discussed below. In any case, the histones are still there, at least to some extent, in the transcribed 10% of the DNA of interphase chromosomes. The histone to DNA ratio may be less in the transcribed portion, but only a little less. So far as the nonhistone chromosomal proteins are concerned, it is agreed on all sides that the template active 10% of chromatin has a higher concentration of nonhistone chromosomal proteins than does the template inactive. The status of the nonhistone chromosomal proteins is confused and made more so by virtue of the fact that a very small number, a dozen or so, major nonhistone chromosomal proteins dominate the picture. Myosin, actin, tropomyosin, alpha and beta tubulin and the HnRNA packaging proteins make up about half of the total mass of the nonhistone chromosomal proteins and dominate every SDS gel electrophoresis picture of the nonhistone chromosomal proteins with the reservation that in many cases nonhistone chromosomal protein degradation by proteases has been extensive, that is, protein degradation by the chromosomal protease has not been controlled (5). The nonhistone chromosomal protease, which has been purified to homogeneity by Ming Ta Chong (2), is a dimer of molecular weight 200,000; it is a serine protease and attacks histones extremely rapidly and nonhistones also but at a slower rate. Amongst the nonhistones its favorite target is myosin which it converts into degradation products of 24,000, 65,000 and other sizes, peptides which are major chromosomal proteins in many of the SDS electrophoresis profiles which have been published in the past. Anyway, degradation by endogenous chromosomal protease must be controlled if one is to understand the ratio of histones to DNA in chromatin fractions and if one is to understand the composition of nonhistone chromosomal proteins in particular chromatin fractions. In any case, study of the nonhistone chromosomal proteins has not brought us any clear understanding from a chemical point of view of the difference between template active, expressed, chromatin and template inactive, nonexpressed, chromatin. There may very well be differences in protein populations between these two fractions but they are not evident in SDS gels of whole chromosomal protein mixtures. There is a small fraction of nonhistone chromosomal proteins, about 2% of the total mass, which binds sequence specifically to homologous DNA (17). These bind only to repetitive sequences of such DNA. Whether these sequence specific homologous DNA binding proteins bind only to the template active DNA, or whether they bind equally to all portions of the genome, is not as yet known. It is important to find out, but we just do not know yet. The experiments are hard and the logistics are difficult.

Now let us go to a further topic. The template active portion of chromatin contains histones in a histone to DNA ratio not greatly different from that of the nonexpressed portion, but even any difference may be an artifact of proteolysis, to which we shall return below. There are differences, however, in the structure and conformation of the DNA between template active and template inactive chromatin. Gottesfeld et al.(8) have shown that the template active portion of chromatin exhibits the CD spectrum, the quinacrine binding and fluorescence characteristics of pure deproteinized DNA. The CD spectrum and quinacrine binding and fluorescence characteristics of template inactive chromatin are entirely different. The CD spectrum is that of a very different kind of DNA--more like C DNA, not B-form DNA. In addition, as was shown by Tuan and Bonner many years ago (18) the DNA of chromatin of which the majority is in non-expressed portion shows hyperchromicity relative to protein free DNA. It is distorted in some fashion. I would say today that it would appear the DNA in winding around the nubody assumes a very contorted figuration in which base pair stacking is partially eliminated by partial sliding of the base pairs out from one another. This could well be the result of the distortion resulting from the winding of DNA around the histone octamer.

The packing ratio of the DNA in chromatin is the ratio of the length of DNA in a structure to the length of the structure. The packing ratio is different for the DNA of the nonexpressed portion of the genome from that of the expressed portion of the genome. In the nucleosomes the packing ratio of DNA is approximately 7, that is, the DNA's contour length is shortened 7 times by being wrapped around the histone octamer. In expressed DNA the contour ratio is 1 or nearly 1, that is, the DNA appears to be fully extended. This is clearly so for actively transcribed ribosomal genes. And although it is more difficult to measure exactly the packing ratio of expressed nonribosomal genes DNA, there is general agreement that it is close to 1. Apparently, then, in the expressed genes the DNA unwinds from the nucleosomes. The histones remain bound to the DNA in some form but the DNA becomes more extended.

Just as DNase II selectively attacks the expressed portion of the genome so do other nucleases selectively attack this same material. Thus, DNase I attacks expressed portions of the genome much more rapidly than it attacks non-expressed portions as shown by the work of Lacy and Axel (11) and Weintraub and Groudine (20) and others. These workers have shown, using cDNA to various messenger RNA as probes, that the DNA of chromatin which has been attacked with DNase I loses the ability to hybridize to the cDNA of a message expressed in that chromatin. We have found, too, that when chromatin is attacked with DNase I and the remaining DNA deproteinized and melted it has lost its ability to hybridize to the sequences contained in the template active portion of the genome which are excised with DNase II.

We have, therefore, a wide spectrum of differences and properties between expressed and nonexpressed portions of the genome. These differences include: the lower melting temperature of template active DNA; the C DNA spectrum, which in template active DNA resembles that of B-form DNA while in template inactive it does not; differences in packing ratio; differences in susceptibility to nuclease attack; and so on. There are also similarities in that both template active and template inactive chromatins are organized in segments about 200 base pairs long. Their existence is revealed by chromatography of nuclease limit digests of the DNA of each of the two fractions. Both fractions also contain histones. Template active chromatin, so far as our preliminary electron microscopic investigations go, does not appear to contain nucleosomes; neither are they evidenced in the pictures of Hamkalo and Miller. Template inactive chromatin appears to be always in the nucleosomal structure.

We may now ask the question: What chemical, physical, biological reasons or mechanisms are involved in maintenance of the difference in properties between expressed and non-expressed portions of the genome? Many individuals have thought about this subject, but so far as I am aware, no one has come up with a good answer to the question. We have attacked this subject in a new and different way during the past few months. It has been shown by Keiji Marushige (12) that when chromatin is acetylated with acetic anhydride, the latter in extremely low concentration, as 0.7 mM, histones, although acetylated, do not fall off the DNA. At the same time, however, the template activity of the chromatin, in the case of calf thymus, increases about 5-fold as a result of such acetylation (from about 15% of the template activity of deporteinized DNA to about 75% of same). Under these conditions histone 1 is modified at an average of 8 sites per molecule, histone II about 2 1/2 sites per molecule, histone 3,1.5 sites per molecule and histone 4,1 site per molecule. The acetylation is thus not very dramatic but it causes a dramatic change in template activity. We (Tom Sargent, Bruce Wallace and myself) have expanded on these findings. Acetylation of chromatin or of nucleosomes from inactive chromatin causes the melting temperature of their DNA to be dramatically lowed. It causes their sedimentation coefficient to be decreased from the 11.5 S characteristic of nucleosomes to 8 S with 0.7 mM acetic anhydride and to about 6 S with 1.4 mM acetic anhydride. Acetylation with acetic anhydride causes DNA of chromatin or of nucleosomes to become available for rapid attack by DNase I or DNase II. By all of these measures acetylation of chromatin transforms the treated material from the template inactive form to a something with the properties of template active chromatin. Since we know that the acetylation is mainly in the ε-amino groups of the lysines, we suppose that the acetylation is mostly on the lysines of the N-terminal portion of histones II, 3 and 4, and that the acetylation causes the N-terminal portions of these histones to become less tightly bound to the DNA. This supposition is supported by the findings of Marushige (12), which showed that acetylation reduces the ionic strength necessary to remove a histone from DNA. For example, histone 1 is removed from DNA to template inactive chromatin at a concentration of 0.5 to 0.6 M NaCl. After acetylation with .7 mM acetic anhydride, histone 1 is removed to a concentration of .15 M NaCl. Similar findings have been described for all of the other histones. They are quite evidently less tightly bound to DNA after acetylation.

That acetylation may really and truly be an agent causing the conversion from the template inactive structure to the template active structure is indicated by preliminary experiments of Bruce Wallace and Tom Sargent. These workers have incubated tissue cultured mammalian cells with labeled acetate. Chromatin was then harvested, fractionated by DNase II fractionation method and the histones separated and their acetyl content measured. The histones of the template active portion of the chromosome had about 4 times the specific activity in acetyl groups of the template inactive histones.

Let us suppose that acetylation is indeed the agency which transforms chromatin from the active to the inactive state. This is not a new suggestion. It was suggested as early as 1965 by Vincent Allfrey (1). Today we have more biochemical evidence that Vincent Allfrey was right about acetylation. That we can say more about the role of acetylation today than we could in 1964 is called progress.

But anyway, let us suppose that acetylation is the mechanism for the conversion about which we are speaking. This only staves off our problem, our ultimate problem, by one step. The next question is how does the histone acetylase, known to be a chromosomal protein (Candido and others), know which histones to acetylate? The problem is even worst than it appears at first glance. Expressed genes, or expressed portions of the genome, consist of single copy sequences interspersed between repeated sequences. It is widely believed that the repeated sequences constitute control elements and there is some basis for this belief. Thus, as described earlier, the single copy sequences which are turned on in the rat liver genome are all attended by repeated sequences and the repeated sequences constitute a subset of the families of repeated sequences of the rat genome. Let us suppose that the agent which is responsible for saying this set of genes shall be turned on (the agent being either a sequence specific DNA binding protein or an RNA) the turning on agent binds to a repeated sequence. Let us suppose that the turning on agent is an agent that says to histone acetylase, acetylate the histones of the adjacent nubody on my right. But the problem is how can this phenomenon translate itself along the chromatin from one nucleosome to the next? The repetitive sequences are interspersed in the rat genome an average of about 2000 base pairs apart. Is the transformation by acetylation somehow cooperative or contagious so that it can spread down the single copy gene? We have already seen above that the average stretch of template active DNA is about 10,000 base pairs long. This whole stretch, which codes principally for HnRNA, must have its histones acetylated simultaneously. It is not enough to imagine that acetylation turns on the end of the gene and moves a way down the gene leaving the chromatin restored to its nucleosomal configuration after transcription. The entire stretch must be transformed simultaneously and in a semi-stable way. So, the translational properties of the effector which turns on acetylation are of great interest. We do not know what it is yet, but we are working hard to try and find out.

Finally, let us go to the effectors which determine what stretches of DNA are in fact turned on. There is a considerable body of evidence by a large number of investigators that some component of the nonhistone chromosomal proteins plays the role of inducer of template activity. These experiments have all been done by reconstitution, that is, dissolving the chromosomal components in 2 M salt, 5 M urea, adding heterologous nonhistone chromosomal proteins to the mixture, dialyzing the salt away and dialyzing the urea away, transcribing the reconstituted chromatin and finding, generally with cDNA probes, that genes that are turned on in the tissue that donated the heterologous nonhistone chromosomal proteins are turned on in the reconstituted chromatin. No one has isolated a protein which has the properties attributed to the mixture. We have already reported that about 2% of the nonhistone chromosomal protein fraction consists of proteins which bind sequence specifically only to homologous DNA. We have, therefore, purified a fraction which could be a candidate. Interestingly enough, this subset of chromosomal proteins binds only to repeated sequences. The homologous DNA sequence specific binding proteins of rat liver chromatin binds specifically to about 20% of the repeated sequences and preliminary evidence indicates the 20% is 20% of the families of repeated sequences. This again, is a property that would be expected of a turning on agent. Even when we are able to prepare large amounts of homologous DNA sequence specific binding proteins and even though we discover how to reconstitute them to DNA, then add histones or histone octamers, as well as chromosomal histone acetylase and acetyl CoA, etc., etc.; even though we are possessed of logistic ability to perform such a reconstitution, we would still be faced with a mystery of how does the sequence specific homologous DNA binding proteins cause the acetylase to acetylate only the histones of the nucleosomes which occupy the sequences which are to be turned on. But like all biological problems, I think this too can be solved. It is just a gigantic problem in enzymology to be solved by test tubization; study of the problem at the *in vitro* test tube level.

ACKNOWLEDGEMENTS

The author wishes to acknowledge the support and counsel of his colleagues, particularly Dr. Bruce Wallace, Dr. Angeline Douvas, Dr. J.R. Wu, Dr. William Pearson, Michael Savage, Tom Sargent, Jim Posakony, Tony Bakke, and Ingelore Bonner. Report of work supported in part by U.S. Public Health Service Grant GM13762.

REFERENCES

1. Allfrey, V.G., Faulkner, R. and Mirsky, A.E. (1964a) Proc. Nat. Acad. Sci. 51, 786.
2. Chong, M.T., Garrard, W. and Bonner, J. (1974) Biochemistry 13, 5128.
3. DeLange, R.J., Smith, E.L., Fambrough, D.M. and Bonner, J. (1968) Proc. Nat. Acad. Sci. 61, 7-8 (Abstract).
4. DiAnna, J. and Chalkley, R. (1969) Biochemistry 8, 3972.
5. Douvas, A.S., Harrington, C.A. and Bonner, J. (1975) Proc. Nat. Acad. Sci. 72, 3902.
6. Fambrough, D.M. and Bonner, J. (1966) Biochemistry 5, 2563.
7. Gottesfeld, J.M., Bonner, J., Radda, G.K. and Walker, I.O. (1974) Biochemistry 13, 2937.
8. Gottesfeld, J.M., Garrard, W.R., Bagi, G., Wilson, R.F. and Bonner, J. (1974) Proc. Nat. Acad. Sci. 71, 2193.
9. Huang, R.C. and Bonner, J. (1962) Proc. Nat. Acad. Sci. 48, 1216.
10. Huang, R.C., Maheshwari, N. and Bonner, J. (1960) Biochem. Biophys. Res. Comm. 3, 689.
11. Lacy, E. and Axel, R. (1975) Proc. Nat. Acad. Sci. 72, 3978.
12. Marushige, K. (1976) Proc. Nat. Acad. Sci. 73, 3937.
13. Marushige, K. and Bonner, J. (1971) Proc. Nat. Acad. Sci. 68, 2941.
14. Noll, M. (1974) Nature 251, 249.
15. Olins, A. and Olins, D. (1974) Science 181, 330.
16. Panyim, S. and Chalkley, R. (1969) Biochemistry 8, 3972.
17. Sevall, J.S., Cockburn, A., Savage, M. and Bonner, J. (1975) Biochemistry 14, 782.
18. Tuan, D. and Bonner, J. (1968) In: "Structural Chemistry and Molecular Biology", Rich, A. and Davidson, N. Eds. W.H. Freeman & Co., San Francisco pp. 412-421.
19. Wallace, R.B., Dube, S. and Bonner, J. (1977) Science, in press.
20. Weintraub, H. and Groudine, M. (1976) Science 193, 848.

CHROMOSOME BANDING AND CHROMOSOMAL PROTEINS

David E. Comings, M.D.

Department of Medical Genetics
City of Hope National Medical Center
Duarte, California 91010

Supported by NIH Grants GM 15886 and GM 23199

ABSTRACT. Chromomeres observed in pachytene cells represent a basic subdivision of the chromosomes in higher organisms. There is an excellent correlation between pachytene chromomeres and the C- and G-bands of mitotic chromosomes. G-band techniques bring out the latent chromomere pattern which is obscured by the high degree of condensation of metaphase chromosomes. The major factors in G-banding appear to be (a) some rearrangement of the chromatin fibers, and (b) the inhibition by non-histone proteins (NHP) of the binding of Giemsa dyes to the DNA of R-band regions. Studies of isolated heterochromatin of *Drosophila virilis* and mouse suggest C-band, satellite rich, heterochromatin is predominately a histone-DNA complex with little NHP. The AT-richness of Q-bands and the GC-richness of R-bands are the major factors in Q- and R-banding. EM studies show that chromatin is attached to the nuclear matrix at multiple sites throughout the nucleus, not just at the nuclear membrane. DNA binding studies indicate the nuclear matrix has preference for the dT-strand of AT-rich DNA. This DNA-matrix interaction may play a role in organizing the chromosome into chromomeres.

The chromosome banding techniques have enormously increased the power and sensitivity of chromosome analysis. They have allowed detection of subtle chromosome aberrations which are of great importance in studies of malformation syndromes and malignancies, and have been responsible for the rapid expansion of human chromosome mapping with somatic cell genetics. While the mechanisms of chromosome banding are not entirely understood, enough studies have been completed to provide a reasonably clear idea of how many of them work. This is an examination of some of these mechanisms.

This paper was also presented at Aspecto de la Organizacion y Function Chromosomica, Montevideo, Uruguay, Feb. 2-5, 1977.

CHROMOMERES = C-, G- AND Q-BANDS

After many years during which cytogeneticists simply examined uniformly stained chromosomes, it was impressive to observe how rapidly the numerous banding techniques were developed once the initial conceptual barrier that banding was possible was broken. Although some early suggestions of banding had previously been reported (1, 2), the true barrier was broken by reports of Caspersson and colleagues (3, 4) by banding chromosomes with quinacrine. The observation that Q-banding could be mimicked by various salts (5, 6) or enzymes (7, 8, 9) followed by Giemsa staining led to the question of whether the banding procedures were inducing bands or simply allowing the ones already present to be visualized.

A number of clues that the bands were already present came from studies showing that they could be visualized by various gentle, non-destructive procedures such as whole mount electron microscopy (10, 11, 12), staining with dilute Giemsa (13), and examining the chromosomes by UV light or phase contrast microscopy (14). The validity of these observations was confirmed by the finding that there was a precise correlation between the pattern of pachytene chromomeres and G-banded metaphase chromosomes (15, 16). The condensed chromatin of the chromomeres probably corresponds to regions of condensed chromatin in interphase nuclei (17).

These observations indicate that the chromomeres represent the major structural subdivision of the chromosome and this structure persists in the pachytene chromosomes, metaphase chromosomes and interphase nuclei.

When elongated prometaphase chromosomes are examined the chromomeres can be seen to be composed of 2-5 smaller chromomeres (18). It is most likely that if it were possible to polytenize mammalian chromosomes the ultimate chromomere pattern might resemble that seen in *Drosophila* salivary gland chromosomes.

It has recently been suggested that the chromomeres are not a static defined unit of the chromosome since their number varies under different conditions (19). The studies of extended chromosomes suggest this variation is simply the result of different degrees of chromosome condensation rather than a true variation in the position or number of chromomeres.

Q-BANDING AND BASE COMPOSITION

Caspersson's original rationale for using quinacrine mustard was that it would be a fluorochrome sensitive to the

GC-rich regions of the chromosome. However, quinacrine without the mustard worked equally well and Weisblum and de Haseth (20, 21) and Ellison and Barr (22) showed that quinacrine was responding to the AT-rich regions of the chromosome. In *in vitro* studies very AT-rich DNA enhanced fluorescence and GC-rich DNA quenched fluorescence (20, 21, 23, 24, 25, 26, 27).

The observation that chromatin, chromatin subfractions, and various protein DNA complexes show a fluorescence intensity different from that compared to the purified DNA (28, 27) raised the question of how much a role proteins played in Q-banding. An additional disturbing feature was the observation that some AT-rich satellite fractions showed the same degree of fluorescence *in vitro* as did main band DNA (27, 29, 30). This raised the question of whether the base composition along the chromosome alone was adequate to account for Q-banding.

In an effort to answer these questions we determined the ratio of quinacrine fluorescence of DNA of varying base composition to DNA of 40% base composition (31). This indicated it required a change in base content of only 6.3% to give a 50% change in quinacrine fluorescence. This agreed with the change of quinacrine fluorescence with the 4th power of the base composition (23). Biochemical evidence (31) indicates there is a difference of up to 10% in the base composition of early replicating R-bands versus late replicating G- or Q-bands (32, 33, 34).

Studies in the change in relative fluorescence along Q-banded chromosomes suggested that these changes in base composition are adequate to account for most Q-banding. However, in some of the most intensely fluorescing heterochromatic regions a deficiency in nonhistone proteins (see below) may allow increased dye binding to account for a degree of brightness which seems greater than can be accounted for on base composition alone.

R-BANDING AND BASE COMPOSITION

The one feature that all R-banding techniques have in common is the exposure to conditions which are capable of denaturing DNA (35). Most utilize heat at about 89° C (36), some utilize alkali (37, 38). Chromatin *in situ* shows a broad range of thermal denaturation (39, 40). Thus, differential melting of relatively AT-rich G-band regions with no melting of the GC-rich R-band regions is possible. The AT-rich regions stain red with acridine orange while the GC-rich regions stain green. Regions of the chromosome containing

AT-rich or GC-rich satellite DNA behave as expected (41, 33, 42, 43). The studies of Schreck et al (44), using anti-cytosine antibodies, also give R-banding of chromosomes.

A recent study by van de Sande, Lin and Jorgeson (45) has provided further strong evidence that R-banding primarily depends upon base composition. They showed that olivomycin fluorescence is enhanced in the presence of GC-rich DNA and unaffected by poly d(A-T). This compound gave R-banding of human chromosomes.

The one aspect of R-banding that is difficult to understand is the fact that Giemsa can also be used to stain R-bands in appropriately treated chromosomes (46) despite the fact that it shows little preferential staining for double versus single stranded DNA *in vitro* (47). Whether the temperature induces the preferential masking of G-band DNA by denaturation of proteins, or whether the thiazin dyes bind more readily to double stranded DNA than the *in vitro* studies indicate, remains to be determined.

Correlation of the R-bands with human aneuploid states is consistent with an enrichment of euchromatin in R-bands (48).

C-BANDING AND DIFFERENTIAL CHROMOSOME CONDENSATION

C-banding seems to be the easiest of the banding techniques to understand. When chromosomes are exposed to sodium hydroxide and warm salts (49), DNA is preferentially extracted from the non C-band regions (37, 50). Subsequent exposure to Giemsa leaves the C-band areas intensely stained. Since pretreatment of the chromosomes with hydrochloric acid does not affect C-banding, we initially felt that the C-band DNA was protected by tightly bound nonhistone proteins (37). However, in recent studies of *Drosophila virilis* and mouse heterochromatin isolated in isotonic buffers, we have found no nonhistone proteins unique to constitutive heterochromatin and found constitutive heterochromatin to be markedly deficient in all types of nonhistone proteins (51). This finding agrees well with alkaline acid fast staining of condensed and decondensed chromatin in individual nuclei (51). Constitutive heterochromatin appears to be largely a DNA histone complex. As such its degree of condensation even after acid extraction may be sufficiently great to provide enough protection to allow C-banding.

N-BANDING AND NUCLEOLAR PROTEINS

A number of procedures have been described which specifically stain the nucleolus organizer regions of chromosomes.

These include N-banding (52, 53, 54, 55) and silver staining (56). The observation that this procedure usually does not stain all of the sites containing ribosomal DNA indicates that only genetically active ribosomal DNA regions give a positive stain. This assumption is confirmed by the finding that with mouse human hybrids none of the human chromosomes are stained (55). This demonstrates in a mammalian hybrid the phenomenon of differential amphiplasty first described by Navashin in plant hybrids in 1928. Navashin noted (57) that in different hybrids there was suppression of nucleolus organizers in the chromosomes of one paternal set by the chromosomes of another paternal set. It was possible to arrange a hierarchical order of the ability of given plants to suppress nucleolar organizers of other species.

Electron microscopy studies show a distinct nucleolar matrix consisting of 50 - 70 Å protein fibers associated with nucleolus organizing regions of metaphase chromosomes (58). It is presumably the protein of this complex which is protecting the DNA of this region to allow it to stain well in the N-banding technique and which is sensitive to silver stain for the silver staining techniques. These protein fibers are probably the same as seen in whole mount electron microscopy of the nuclear matrix (see below).

G-BANDING

The mechanism of G-banding is the least understood of the banding techniques. Basically the procedure allows a chromomere pattern already present and obscured by the degree of condensation of metaphase chromosomes to become visible. Whole mount electron microscopy studies strongly suggest a role of chromatin rearrangement in G-banding (59, 60). These studies show that the untreated chromosomes are uniformly dense to the electron beam while brief treatment with trypsin results in a visible banding pattern. Since no stain is used in these procedures and the electron beam is sensitive only to relative density of the chromatin, the trypsin must have rearranged some of the chromatin out of the R-band region and into the G-band regions.

The following facts indicate the dye itself is also playing a role:

1. Giemsa staining significantly enhances the banding pattern compared to Feulgen (37).

2. Dyes like thianin which show little metachromasia and stacking give poor banding while dyes like methylene blue which show a lot of metachromasia and stacking produce good banding (61, 47, 62).

Scanning EM (63) and Nomarski optics (64, 65) show a large build up of dye chromatin complex in the C- and G-bands and with little build up in the inner bands. Since Giemsa dyes interact with chromatin primarily by side stacking onto the DNA, this implies there is something inhibiting the binding and side stacking of the dye in the interband region. In an attempt to examine this phenomenon in further detail we examined the binding of methylene blue to DNA and various chromatin preparations by equilibrium dialysis (47). The ratio of moles of dye bound to moles of DNA nucleotides was 1.0 for free DNA, 0.5 for chromatin, and 0.85 for acetic acid methanol fixed chromatin. However, when the fixed chromatin was treated with 2 x SSC at 60° for 3 hours, as in the ASG technique, the ratio dropped to 0.5. This indicated that after ASG treatment fewer DNA sites were available for binding and side stacking of dyes. The one feature that all the G-banding techniques have in common is their ability to denature protein. We interpret these results to indicate that before treatment, while the proteins are still in a reasonably native state, they cover lesser DNA and the dye has uniform access to the DNA on the chromosomes. After denaturation the proteins are capable of more efficient covering the DNA in the R-band regions and these regions will stain more poorly. Since removal of the histones with acid does not affect the G-banding (66) and since the histones are uniformly distributed all along the chromatin while the non-histone proteins are enriched in euchromatic regions, the proteins involved are presumably the nonhistone proteins (67).

NUCLEAR MATRIX AND THE ORGANIZATION OF INTERPHASE CHROMATIN

The nucleus is generally considered to consist of a nucleolus and a mass of chromatin surrounded by a nuclear membrane to which some of the chromatin is attached. Recent studies have shown, however, that if all of the chromatin is removed from the nucleus by a combination of high salt and DNase treatment, by phase microscopy the nucleus still appears much as it did originally (68, 69, 70). This is due to the presence of a structure termed the nuclear matrix (68). By electron microscopy it is seen to be composed of a nuclear pore complex (71), a nucleolar matrix, and an intra-nuclear matrix (69, 70). By whole mount electron microscopy the matrix is seen to have a fibrillar structure composed of 20 - 30 Å fibers which aggregate to form still larger fibers (70). By SDS gel electrophoresis the matrix is composed primarily of three polypeptides of 65,000, 67,000 and 68,000 daltons. When 90% of the chromatin is removed from the nucleus by treatment with 2 M NaCl, the remaining chromatin is seen to be attached at many sites to all three portions of the matrix. Thus, in addition to being attached to the nuclear membrane, chromatin is also attached to the nucleolar and intranuclear matrix. When these nuclei are examined by whole mount electron microscopy the DNA is seen to be attached as multiple supercoiled loops to the nuclear matrix (70). The presence of a supercoiled intranuclear DNA has recently been described in mouse (72), human (73) and *Drosophila* (74). Ide et al (72) presented evidence that the supercoiled structures were attached to a nonhistone protein of 68,000 daltons by SDS gel electrophoresis. This is presumably nuclear matrix. Amino acid analysis of the nuclear matrix protein shows it is very similar in composition to human red cell membrane proteins (Comings, unpublished). Berezney and Coffey (75) have suggested that recently replicated DNA is preferentially associated with the nuclear matrix. Although it is clear now that the replication fork is not associated with the inner nuclear membrane in eukaryotes, it may be associated with the membrane-like intranuclear matrix protein. Studies with fluorescent tagged antimatrix antibodies show it is distributed throughout the cell, not just in the nucleus (Comings and Kovacs, unpublished). This is consistent with the gel electrophoresis studies which show similar polypeptides in the cytoplasm (76).

Because of its intimate association with chromatin, we have recently examined the DNA binding properties of nuclear matrix protein (77). The matrix protein has a high affinity for DNA, preferentially binds to AT-rich DNA, to single stranded DNA, and shows a marked preference in binding to poly dT compared to poly dA, dG or dC. These properties suggest that the binding of AT-rich sequences in the G-bands may be responsible for the chromomere organization of the chromosomes. Why there is such a non-random arrangement of GC- and AT-rich DNA along the chromosome with preferential clustering of this DNA into R-bands and G-bands respectively, remains to be elucidated.

REFERENCES

1. Levan, A. (1946) *Hereditas* 32, 449-468.
2. Stubblefield, E.P. (1964) *Cytogenetics of Cells in Culture* (Academic Press) 3, 223-248.
3. Caspersson, T., Farber, S., Foley, G.E., Kudynowski, J., Modest, E.F., Simonsson, E., Wagh, U. and Zech, L.

(1968) *Exp. Cell Res.* 49, 219-222.
4. Caspersson, T., Zech, L., Modest, E.J., Foley, G.E., Wagh, U. and Simonsson, E. (1969) *Exp. Cell Res.* 58, 128-140.
5. Sumner, A.T., Evans, H.J. and Buckland, R.A. (1971) *Nature New Biol.* 232, 31-32.
6. Drets, M.E. and Shaw, M.W. (1971) *Proc. Nat. Acad. Sci.* 68, 2073-2077.
7. Dutrillaux, B., de Grouchy, J., Finaz, C., Lejeune, J. (1971) *C. R. Acad. Sci.* 273, 587-588.
8. Seabright, M. (1971) *Lancet* 2, 971-972.
9. Wang, H.C. and Federoff, S. (1972) *Nature New Biol.* 235, 52-53.
10. Bahr, G.F., Mikel, U. and Engler, W.F. (1973) *Chromosome Identification - Technique and Applications in Biology and Medicine* (Academic Press), 280-289.
11. Comings, D.E. and Okada, T.A. (1975) *Exp. Cell Res.* 93, 267-274.
12. Bahr, G.F. and Larsen, P.M. (1974) *Adv. Cell and Molecular Biol.* 3, 191-212.
13. Sanchez, O., Escobar, J.I. and Yunis, J.J. (1973) *Lancet* 2, 269.
14. McKay, R.D.G. (1973) *Chromosoma* 44, 1-14.
15. Okada, T.A. and Comings, D.E. (1974) *Chromosoma* 48, 65-71.
16. Ferguson-Smith, M.A. and Page, B.M. (1973) *J. Med. Genet.* 10, 282-286.
17. Comings, D.E. (1974) *Birth Defects* (Excerpta Medica, Amsterdam), 44-52.
18. Yunis, J.J. (1976) *Science* 191, 1268-1270.
19. Lima-de-faria, A. (1975) *Hereditas* 81, 249-284.
20. Weisblum, B. and de Haseth, P.L. (1972) *Proc. U.S. Nat. Acad. Sci.* 69, 629-632.
21. Weisblum, B. and de Haseth, P.L. (1973) *Chromosomes Today* Plenum Press, volume 4, 35-51.
22. Ellison, J.R. and Barr, H.J. (1972) *Chromosoma* 36, 375-390.
23. Pachmann, U. and Rigler, R. (1972) *Exp. Cell Res.* 72, 602-608.
24. Selander, R.K. and de la Chapelle, A. (1973) *Nature New Biol.* 245, 240-243.
25. Michelson, A.M., Monny, C. and Kovoor, A. (1972) *Biochimie* 54, 1129-1136.
26. Latt, S.A., Brodie, S. and Munroe, S.H. (1974) *Chromosoma* 49, 17-40.
27. Comings, D.E., Kovacs, B.W., Avelino, E. and Harris, D.C. (1975) *Chromosoma* 50, 111-145.

28. Gottesfeld, J.M., Bonner, J., Radda, G.K. and Walker, I.O. (1974) *Biochemistry* 13, 2937-2945.
29. Simola, K., Selander, R.K., de la Chapelle, A., Corneo, G. and Ginelli, E. (1975) *Chromosoma* 51, 199-205.
30. Bostock, C.J. and Christie, S. (1974) *Exp. Cell Res.* 86, 157-161.
31. Comings, D.E. and Drets, M.E. (1976) *Chromosoma* 56, 199-211.
32. Ganner, E. and Evans, H.J. (1971) *Chromosoma* 35, 326-341.
33. Dutrillaux, B. (1975) *Chromosoma* 52, 261-273.
34. Dutrillaux, B., Couturier, J., Richer, C.L. and Viegas-Pequignot, E. (1976) *Chromosoma* 58, 51-61.
35. Verma, R.S. and Lubs, H.A. (1975) *Am. J. Human Genetics* 27, 110-117.
36. Dutrillaux, B. and Covic, M. (1974) *Exp. Cell Res.* 85, 143-153.
37. Comings, D.E., Avelino, E., Okada, T.A. and Wyandt, H.E. (1973) *Exp. Cell Res.* 77, 469-493.
38. Wyandt, H.E., Vlietinck, R.F., Magenis, R.E. and Hecht, F. (1974) *Humangenetik* 23, 119-130.
39. Darzynkiewicz, Z., Traganos, F., Arlin, Z.A., Sharpless, T. and Melamed, M.R. (1976) *J. Histochem, Cytochem.* 24, 49-58.
40. Rigler, R., Killander, D., Bolund, L. and Ringertz, N.R. (1969) *Exp. Cell Res.* 55, 215-224.
41. Comings, D.E. and Wyandt, H.E. (1976) *Exp. Cell Res.* 99, 183-185.
42. de la Chapelle, A., Schroder, J. and Selander, R.K. (1973) *Chromosoma* 40, 347-360.
43. de la Chapelle, A., Schroder, J., Selander, R.K. and Stenstrand, K. (1973) *Chromosoma* 42, 365-382.
44. Schreck, R.R., Warburton, D., Miller, O.J., Beiser, S.M. and Erlanger, B.F. (1973) *Proc. Nat. Acad. Sci.* 70, 804-807.
45. van de Sande, J.H., Lin, C.C. and Jorgenson, K.F. (1977) *Science* 195, 400-402.
46. Dutrillaux, B. and Lejeune, J. (1971) *C. R. Acad. Sci.* 272, 2638-2640.
47. Comings, D.E. and Avelino, E. (1975) *Chromosoma* 51, 365-381.
48. Hoehn, H. (1975) *Am. J. Human Genet.* 27, 676-686.
49. Arrighi, F.E. and Hsu, T.C. (1971) *Cytogenetics* 10, 81-86.
50. Pathak, S. and Arrighi, F.E. (1973) *Cytogenet. Cell Genet.* 12, 414-422.
51. Comings, D.E., Harris, D.C., Okada, T.A. and Holmquist,

G.P. (1977) *Exp. Cell Res.* (in press).
52. Matsui, S. and Sasaki, M. (1973) *Nature* 246, 148-150.
53. Matsui, S. (1974) *Exp. Cell Res.* 88, 88-94.
54. Funaki, K., Matsui, S. and Sasaki, M. (1975) *Chromosoma* 49, 357-370.
55. Tantravahi, R., Dev, V.G., Miller, D.A. and Miller, O.J. (1975) *1st International Congress Cell Biol.* 79a.
56. Goodpasture, C. and Bloom, S.E. (1975) *Chromosoma* 53, 37-50.
57. Navashin, M. (1934) *Cytolgia* 5, 169-203.
58. Hsu, T.C., Brinkley, B.R. and Arrighi, F.E. (1967) *Chromosoma* 23, 137-153.
59. Burkholder, G.D. (1974) *Nature* 247, 292-294.
60. Burkholder, G.D. (1975) *Exp. Cell Res.* 90, 269-278.
61. Comings, D.E. (1975) *Chromosoma* 50, 89-110.
62. Lober, G., Zimmer, C., Sarfert, E., Dobel, P. and Riegler, R. (1973) *Studia Biophysica* 40, 141-150.
63. Ross, A. and Gormley, I.P. (1973) *Exp. Cell Res.* 81, 79-86.
64. Cervenka, J., Thorn, H.L. and Gorlin, R.J. (1973) *Cytogenet. Cell Genet.* 12, 81-86.
65. Schuh, B.E., Korf, B.R. and Salwen, M.J. (1975) *Humangenetik* 28, 233-237.
66. Comings, D.E. and Avelino, E. (1974) *Exp. Cell Res.* 86, 202-206.
67. Holmquist, G.P. and Comings, D.E. (1976) *Science* 193, 599-602.
68. Berezney, R. and Coffey, D.S. (1974) *Biochemical Biophys. Res. Comm.* 60, 1410-1417.
69. Berezney, R. and Coffey, D.S. (1976) *Adv. Enzyme Regulation* 14, 63-100.
70. Comings, D.E. and Okada, T.A. (1976) *Exp. Cell Res.* 103, 341-360.
71. Aaronson, R.P. and Blobel, G. (1975) *Proc. Nat. Acad. Sci.* 72, 1007-1011.
72. Ide, T., Nakane, M., Anzai, K. and Andoh, T. (1975) *Nature* 258, 445-447.
73. Cooke, P.R. and Brazell, I.A. (1975) *J. Cell Sci.* 19, 261-279.
74. Benyajati, C. and Worcel, A. (1976) *Cell* 9, 393-407.
75. Berezney, R. and Coffey, D.S. (1975) *Science* 189, 291-293.
76. Comings, D.E. and Harris, D.C. (1976) *J. Cell Biol.* 70, 440-452.
77. Comings, D.E. and Wallack, A.S. (1977) submitted.

LOCALISATION OF SEQUENCES COMPLEMENTARY TO HUMAN SATELLITE DNAs IN MAN AND THE HOMINOID APES

John R. Gosden, Arthur R. Mitchell, Hector N. Seuanez

M.R.C. Clinical and Population Cytogenetics Unit,
Western General Hospital, Edinburgh, Scotland

ABSTRACT. The location of four human satellite DNAs in human chromosomes has been determined by in situ hybridisation (6). The method used for analysing the results has permitted a quantitative distribution of these DNA sequences to be calculated.

DNA sequences which hybridise with RNAs complementary to human satellite DNAs (satellite cRNAs) are present in the hominoid apes, and the location of these sequences in the chimpanzee (Pan troglodytes) gorilla (Gorilla gorilla) and orang utan (Pongo pygmaeus) has been investigated by the same method.

The distribution of these human equivalent DNA sequences in the hominoid apes confirms that the sequences of the four human satellites are quite distinct, and that their common hybridisation at many sites in man does not arise from sequence contamination or from cross reaction. The relationship between the distribution of these sequences and chromosome homology determined by banding patterns in the apes (15) suggests that such homology does not necessarily include repeated DNA sequences. In several cases homologous chromosome sites do contain the same satellite DNA, but other sites show differences which cannot always be explained by chromosome rearrangement. The distribution of human satellite DNAs in relation to the phylogeny of the Hominidae allows us to speculate about the mode of evolution of these satellites.

INTRODUCTION

Evolution proceeds by the occurrence and fixation of quantitative or qualitative changes in DNA: this is a basic tenet of biology. Examination of homologous macromolecules in man and his closest relatives, the great apes, however, reveals surprisingly little difference between them, either in immunological reaction (12) or at the level of the protein sequence (3). This is in contrast to the very obvious morphological differences.

The chromosomes of the four species we have studied (man, chimpanzee (Pan troglodytes), gorilla (Gorilla gorilla) and orang utan (Pongo pygmaeus) have been compared, and the majority have been assigned a homology between species, on the basis of morphology and Giemsa banding pattern (20). These homologies have been published in the Paris Conference (1971) Supplement (1975) from which the chromosome diagrams in this paper are taken. A number of gene assignments have now been made, and the great majority are found on homologous chromosomes in the different species (1).

It seems, therefore, that the DNA changes expected to produce evolutionary modifications may not be found at the level of the structural genes, and that many gene linkage groups have not been affected by chromosome rearrangements (25). With the finding that a highly repeated human DNA sequence would react with the chromosomes of chimpanzee and orang utan (11) a further aspect of the man/ape relationship became accessible. We have developed a system for quantitative analysis of the results of in situ hybridisation (4) and have used this method to determine the distribution of the four major human satellite DNAs in human chromosomes (6). We have extended this work to an analysis of the distribution of sequences which cross-react with these four human DNAs in the chromosomes of chimpanzee, gorilla and orang utan, and have examined the distribution of these human-homologous repeated DNA sequences (HHR-DNA) in the context of chromosome homologies defined by Giemsa banding patterns (7,14,17). In the present paper we examine in detail some of the anomalous sites of hybridisation of RNAs complementary to human satellite DNAs. These are homologous sites in which hybridisation is detected in some species, but not in all four, and we believe that they provide information both about the mode of evolution of these HHR-DNA sequences, and the evolution of the contemporary hominoid karyotypes.

MATERIALS AND METHODS

a) Chromosomes of Gorilla, Pongo and Homo were prepared from peripheral blood leukocyte cultures, and those of Pan from cultured fibroblasts as described previously (14).
b) Human DNA was prepared, satellite DNAs I, II, III and IV purified and their complementary RNAs (cRNAs) synthesised as before (5,6). E.coli RNA polymerase was supplied by the Boehringer Corporation (London) Ltd., and ^{3}H-nucleoside triphosphates by the Radiochemical Centre, Amersham. All cRNA preparations had the same specific radioactivity.
c) Cell selection and quinacrine fluorescence photography, in situ hybridisation and autoradiography, and grain distri-

bution analysis were carried out as before(4,6,14). Each set of *in situ* hybridisation experiments used aliquots from the same preparation of cRNA, hybridised to identically treated chromosome preparations under exactly the same conditions. All autoradiographs were exposed for 21 days. This ensures that the quantitative data are directly comparable between species and between experiments.

RESULTS AND DISCUSSION

The detailed results of these experiments have been described elsewhere (7,14), and are summarised in Table 1.

Species	Hybridisation of cRNA I	II	III	IV
Homo	+	+	+	+
Pan	++	-	+++	+
Gorilla	++	++	+++	++
Pongo	++	++	++	±

Table 1

Hybridisation of the cRNA from human satellite II is not detectable on chimpanzee chromosomes, and cRNAIV hybridises almost exclusively to the Y chromosome of orang utan. This alone is sufficient to show that these two sequences are distinct both from each other and from the cRNAs to human satellite DNAs I and III.

Analysis of the distribution of these sequences shows that when hybridisation to the chromosomes of a given species is detected, the majority of the sites of hybridisation are homologous to the sites of hybridisation of the same cRNA in the other species. There are quantitative differences in the amount of hybridisation at homologous sites in the different species, but in many cases these differences are no greater than those found between phenotypically normal individuals in man. However, there are several sites of hybridisation where these general observations do not hold true, and we wish to discuss five of these in the current paper.

Human chromosome 2(HSA*2 in the nomenclature of the Paris Conference (1971) Supplement (1975)) is a metacentric chromosome, to which no hybridisation is detectable with any

---

* HSA = *Homo sapiens* GGO = *Gorilla gorilla*
PTR = *Pan troglodytes* PPY = *Pongo pygmaeus*

of the four cRNAs. The homologues to this chromosome consist of two pairs of chromosomes in each of the apes: in chimpanzee they are an acrocentric (PTR12) and a submetacentric (PTR13); in gorilla they are also an acrocentric (GGO12) and a submetacentric with a very small short arm (GGO11) and in orang utan two pairs of acrocentric chromosomes (PPY11 and 12).

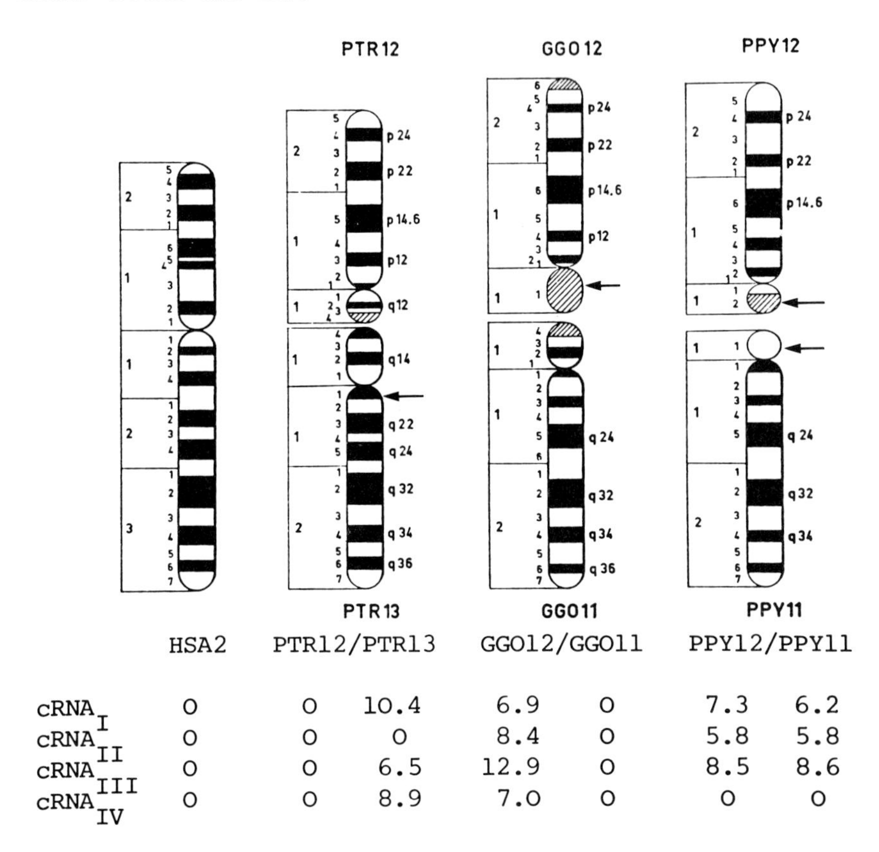

	HSA2	PTR12/PTR13		GGO12/GGO11		PPY12/PPY11	
$cRNA_{I}$	0	0	10.4	6.9	0	7.3	6.2
$cRNA_{II}$	0	0	0	8.4	0	5.8	5.8
$cRNA_{III}$	0	0	6.5	12.9	0	8.5	8.6
$cRNA_{IV}$	0	0	8.9	7.0	0	0	0

Table 2 Hybridisation of Human Satellite cRNAs to HSA2 and its Ape Homologues
Arrows indicate sites of hybridisation. Figures indicate mean number of grains/site.

Table 2 shows the distribution of hybridisation of each of the four human satellite cRNAs in grains per chromosome on HSA2 and its homologues. HSA2 shows no hybridisation whatever, but PTR13 (homologous to most of the long arm of HSA2) shows a lot of hybridisation with cRNAs I, III and IV, at a centromeric band which is absent in the human chromosome. The gorilla homologue to the human long arm (GGO11) shows no hybridisation, but GGO12 is homologous to the human short arm, and this shows hybridisation to its centromere and short arm (a variable region absent from HSA2) with all four cRNAs. In orang utan, the situation is different again, for cRNAs I, II and III hybridise in approximately equal amounts to the short arm/centromere regions of both PPY11 and PPY12, both of which are regions lacking homology with HSA2. Thus these homologous chromosomes with many apparently identical Giemsa bands, show complete variation in their capacity to hybridise with cRNA to human satellite DNAs. The presence of repeated sequences in the short arms of the orang utan homologues, and their absence from the human chromosome can be explained as follows: loss of satellite DNA following telomeric fusion of two acrocentric chromosomes has been observed in man (Gosden, unpublished results) and it is reasonable to assume that the same sort of event has taken place in the evolution of the Hominidae. In a similar way, the absence of detectable repeated sequences from the short arm of GGO11 may be ascribed to the same sort of rearrangement during speciation. The chimpanzee, however, is a more difficult case, as the acrocentric homologue to HSA2p has no detectable hybridisation, but the submetacentric (PTR13) hybridises at its centromere, a region which must now be included in the human long arm (HSA2q) between bands q14 and q22, both of which are represented in the chimpanzee chromosome. It seems therefore, that this centromere has been inactivated in man (2,23), with a concomitant loss of repeated DNA.

Table 3 shows the results of hybridisation to HSA7 and its homologues. Here the human chromosome shows very little hybridisation with any of the cRNAs, though with all four the level is above background, and this hybridisation is pericentric. Neither the gorilla nor the orang utan homologues (GGO6 and PPY10) show any hybridisation with cRNAs I, II and IV, and the grains found with cRNAIII are barely above background (they represent less than 0.5% of total hybridisation), and appear to be randomly distributed. The chimpanzee homologue, PTR6, however, forms a major site of hybridisation of cRNAs I, III and IV, and this is also located pericentrically. The Giemsa banding pattern of this region is almost identical to that of HSA7, so that the presence of this large amount of human-homologous repeated DNA (HHR-DNA) is in no way related to the presence of additional bands. Moreover, this chimpanzee chromosome has no C band at the site of hybridisation (17) which again confirms the lack of correspondence between banding patterns and specific DNA content (19,21). In this set of homologous chromosomes, therefore, repeated DNA is present in one species, absent from two, and barely detectable in the fourth.

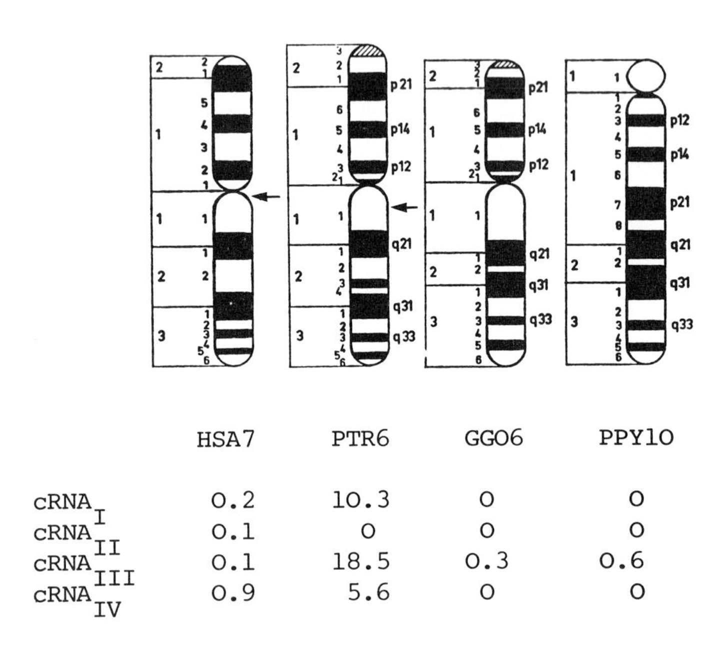

	HSA7	PTR6	GGO6	PPY10
$cRNA_{I}$	0.2	10.3	0	0
$cRNA_{II}$	0.1	0	0	0
$cRNA_{III}$	0.1	18.5	0.3	0.6
$cRNA_{IV}$	0.9	5.6	0	0

Table 3 Hybridisation of Human Satellite cRNAs to HSA7 and its Ape Homologues
Arrows indicate sites of hybridisation. Figures indicate mean number of grains/site.

Finally let us examine the human E group chrosomes (HSA16, 17 and 18) and their homologues, shown in Tables 4,5 and 6. In man, this is a group of small meta- and submetacentric chromosomes, of which HSA16 hybridises only cRNAII, HSA17 hybridises cRNAII and cRNAIV and HSA18 hybridises no cRNA. If we examine the homologues to each of these chromosomes individually, those to HSA16 (Table 4) are all metacentric chromosomes. PTR18 shows a great deal of hybridisation of cRNAIII, and about half of this amount of cRNAI

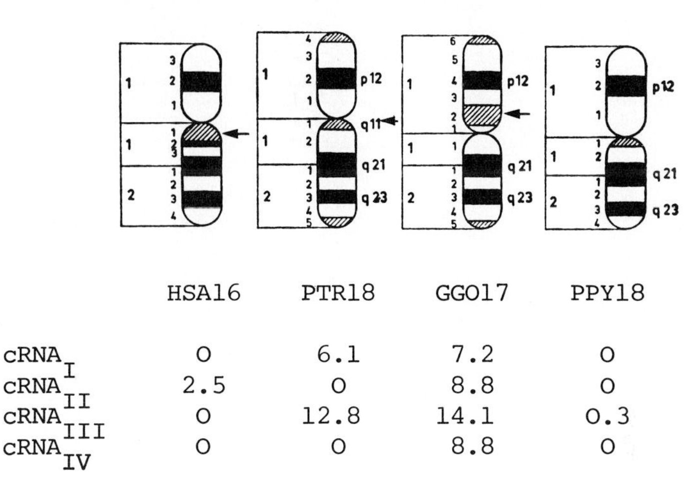

	HSA16	PTR18	GGO17	PPY18
$cRNA_{I}$	0	6.1	7.2	0
$cRNA_{II}$	2.5	0	8.8	0
$cRNA_{III}$	0	12.8	14.1	0.3
$cRNA_{IV}$	0	0	8.8	0

Table 4 Hybridisation of Human Satellite cRNAs to HSA16 and its Ape Homologues
Arrows indicate sites of hybridisation. Figures indicate mean number of grains/site.

to a region around the centromere. Neither of these cRNAs is detected in the human homologue. GGO17 has a secondary constriction in the short arm to which all four cRNAs hybridise, and PPY18 hybridises with none of them. In this case, the only sequence which is detectable in HSA16 is absent from all chimpanzee chromosomes, and two sequences are present in chimpanzee, which are also present in GGO17, but at a different site, though PTR18 can be reconstructed from GGO17 by postulating a pericentric inversion. PPY18 hybridises none of the cRNAs.

HSA17 (Table 5) is a submetacentric, which hybridises cRNAsII and IV near the centromere; PTR19 is more nearly metacentric, and hybridises cRNAsI, III and IV to the pericentric region, while GGO19, a true metacentric, hybridises none of the cRNAs. Pongo lacks a homologue for this chromosome.

HSA18 (Table 6) is a submetacentric, with a very small short arm and hybridises none of the cRNAs, but its ape homologues, which are all acrocentric, all hybridise at least some of them to the short arm/centromere region. Here again, we can postulate changing from the ape to the human chromosome by a pericentric inversion, and a concomitant loss of satellite DNA.

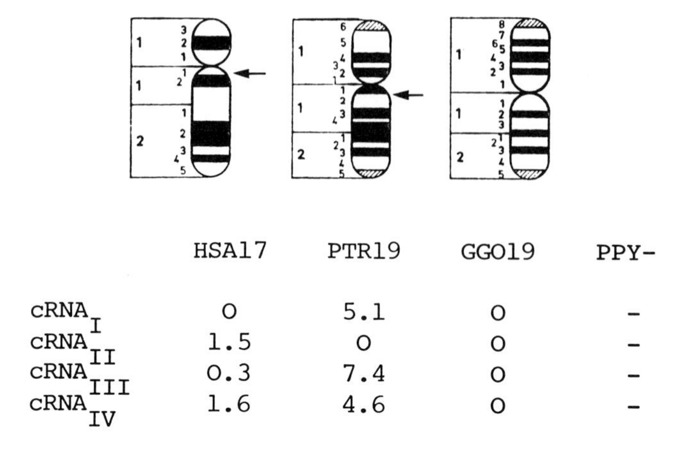

	HSA17	PTR19	GGO19	PPY-
$cRNA_{I}$	0	5.1	0	-
$cRNA_{II}$	1.5	0	0	-
$cRNA_{III}$	0.3	7.4	0	-
$cRNA_{IV}$	1.6	4.6	0	-

Table 5 Hybridisation of Human Satellite cRNAs to HSA17 and its Ape Homologues
Arrows indicate sites of hybridisation. Figures indicate mean number of grains/site.

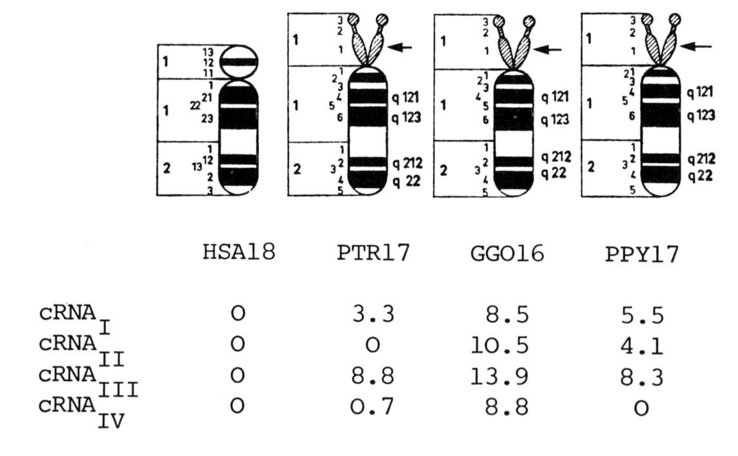

	HSA18	PTR17	GGO16	PPY17
$cRNA_{I}$	0	3.3	8.5	5.5
$cRNA_{II}$	0	0	10.5	4.1
$cRNA_{III}$	0	8.8	13.9	8.3
$cRNA_{IV}$	0	0.7	8.8	0

Table 6 Hybridisation of Human Satellite cRNAs to HSA18 and its Ape Homologues
Arrows indicate sites of hybridisation. Figures indicate mean number of grains/site.

The majority of known gene assignments are to homologous chromosomes in each species. Those repeated DNAs which are known to be transcribed (e.g. the 5s and the large ribosomal RNA cistrons) are also found on homologous chromosome in man and the great apes (9,10,18). However, as we have seen, there are several sites where these highly repeated human-homologous DNAs, which are not, apparently, transcribed into RNA, are found on a chromosome pair in one species, but not on its homologues in the other species.

In some of these cases (e.g. HSA2, HSA18) differences in distribution of HHR-DNA can be explained by loss of such sequences following, or in association with, chromosomal rearrangement. This, however, is in contrast to the results of Hatch et al (8) who found that changes from acrocentric to metacentric chromosome morphology in species of the Kangaroo rat, Dipodomys, were associated with an increase in satellite DNA content. They suggested that this DNA was itself implicated in these chromosome rearrangements, and thus played a role in the evolutionary changes resulting from them.

Moreover, in other cases (e.g. HSA7, 16, 17) HHR-DNAs are present in one or more species and absent in others, with no apparently associated chromosome rearrangement, suggesting that the amplification of these DNAs may have occurred de novo, rather than been inherited from some common ancestor, and that the distribution of these sequences in the karyotype is controlled by a different process from that governing the evolution of the rest of the genome.

Despite the number of hypotheses advanced about the function of satellite DNAs (8,13,16,22,24,26) we appear to be no nearer finding a model which is generally valid. We have suggested that the absence of detectable satellite DNA sequences from certain human chromosomes rules out models of satellite function in man requiring the presence of some satellite sequences in every chromosome (6). Our results with the hominoid apes support this conclusion, but the fact that these apparently inert sequences are sufficiently conserved during the evolution of the Hominidae to cross react between the contemporary species indicates that the sequences themselves may be of some functional importance. We trust that the steady accummulation of information about satellite DNAs will produce a solution to the mystery.

ACKNOWLEDGEMENTS

We would like to thank Sandra Lawrie for her labours in photomicrography, and Dr. Christine Gosden for helpful discussion and for the chimpanzee cell culture without which the work would not have been complete.

REFERENCES

1. Baltimore Conference (1975): *Third International Workshop on Human Gene Mapping*. Birth Defects:Original Article Series, XII, 7, 1976 The National Foundation, New York.
2. Dutrillaux, B. (1975) *Sur la nature et l'origine des chromosomes humains*. L'Expansion Scientifique, Paris.
3. Goodman, M. (1974) In *Phylogeny of the Primates*, Eds W.P. Lackett and J.S. Szalay p. 219.
4. Gosden, J.R., Buckland, R.A., Clayton, R.P. and Evans, H.J. (1975) *Exp. Cell Res.* 92, 138.
5. Gosden, J.R. and Mitchell, A.R. (1975) *Exp. Cell Res.* 92, 131.
6. Gosden, J.R., Mitchell, A.R., Buckland, R.A., Clayton, R.P. and Evans, H.J. (1975) *Exp. Cell Res.* 92, 148.
7. Gosden, J.R., Mitchell, A.R., Seuanez, H.N. and Gosden, C.M. Submitted to *Chromosoma (Berl.)*.
8. Hatch, F.T., Bodmer, A.J., Mazrimas, J.A. and Moore, D.H. II (1976) *Chromosoma (Berl.)* 58, 155.
9. Henderson, A.S., Atwood, K.C., Yu, M.T. and Warburton, D. (1976) *Chromosoma (Berl.)* 56, 29.
10. Henderson, A.S., Warburton, D. and Atwood, K.C. (1974) *Chromosoma (Berl.)* 46, 435.
11. Jones, K.W., Prosser, J., Corneo, G., Ginelli, E. and Bobrow, M. (1973) In: *Modern Aspects of Cytogenetics:* Symposium Media Heochst, 643.
12. King, M.C. and Wilson, A.C. (1975) *Science* 188, 107.
13. Miklos, G.L. Gaber and Nankivell, N.R. (1976) *Chromosoma (Berl.)* 56, 143.
14. Mitchell, A.R., Seuanez, H.N., Lawrie, S.S., Martin, D.E. and Gosden, J.R. *Chromosoma (Berl.)* in press.
15. Paris Conference (1971), Supplement (1975): *Standardization in Human Cytogenetics*. Birth Defects:Original Article Series, XI, 9 1975. The National Foundation, New York.
16. Paul, J. (1972) *Nature* 238 444.
17. Seuanez, H.N., Mitchell, A.R., Fletcher, J., Martin, D.E. and Gosden, J.R. Submitted to *Chromosoma (Berl.)*.
18. Steffensen, D.M., Prensky, W. and Dufy, P. (1974) *Cytogenet. Cell Genet.* 13, 153.
19. Sumner, A.T. (1972) in *Current Chromosome Research* Eds. Jones K. and P.E. Brandham p. 17.
20. Sumner, A.T., Evans, H.J. and Buckland, R.A. (1971) *Nature New Biol.* 232, 31.

21. Sumner, A.T., Mitchell, A.R. and Gosden, J.R. (1975) *Cytobios* 13,151.
22. Sutton, W.D. (1972) *Nature New Biol.* 237,70.
23. Turleau, C. and de Grouchy, J. (1973) *Humangenetik* 20,151.
24. Walker, P.M.B. (1971) *Prog. in Biophys. and Mol. Biol.* 23,145.
25. Wilson, A.C., Sarich, V.M. and Maxson, L.R. (1974) *Proc. Nat. Acad. Sci. U.S.A.* 71,3028.
26. Yunis, J.J. and Yasmineh, W.G. (1971) *Science* 174, 1200.

IMMUNOLOGICAL APPROACHES TO CHROMOSOME BANDING

Orlando J. Miller and Bernard F. Erlanger

Departments of Human Genetics and Development, Obstetrics and Gynecology, and Microbiology, College of Physicians and Surgeons, Columbia University, New York, N.Y. 10032

ABSTRACT. Antibodies to macromolecular components of chromatin have been used to probe the structure of human and other mammalian metaphase chromosomes. Antibodies that react with specific nucleotides or oligonucleotides in single stranded DNA have been particularly useful because the specificity of the antibodies, when used in conjunction with methods for selective denaturation of DNA, permit inferences to be made regarding sequence organization in the DNA of chromosomes. UV irradiation exposes nucleotide sequences that bind anti-adenosine and anti-thymine but not anti-cytidine or anti-5-methylcytidine. These sequences are concentrated in Q- (G-) bands, with their abundance paralleling the intensity of staining of the chromosome bands by Q- or G-banding procedures. Photooxidation exposes sequences that bind all four antibodies. These sequences are concentrated in R-bands, with their abundance matching the intensity of staining of each chromosome segment by R-banding procedures. It is likely, therefore, that Q-, G- and R-banding patterns, whether produced by immunological procedures or by standard banding methods, reflect reactions with the same families of nucleotide sequences, which have a highly non-uniform and characteristic distribution throughout the genome.

The location of some of the highly reiterated human satellite DNAs can be detected by immunological methods. UV irradiation appears to generate longer single stranded regions in them than photooxidation, since only anti-5-methylcytidine (M) binds to the relevant regions after photooxidation while antibodies to M, A, C and AAA bind after UV. Specific identification of a small number of chromosome bands in the human genome has been achieved and put to clinical use.

INTRODUCTION

Antinucleoside antibodies.

Some years ago, one of us (BFE) in collaboration with Dr. Sam Beiser, developed a general method by which nucleosides, oligonucleotides or certain other haptens could be covalently bound to an immunogenic carrier protein such as bovine serum albumin (BSA) and used to prepare antibodies specific for the haptenic group (1). Since that time, antibodies to more than 25 nucleosides or nucleotide sequences have been prepared (2, 3) and there is every reason to believe that antibodies to a very large number of additional specific oligonucleotides can be prepared.

Antinucleoside and antinucleotide antibodies of high specificity are easily obtained directly from the immunized animal or after suitable absorption techniques that remove cross-reacting antibodies. Thus, for example, antibodies essentially monospecific for 5-methylcytosine can be obtained from rabbits immunized with 5-methylcytidine conjugates if absorption is carried out with a cytidine-protein conjugate (4). Antisera to 5-bromodeoxyuridine (BrdU) cross-react extensively with thymine, and exhaustive absorption with a thymine conjugate removes all BrdU antibodies (5). However, anti-IdU cross-reacts strongly with BrdU and to a lesser extent with thymine and can be used to detect BrdU in DNA (5) and on chromosomes (6). Antisera can be specifically purified also be binding to an insolubilized hapten, washing free of other immunoglobulins and eluting by low pH or free hapten (7).

Reaction with denatured DNA.

Antinucleoside antibodies will not react with native DNA but will react with specific bases in single-stranded or denatured DNA (1). Consequently, they provide a powerful tool for the study of the organization of DNA in chromatin and chromosomes. For example, one can test the hypothesis that the average base composition of the DNA varies along chromosome arms, providing a chemical basis for chromosome banding patterns (8, 9, 10, 11).

RESULTS

We first used chromosomes whose DNA had been denatured by NaOH (12). Unfortunately, the chromosome morphology was so distorted by this treatment that only generalized binding of anti-adenosine (anti-A) was observed. When we turned to a denaturing method used for *in situ* hybridization, where retention of chromosome morphology is important, we obtained more interesting results. Using 95% formamide in 2 X SSC at 65° for 30-60 minutes, we observed highly non-uniform binding of anti-A, as visualized by indirect immunofluorescence (13). The resultant banding pattern was an exact Q-banding pattern, with the exception of the distal quinacrine-bright portion of the Y chromosome, which showed minimal binding of anti-A. This led us to conclude that Q-banding does reflect regional differences in base composition. However, while anti-A gave the most sharply differentiated binding to Q-bright bands of formamide-denatured chromosomes, antibodies of other specificities, e.g., anti-C and anti-G, also bound, though with much less specificity than anti-A, indicating nevertheless that the DNA made available for antibody binding in the Q-bright regions by formamide denaturation contains detectable G-C base pairs as well as A-T base pairs. The limited binding of antibodies in the Q-dull bands (R-bands) remains unexplained. It is possible that DNA was removed from the chromosome, perhaps especially in the R-bands, by the denaturing treatment. Formamide in 2 X SSC at 65° for 2.5 hours extracts a great deal of DNA from chromosomes, and some of the other banding techniques extract a variable amount of DNA, up to 80% or so from chromosome arms and much less from C-band regions (14, 15). If this is the explanation of our finding, it means that there has been rather selective denaturation along the arms, removing DNA primarily from the R-bands. Another possibility was suggested by David Comings (this volume), who noted that treatment of fixed chromatin by 2 X SSC at 60° reduces the availability of DNA for dye binding to the level found in native chromatin (it is higher in fixed chromatin). He suggested this might be due to conformational changes in non-histone proteins, which are concentrated in euchromatin of the R-band regions. Another explanation is that the DNA in the R-band regions was simply not denatured sufficiently, or renatured to such an extent that antibody binding was very limited. Whatever the explanation, our results support the idea that Q-bright bands are relatively A-T rich when compared to Q-dull bands. The use of different methods for denaturing DNA has provided a means of obtaining further information about the organization of DNA in chromosomes.

Selective denaturation of DNA.

Antinucleoside antibodies react only with bases in single stranded regions of DNA; therefore selective denaturing techniques can be used to generate single-stranded regions in DNA with a certain nucleotide sequence while leaving other sequences denatured (Table 1). For example, G-C rich sequences have greater thermal stability than A-T rich sequences because of the 3 hydrogen bonds per GC base pair and only 2 per A-T base pair. Therefore, mild heat will denature A-T rich sequences, e.g., ATAT or TTTT, although such sequences may also re-associate very quickly. Ultraviolet irradiation (UV) leads to the production of pyrimidine dimers, TT, TC and CC, in decreasing order of abundance (16) and will denature sequences in which there are adjacent pyrimidines in one polynucleotide strand, e.g., TTTT, TCTC or CCCC.

TABLE 1

ABILITY OF VARIOUS TREATMENTS TO GENERATE SINGLE-STRANDED REGIONS IN REPETITIVE DNA

	Base sequence		
Treatment	ATATAT	TTTTTT	GCGCGC
Mild heat	+	+	-
UV irradiation	-	+	-
Photooxidation	-	-	+

Pyrimidine dimers are relatively stable, although they can be abolished by an enzymatic photoreactivating system. A third method, photooxidation, selectively modifies guanine residues in DNA (17, 18) and will denature G-C rich sequences, e.g., GCGC or GGGG. Thus an alternating A-T sequence (poly d(A-T): poly d(A-T) could be denatured by mild heat, which denatures A-T rich DNA more readily than the more stable DNA rich in GC base pairs but would not be denatured by UV-irradiation, which requires adjacent pyrimidines for dimerization and denaturation, or by photooxidation. On the other hand, a (poly d(A): poly d(T) sequence would be denatured by either UV-irradiation or mild heat but not by photooxidation, while a poly d(G-C): poly d(G-C) sequence would be denatured by photooxidation but not by either of the other two treatments.

Distribution of sequence families along chromosomes.

From these examples, it is clear that there may be many nucleotide sequences in DNA which can be made single-stranded by one denaturation method but not by another. The question we have asked is whether such sequences are distributed uniformly along the lengths of chromosomes or whether they have a more restricted distribution. Antibodies to oligonucleotides of lengths 2, 3, 4 etc. would permit accurate detection of these specific sequences in denatured DNA but one would then obtain information only about a specific sequence with each antiserum. Antibodies to single bases and the combined use of antisera of different specificities should provide more general information about families of denatured sequences.

Antibody binding after photooxidation.

After photooxidation of human chromosomes, anti-C binds in a highly non-uniform fashion, producing a typical R-banding pattern (19). The same is true in the mouse (20) and probably in several other mammalian species: rat, guinea pig and Chinese hamster (21). One rather unexpected finding in the human is that the centromeric heterochromatin (C-band) region of chromosome 9, which is known to be the location of the relatively G-C rich satellite III DNA (22, 23) does not bind anti-C intensely after photooxidation although it does after UV-irradiation. Similarly, the secondary constrictions, or nucleolus organizer regions, of chromosomes 13, 14, 15, 21 and 22, sites of the very G-C rich rRNA genes, fail to bind anti-C to any extent after photooxidation. These paradoxical results are still unexplained although they may reflect interference by proteins. Another surprising finding with photooxidation of human chromosomes is that anti-A binds to about the same extent as anti-C and produces an identical R-banding pattern. One explanation of this finding is that photooxidative modification of guanine residues leads to denaturation of sequences that contain A-T base pairs as well as G-C base pairs. Alternatively, one might suspect that photooxidation is not totally selective but also modifies some adenosine or thymine residues. However, if this were the case, the relatively A-T rich satellite DNAs in the C-band regions of chromosomes 1, 9, 16 and the Y should show intense binding, and they don't. A third explanation, that neither the G-C rich nor the A-T rich satellite DNA are available for antibody binding because of interfering chromosomal proteins or because of their organization within the C-band heterochromatin, is ruled out by the results obtained with antisera to 5-methylcytidine (anti-M).

After photooxidation, anti-M shows relatively intense binding to the C-band regions of chromosomes 1, 9, 15, 16 and the Y (21). Since the only satellite DNA present in abundance on all these 5 pairs of chromosomes is satellite II (22) this satellite, or at least sequences within it containing 5-methylcytidine residues, must be available for antibody binding. Anti-M also produces an R-banding pattern of approximately the same intensity, which is surprising considering that 5-methylcytidine makes up only about 1% of the bases in human DNA (24). The organization of the DNA may tend to make its 5-methylcytidine residues more readily available for interaction with molecules coming from outside the chromosome.

Antibody binding after ultraviolet irradiation.

After UV irradiation, anti-A binds in a highly non-uniform fashion, producing a Q-banding pattern along the chromosome arms. In addition, there is extensive binding to the C-band regions of chromosomes 1, 9, 16 and to a lesser extent the Y, and minimally to that of chromosome 15 (25), confirming the availability of denatured bases in one or more repetitive DNAs. Anti-T, anti-thymine dimers (kindly provided by Dr. L. Levine, Brandeis Univ.), and anti-ApA (26, 27) produces exactly the same pattern of binding. UV followed by anti-C on the other hand, produces a unique pattern. In some individuals the C-band of a single chromosome pair, number 9, shows extensive binding. In other individuals the C-band on a chromosome 15 or 22 also shows binding of anti-C, though to a smaller extent. All chromosome arms show minimal antibody binding. This pattern of binding is highly reminiscent of the locations of satellite III DNA as revealed by in situ hybridization (22, 23). Because of the generally low level of antibody binding throughout most of the genome, UV-anti-C provides an almost unique probe for the C-band regions of chromosome 9 and in some individuals and families probably also of the C-band regions of one or two other chromosomes.

It is interesting that Q-bands contain cytidine residues that are available for antibody binding after the formamide denaturation procedure but not after the more restricted denaturation provided by either photooxidation or UV-irradiation.

After UV-irradiation, anti-M binding, other than minimal generalized binding, is restricted to the C-band regions especially of chromosomes 1, 9, 15, 16 and the Y. Thus it resembles C-band binding of anti-A, except for chromosome 15, which shows moderately extensive anti-M binding (28). Since in most individuals the other acrocentric chromosomes show minimal or no binding of anti-M, it provides a unique tag for chromosome 15, and especially for its C-band region. This has already found application in clinical cytogenetics. In one family its use clearly demonstrated that a 47th chromosome, the size of a G-group-chromosome, present in first cousins with mental retardation and several malformations, was derived from a chromosome 15 (29). In six of eleven unrelated individuals with a bisatellited G-group sized chromosome, the extra chromosome was derived from number 15 and had a 5-methylcytidine-rich band near each end. This chromosome arose as a duplication of the short arm and an indeterminate though short segment of the long arm (30). The bisatellited chromosome in the case presented by Van Dyke and his colleagues (this volume), is of this type, with a 5-methylcytidine-rich band near each end (unpublished observations).

Additional information has been obtained by the use of antibodies to the trinucleotide, ApApA (AAA). These antibodies were produced and characterized by Rose D'Alisa who found that they did not cross-react with A, only slightly with AA, and were highly specific for AAA (31). When she applied them to UV-irradiated human chromosomes, she observed extensive binding only to the C-band regions of chromosomes 1, 9, 16 and the Y and probably 22 (32).

DISCUSSION

Nature of the denatured sequences.

The restricted availability of DNA bases for binding specific antibodies after selective denaturation procedures provides a basis for predicting the nature of the sequences that are rendered single-stranded by each of these procedures. Let us first consider the sites of the most extensive antibody binding, the C-band regions of 1, 9, 15, 16 and Y (Table 2). After photooxidation, 5-methylcytidine is available in all these C-band regions, and it is the only base available. Methylation occurs mainly in the CpG doublet in mammalian DNA and symmetrically in the two strands (33, 34) giving adjacent GM base pairs. Photooxidation should thus free up the M in each strand, making it available for antibody binding if the GM pairs are sufficiently clustered. Since A, T and C are not available for antibody binding in the C-bands after photooxidation, this single-stranded region cannot extend much beyond the photoreactive GM pairs. Specifically, there appear to be no GC pairs in the denatured sequence, which is thus different from that present in kangaroo rat HS-β satellite DNA. The latter sequence reacts with anti-C and anti-A, as well as anti-M, after photooxidation.

The unavailability of C residues in the C-band region of chromosome 9 is difficult to understand since the relatively GC rich satellite III DNA is concentrated in this region (22, 23). Either satellite III DNA has a nucleotide sequence that is not denatured by photooxidation, which appears very unlikely, or the DNA is complexed to protein or involved in other conformational changes that make it unavailable to antibody. Photooxidation can quickly alter cysteine, methionine, histidine and tryptophan residues (35) and might exert part of its effect because of this. In either case, it is unlikely that the M residues detected in such abundance in this region are in satellite III. They could be in satellite I, II or IV, which are also present in abundance in this region of chromosome 9 (22). Satellite II DNA appears to be the strongest candidate.

After UV-irradiation, M residues are available in great abundance in the C-band regions of the same five pairs of chromosomes, 1, 9, 15, 16 and the Y. Since this region of chromosome 16 contains satellite II DNA but none of the

RESTRICTED AVAILABILITY OF DNA BASES IN C-BANDS
AFTER SELECTIVE DENATURATION,
SATELLITE DNAs PRESENT, AND POSSIBLE BASE SEQUENCES EXPOSED

C-band of	Bases available after			Possible
chromosome	UV-irradiation	Photooxidation	Satellite DNAs*	nucleotide sequence
1 and 16	A AAA M	M	II	A-T-T-T-M-G
9	A AAA M C	M	I, II, III, IV	C-T-T-T-M-G
Y	(A) AAA M	M	I, II, III, IV	T-T-T-T-M-G
15	(A) M	M	I, II, III, IV	T-A-T-T-M-G

* Gosden et al., 1975.

other 3 well characterized human satellite DNAs, we have concluded that satellite II is methylated. Some of the other satellite DNAs may also be methylated, but our data provide no critical evidence of this point. They do, however, match in situ hybridization data (Gosden, this volume) on the widespread presence of satellite II DNA in gorilla and its virtual absence in chimpanzee (36). The C-band of chromosome 16 also contains available A, T and AAA residues after UV-irradiation. The results with both denaturing procedures could be accounted for if satellite II DNA contains abundant copies of the sequence ATTTMG. Dimerization of the two Ts or one T and the adjacent M would make A, T, AAA and M residues available. The results with chromosome 1 could be accounted for in the same way, and so could most of the results with chromosome 9, which contain quite a lot of satellite II DNA.

The availability of C residues on chromosome 9 after UV irradiation is compatible with the abundance of the relatively GC rich satellite III DNA on this chromosome. However, in view of the lack of available C residues in this region after photo oxidation, it is uncertain whether the DNA in the satellite fraction has been made single-stranded; it is clearly not available for anti-C binding. An alternative explanation should perhaps be considered, i.e., that one of the other satellite DNAs is different on chromosome 9 than it is on the other chromosomes, as a result of evolutionary change. A single nucleotide substitution leading to the sequence CTTTMG would account for all the observations except the continued availability of A residues. However, if this sequence was present in satellite I or IV, the presence of satellite II as well would account for the findings.

Although the C-bands of chromosomes Y and 15 have been shown to contain some of the same satellite DNAs as those of chromosome 1, 9 and 16, they present characteristic differences, especially the restricted availability of A residues in both after UV-irradiation and the lack of available AAA residues on chromosome 15. Here, too, various possible explanations of these discrepancies could be provided. The one we want to stress is change in nucleotide sequence, e.g., to TTTTMG on the Y and TATTMG on chromosome 15.

How long must a single-stranded region be for antibody binding to occur? The antibodies we are using are 7S immunoglobulin molecules, whose overall dimensions are about 80 by 110 Å and whose combining sites, are perhaps only 10 by 20 or 30 Å. Evidence from studies carried out by Rhona Schreck (21) with the kangaroo rat, whose HS-β satellite DNA has a basic repeating sequence 10 base pairs long, indicates that antibodies can bind to single-stranded regions no larger than 5, or at most 6, unpaired bases long. The evidence presented here indicates that a region only 2 or at most perhaps 4 nucleotides long can accomodate a 5-methylcytidine antibody molecule.

Considering now the chromosome arms rather than the C-bands, there are striking restrictions on the availability of bases for antibody binding after selective denaturation procedures (Table 3). The bases available in DNA in the Q-bands after UV-irradiation are A and T, not C or M, and after photooxidation none of these four bases is available; that is, antibody binding is minimal. The DNA in the R-bands shows even more striking restrictions, with none of the four bases available after UV-irradiation although all four are available after photooxidation. We must conclude that Q-bands and R-bands contain quite different proportions of particular nucleotide sequences that are denatured by one or the other of these procedures. The R-bands must contain sequences such as GCGCG or GTGCA, which could be denatured by photooxidation, but cannot contain very many sequences such as CCCCC or CTCCT since UV-irradiation, too, could denature these. The Q-bands must contain sequences such as TTTTT that can be denatured by UV-irradiation but not by photooxidation, as well as C residues in other sequences that are not denatured by either UV-irradiation or photooxidation but are denatured by 95% formamide in 2 X SSC at 65°. Bright quinacrine fluorescence of the Q-bands could be due to intercalation of quinacrine into the first type of sequences but not the second, since interspersion of GC pairs quenches quinacrine fluorescence (9).

TABLE 3

RESTRICTED AVAILABILITY OF DNA BASES AFTER SELECTIVE DENATURATION OF CHROMOSOMES

Regions	Bases available after UV-irradiation	Bases available after Photooxidation
Q-bands	A T	
R-bands		A T C M

Are the nucleotide sequences whose non-uniform distribution along the chromosomes has been revealed by these immunochemical methods responsible for the banding seen by Q-, G-, R-, C- and G11 banding techniques? It seems almost certain that they are. This conclusion is supported by the evidence summarized so well by David Comings (37 and this volume) and also by the work of Vande Sande and Lin (38 and this volume), who studied the binding of olivomycin to various polynucleotides and found a specificity for G-C pairs. This compound produces a very clear fluorescent R-banding pattern along human chromosomes. The basis for the R-banding appears to be the much stronger binding of this compound to DNA in R-bands, which must be rich in poly d(G-C):poly d(G-C) sequences than to DNA in Q-bands, which is poor in these sequences.

Antibodies to histones and histone-histone complexes.

One of our graduate students, N. Mihalakis, has prepared antibodies to salt-extracted calf thymus histone H1 and to the histone-histone dimers, H2A-H2B and H3-H4 (39) and has been studying their reactions with chromosomes. In brief, he has found that antibodies to complexes will react with human HeLa cell chromosomes and Drosophila polytene chromosomes, but only after removal of some of the histones, using increasing concentrations of acetic or propionic acid. Furthermore, binding of anti-H3-H4 occurs only after more stringent acid extraction, as one might expect in terms of the Weintraub-Worcel-Alberts model of the superhelical structure of chromatin (40). It seems likely that antibodies directed against histone dimers would be of more use in unraveling the complexities of chromatin structure than antibodies to individual histones, since some of these would be directed against antigenic determinants on parts of the molecule that would be buried in the naturally occuring dimers. In addition, under no circumstances did Mihalakis observe binding to chromosomes of antibodies to salt -extracted histone H1. The explanation we favor is that the strong antigenic determinants were in the hydrophobic portion of the molecule, and in chromosomes this portion in involved in protein-protein interactions and is thus unavailable for antibody binding.

ACKNOWLEDGEMENTS

This work was supported by grants from the NIH (AI 06860, CA 12504 and GM 22966) and the National Foundation-March of Dimes.

REFERENCES

(1) Erlanger, B.F. and Beiser, S.M. (1964) Proc. Natl. Acad. Sci. U.S. 52, 68.

(2) Stollar, B.D. in THE ANTIGENS Vol. I. ed. M. Sela (1973) Academic Press, N.Y., 1.
(3) Miller, O.J. and Erlanger, B.F. in PATHOBIOLOGY ANNUAL (1975) ed. H. Ioachim, Appleton-Century-Crofts, N.Y.,71.
(4) Lubit, B.W., Pham, T.D., Miller, O.J. and Erlanger, B.F. (1976) Cell 9, 503.
(5) Sawicki, D.L., Erlanger, B.F. and Beiser, S.M. (1971) Science 174, 70.
(6) Gratzner, H.G., Leif, R.C., Ingram, D.J. and Castro, A. (1975) Exp. Cell Res. 95, 88.
(7) Szafran, H., Beiser, S.M. and Erlanger, B.F. (1969) J. Immunol. 103, 1157.
(8) Caspersson, T., Modest, E.J., Foley, G.E., Wagh, U. and Simonsson, E. (1969) Exp. Cell Res. 58, 128.
(9) Pachmann, U. and Rigler, R. (1972) Exp. Cell Res. 72, 602
(10) Weisblum, B. and de Haseth, P. (1972) Proc. Natl. Acad. Acad. Sci. U.S. 69, 629.
(11) Michelson, A.M., Monny, C. and Kovoor, A. (1972) Biochemie 54, 1129.
(12) Freeman, M.V.R., Beiser, S.M., Erlanger, B.F. and Miller, O.J. (1971) Exp. Cell Res. 69, 345.
(13) Dev, V.G., Warburton, D., Miller, O.J., Miller, D.A., Erlanger, B.F. and Beiser, S.M. (1972) Exp. Cell Res. 74, 288.
(14) Hubbell, H.R., Sahasrabuddha, C.G. and Hsu, T.C. (1976) Exp. Cell Res. 102, 385.
(15) Pathak, S. and Arrighi, F.E. (1973) Cytogenet. Cell Genet. 12, 414.
(16) Setlow, R.B. (1966) Science 153, 379.
(17) Simon, M.I. and Van Vunakis, H. (1962) J. Mol. Biol. 4, 488.
(18) Garro, A.G., Erlanger, B.F. and Beiser, S.M. (1968) in NUCLEIC ACIDS IN IMMUNOLOGY, ed. O.J. Plescia and W. Braun, Springer-Verlag, N.Y., 47.
(19) Schreck, R.R., Warburton, D., Miller, O.J., Beiser, S.M. and Erlanger, B.F. (1973) Proc. Natl. Acad. Sci. US 70, 804.
(20) Schreck, R.R., Dev, V.G., Erlanger, B.F. and Miller, O.J. (1977) Chromosoma, in press.
(21) Schreck, R.R. (1976) Thesis, Columbia University.
(22) Gosden, J.R., Mitchell, A.R., Buckland, R.A., Clayton, R.P. and Evans, H.J. (1975) Exp. Cell Res. 92, 148.
(23) Jones, K.W., Prosser, J., Corneo, G. and Ginelli, E. (1973) Chromosoma 42, 445.
(24) Brown, G.M. and Attardi, G. (1965) Biochem. Biophys. Res. Comm. 20, 928.

(25) Schreck, R.R., Erlanger, B.F. and Miller, O.J. (1974) Exp. Cell Res. 88, 31.
(26) Beiser, S.M. and Erlanger, B.F. (1966) Cancer Res. 26, 2012.
(27) Wallace, S.S., Erlanger, B.F. and Beiser, S.M. (1971) Biochemistry 10, 679.
(28) Miller, O.J., Schnedl, W., Allen, J. and Erlanger, B.F. (1974) Nature 251, 636.
(29) Breg, W.R., Schreck, R.R. and Miller, O.J. (1974) Am. J. Hum. Genet. 26, 17a.
(30) Schreck, R.R., Breg, W.R., Erlanger, B.F. and Miller, O.J. (1977) Human Genetics, in press.
(31) D'Alisa, R.M. and Erlanger, B.F. (1974) Biochemistry 13, 3575.
(32) D'Alisa, R.M. (1973) Thesis, Columbia University.
(33) Sinsheimer, R.L. (1955) J. Biol. Chem. 215, 579.
(34) Doskocil, J. and Sorm, F. (1962) Biochim. Biophys. Acta 55, 953.
(35) Ray, W.J. Jr., and Koshland, D.E. Jr. (1960) in Brookhaven Symposia in Biology, No. 13 Protein Structure and Function, 135.
(36) Schnedl, W., Dev, V.G., Tantravahi, R., Miller, D.A., Erlanger, B.F. and Miller, O.J. (1975) Chromosoma 52, 59.
(37) Comings, D.E., Kovacs, B.W., Avelino, E. and Harris, D.C. (1975) Chromosoma 50, 111.
(38) Vande Sande, J.H., Lin, C.C. and Jorgenson, K.F. (1977) Science 195, 400.
(39) Mihalakis, N., Miller, O.J. and Erlanger, B.F. (1976) Science 192, 469.
(40) Weintraub, H., Worcel, A. and Alberts, B. (1976) Cell 9, 409.

THE SYNAPTONEMAL COMPLEX AND MEIOSIS

Montrose J. Moses

Department of Anatomy, Duke University Medical Center
Durham, North Carolina 27710

ABSTRACT.

The synaptonemal complex (SC) forms the proteinaceous axis of pachytene bivalents and appears to be prerequisite for crossing over. Unpaired homologues have single axes that become lateral elements of the SC at synapsis when joined to a central element by transverse filaments. Attachment plaques anchor SC ends to the nuclear envelope . A simple whole mount spreading method yields whole complements of selectively stained SC's in which kinetochores are represented by differentiations of the lateral elements. Bivalents are identifiable by length and kinetochore position, and thus pachytene karyotypes have been obtained. Autosomal SC's change in length during pachytene, but their relative lengths and arm ratios remain constant, ruling out major biological variation and preparative distortions, and implying that autosomal length is under uniform control. A 1:1 relationship between relative lengths of SC's and mitotic autosomes indicates that the control operates in meiotic and somatic cells alike. Axial elements develop from attachments on the nuclear envelope at or just prior to synapsis. Synapsis usually begins at the nuclear envelope and proceeds toward the kinetochore. Although axial elements evidently do not participate directly in chiasma formation, transient "recombination nodules" are observed in occasional sets of SC's. X and Y chromosomes operate on an independent schedule. They may be paired over an SC region that is longest at early pachytene; precocious desynapsis often occurs, although the SC of the paired region may persist until diplotene in some species. Studies of acrocentric-metacentric trivalent formation in lemur hybrids, and translocations in the mouse show that SC analysis in spreads is uniquely applicable to cytogenetic analysis. Axial structures are also observed in mitotic chromosomes, implying a fundamental role in chromosome organization. Intact SC's have been isolated with attached annular sheets at their ends, suggesting common characteristics.

INTRODUCTION

Meiosis has evolved as a highly conserved process in eukaryotes. Its selective advantages derive from the production of haploid gametes for sexual reproduction and from the genetic variability provided by reassortment and recombination of parental genes. Also highly conserved is a chromosomal structure, the synaptonemal complex (SC; reviews: 1, 2, 3, 4) that has evolved in eukaryotes in such a close parallel with meiosis that it may reasonably be asked whether the SC is not to some extent responsible for its evolutionary success.

The exact role of the SC in meiosis is presently not understood. However, the recent development of serial section reconstruction and microspreading whole mount techniques (discussion in 5) provides information about whole complements of SC's that confirms and extends the early views about the nature and behavior of the SC in meiotic processes which were based mainly upon observations on random thin sections (1,2).

Chromosome reduction and crossing over, the two special functions that characterize meiosis, depend on pairing and recombination respectively. These latter events are sometimes encountered in mitotic cells as well, and must thus occur independently of the SC, which has never been found associated with somatic chromosomes. The recognition and pairing of homologous maternal and paternal chromosomes is a well known but poorly understood occurrence in many somatic cells (e.g., in Diptera). However, with the possible exception of the special synaptic register that occurs in polytene chromosomes, it is the process of synapsis and maintenance of the synaptic state leading to crossing over that is unique to meiosis. Recombination, a molecular event involving exchange between homologous non-complementary DNA single strands, occurs also in somatic cells, but at a low frequency. Thus the refinement of pairing, i.e., synapsis, and the facilitation of recombinational events leading to crossing over, are the characteristics of meiosis with which the SC is most likely to be functionally associated.

THE NATURE OF THE SC

The SC forms the axis of the pachytene bivalent, extending continuously from one end to the other. Chromatin radiates from it in lampbrush-like loops. The typical morphology of the SC is shown in Fig. 1. Two parallel dense filamentous lateral elements flank a central element that lies in the same plane, and are joined to it by thin (less than 2.0 nm) transverse filaments (Fig. 2). The available cytochemical evidence indicates that protein is the main component of all

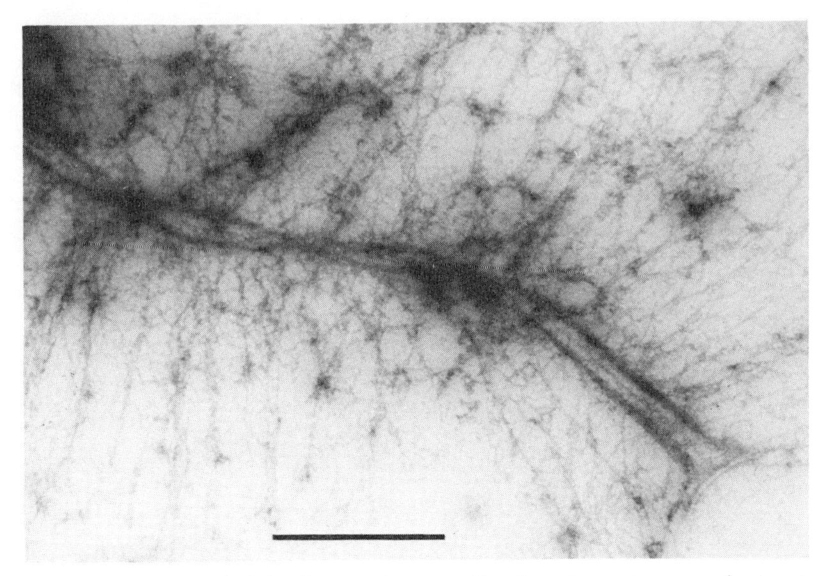

Fig. 1. Autosomal bivalent from a Syrian hamster pachytene spermatocyte spread on 0.45% NaCl, stabilized with uranyl acetate, and protected dried from 0.5% polyvinyl pyrrolidone. Chromatin fibrils radiate as collapsed loops from the lateral elements of the prominent synaptonemal complex (SC), which is straight and thick near its termination on the nuclear envelope, and is twisted elsewhere. Central elements and transverse filaments are also visible. Magnification approximately 23,000 X. (Bar = 1 μm). From Moses and Solari (39), with permission of the Journal of Ultrastructure Research.

of these elements. The lateral elements probably consist of arginine-rich, non-histone proteins (6, 1), and the central element of other protein(s) with different solubility and enzyme sensitivity (7, 8). No DNA is detectable cytochemically (9, 1, 2, 3, 10, 8), but a DNase sensitive 6.5 nm thread is a component of each lateral element (7). Chromatin that is resistant to DNase remains associated with the lateral elements after digestion. Evidence concerning the RNA content of the SC is contradictory, probably because of the equivocality of some of the cytochemical tests employed. RNA has been reported to be a major component of the lateral (11, 12, 13) and central (13) elements on one hand, while according to other studies, no nucleic acid of any sort is detectable in the SC (6, 9, 2). The failure of RNase to alter the structure of SC's in spread preparations (10) and of tritiated uridine either to be incorporated in the SC, or to be carried into it by previously synthesized RNA (14), tend to support the negative conclusions.

The SC usually terminates on the nuclear envelope in an

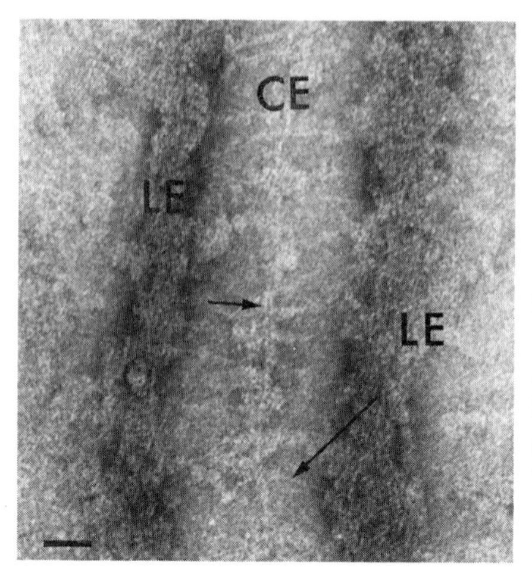

Fig. 2. Autosomal bivalent of the Syrian hamster, spread on 0.45% NaCl, digested with DNase for 2 hours, and negatively stained with uranyl acetate. LE: lateral element; CE: central element of the SC. Transverse filaments (long arrow) join the LE's to the CE. Stretches of longitudinal filaments (short arrow) run along the CE. Magnification approximately 53,000 X. (Bar = 0.1 μm). From Solari and Moses (8).

attachment plaque, and in spread preparations, pieces of the envelope adhere to the ends (Fig. 1). Arrays of packed nuclear annuli often extend from the dense attachment plaques, frequently joining adjacent SC's (Fig. 3).

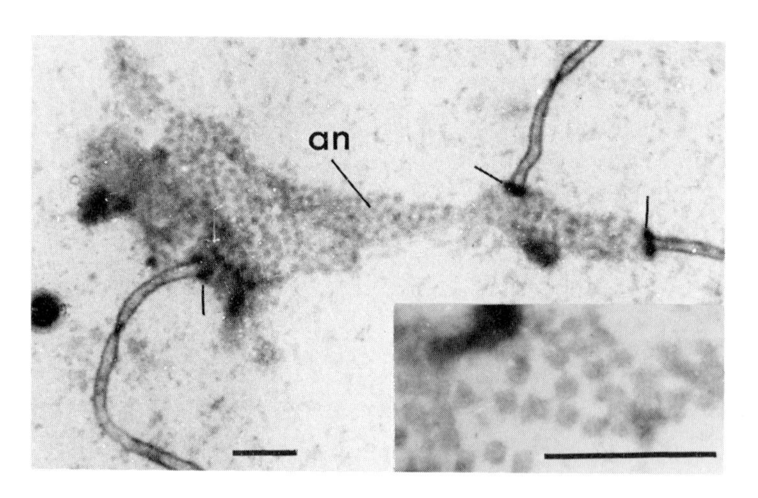

Fig. 3. Chinese hamster pachytene spermatocyte, Counce-Meyer spread. Clusters of nuclear annuli (an), some of which show regular packing, often connect the terminal plaques (-) by which the SC's are attached to the nuclear envelope. Individual annuli, some of which are connected by fine processes, are seen at higher magnification in the inset. Magnification approximately 8,500 X; inset 23,400 X. (Bars = 1 μm). From Moses (5) with permission of Chromosoma.

The unexpected discovery that the kinetochore is represented on the SC by a localized differentiation of the lateral elements was made by Counce and Meyer (15) coincident with their development of a micro-spreading technique that produces selectively stained SC's. The method is simple and rapid: in brief, it consists of spreading a small drop of suspended meiocytes on the surface of 0.5% NaCl, collecting the spread cells by adhesion to formvar-carbon coated grids, fixing with formaldehyde, drying in "Photoflo" (Eastman Kodak: 16) and staining with ethanolic phosphotungstic acid. Chromatin staining is suppressed; the SC, attachment plaques, kinetochore (e.g., Fig. 10), nuclear annuli, and nucleolar components stain strongly. Application of the method to mammalian spermatocytes (described in detail in 5) has yielded full complements of autosomal and sex bivalents as represented by their SC's in the golden hamster (17), man (18), Lemur (19), mouse (20), and the Chinese hamster (5, 21, 22), among other species. The Chinese hamster (Fig. 4) has been investigated most thoroughly and appears to be generally representative of the mammals studied so far.

SYNAPSIS AND DESYNAPSIS

The nature of the SC and its relationship to the chromosome in meiosis is best appreciated by following its formation and dissociation during meiotic prophase using the micro-spreading technique. The subsequent discussion is based largely on observations of the Chinese hamster, and relates primarily to mammals, although it undoubtedly applies more widely as well.

Serial section reconstructions of whole meiotic nuclei have made an important contribution to an understanding of the SC. SC formation in early meiotic prophase has been followed in *Locusta* (23), *Brachystola* (24), and Maize (25). While the method has the advantage of preserving structures *in situ*, it has the disadvantage of being laborious, exacting, and slow. As a consequence, sample sizes are apt to be small. Furthermore, structures such as the kinetochore are often not recognizable. The spreading method, on the other hand, is rapid and direct; it flattens, rather than disrupts nuclei, leaves many structures and relationships intact, and, as will be shown later, is capable of producing surprisingly little distortion.

Synapsis and desynapsis as observed in Counce-Meyer spreads of mammalian spermatocytes is summarized schematically in Fig. 5, where the relation of the axes to the paired chromosomes is also indicated.

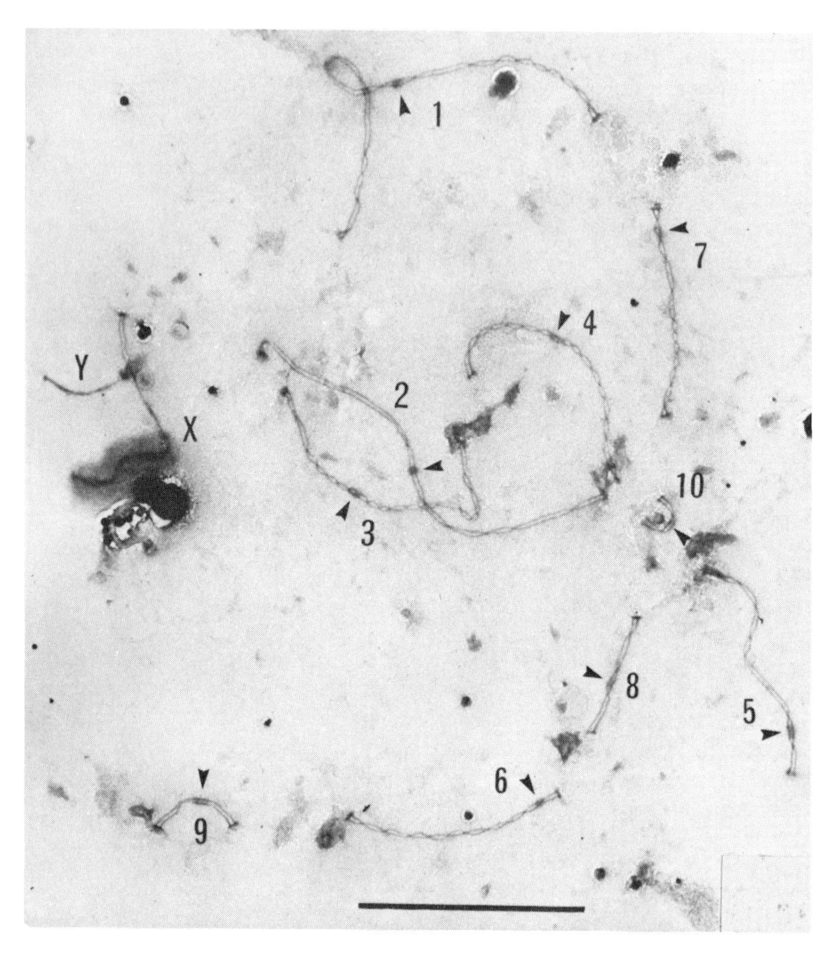

Fig. 4. Complete set of SC's from a Chinese hamster spermatocyte: Counce-Meyer preparation. Autosomal SC's are numbered according to ranked length as measured from one terminal attachment plaque to the other. Differentiation of the XY pair indicate the stage to be mid-late pachytene. Kinetochores (arrowheads) distinguish long and short arms for measurement. Magnification approximately 3,000 X. (Bar = 10 μm). From Moses et al (22) with permission of Chromosoma.

The simple filamentous structures that are the axes of single chromosomes prior to synapsis end in attachment plaques. They form first at the nuclear envelope just before, or at the time of synapsis, depending on genus. They develop as thin tortuous, often discontinuous threads that thicken as they mature. Kinetochores are often represented as thickenings on the unpaired axial elements. Leptotene, defined as the stage

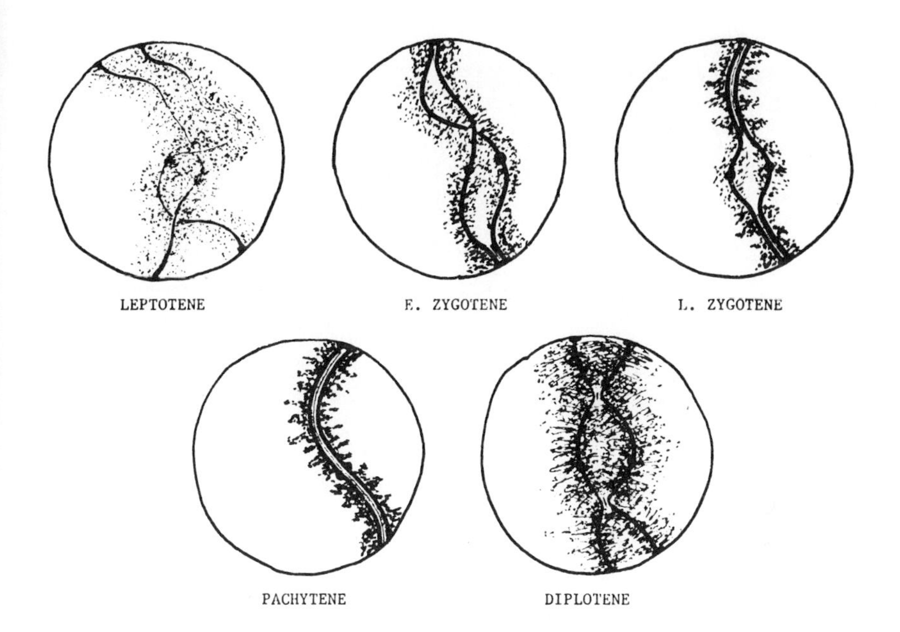

Fig. 5. Schematic diagram of autosomal synapsis and desynapsis in meiosis based on studies of spread preparations from 5 mammals. The essential structural components are shown: nuclear envelope, axial elements (chromosome axes), SC (with lateral and central elements), chromatin extending from the lateral (axial) elements, kinetochores on lateral (axial) elements, terminal plaques by which axial and lateral elements are anchored to the inner face of the nuclear envelope, and nuclear envelope.

at which only single axial elements are observed in the full complement, is seldom encountered, indicating that appearance of the axis is quickly followed by SC formation. Unpaired ends are apt to be widely separated; there is no evidence that ends are paired prior to SC formation. Synapsis, represented by formation of the SC through the coming together of homologous axial elements coincident with the appearance of the transverse filaments and central element material, also initiates at the nuclear envelope (Fig. 6). Completion of the axial elements may not occur until after SC formation is well underway. Neither SC initiation nor termination is synchronous. SC formation progresses proximally toward the kinetochore, which in the Chinese hamster is the last region to synapse (Fig. 7). The stage at which both SC's and unpaired lateral elements exist concomitantly is defined as _zygotene_.

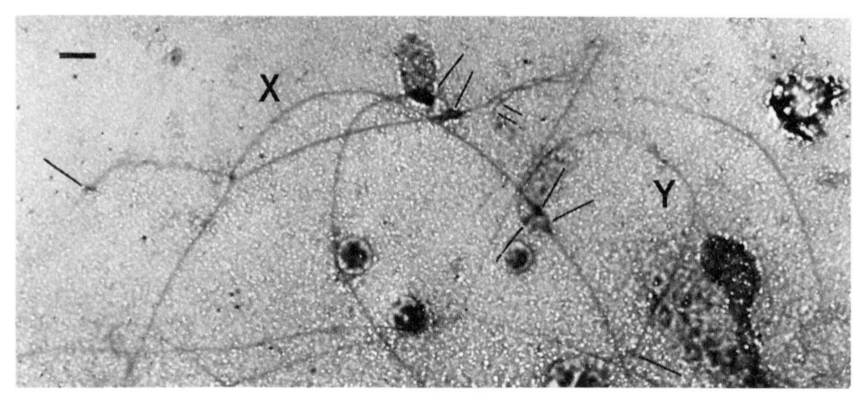

Fig. 6. Early zygotene: portion of a pachytene Chinese hamster spermatocyte from a Counce-Meyer spread. Single axial elements predominate, but a few short lengths of SC have formed, one of which is indicated at its termination (=). Distal to this point the SC branches into two axial elements that diverge to widely separated parts of the nucleus. The X and Y axes, of different lengths, are joined at their attachment plaques but have not yet synapsed to form an SC. Attachment plaques of unpaired axial elements are indicated by (-). Magnification approximately 5,800 X. (Bar = 1 μm). From Moses (5), with permission of Chromosoma.

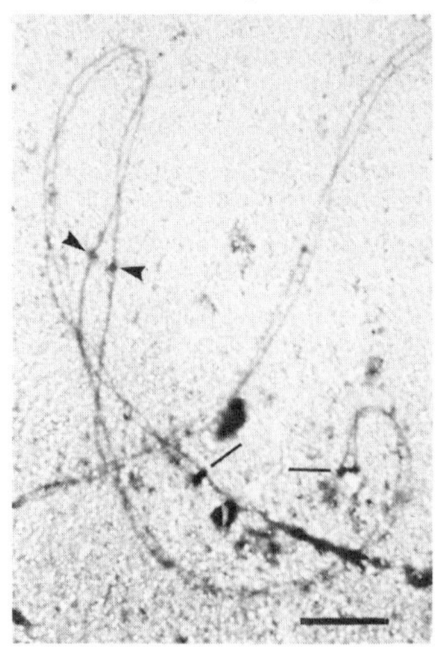

Fig. 7. Late zygotene: Chinese hamster autosomal bivalent (identified as #2). The SC is incomplete in the region of the kinetochore (arrowheads), indicating that synapsis has proceeded from the ends proximally. Each unpaired axis has a distinct kinetochore. Magnification approximately 9,000 X. (Bar = 1 μm). From Moses (5), with permission of Chromosoma.

At pachytene the prophase stage of longest duration, and the one in which recombination probably occurs, each autosomal bivalent is represented by a complete SC (Fig. 4). Ends remain attached to the nuclear envelope via attachment plaques throughout pachytene. Twists along the SC may occur, but although their frequency increases during pachytene, they appear to be random. SC's sometimes hook around each other, possibly as a consequence of the spreading technique, but interlocking is seldom encountered, suggesting that at the time of synapsis, movement of other ends between homologous ends is somehow prevented, or the consequences are corrected by a process that is not yet apparent.

Autosomal desynapsis (diplotene) occurs within a short period of time, but is not entirely synchronous. It may start at the bivalent ends or interstitially. In the mouse, where kinetochores are virtually terminal, the basal heterochromatic knobs often hold the ends together, and several basal knobs may fuse to give a small chromocenter containing a number of ends (Fig. 8). In general, chromatin at late pachytene-diplotene tends to condense and to stain more conspicuously. Axes lengthen and are more susceptible to stretching; subsequently they begin to disassociate and are lost from view. The attachment plaques are the last structures to be seen. At the latest stage of diplotene observed in the mammals studied there is no sign of axial doubling that would reflect chromatid

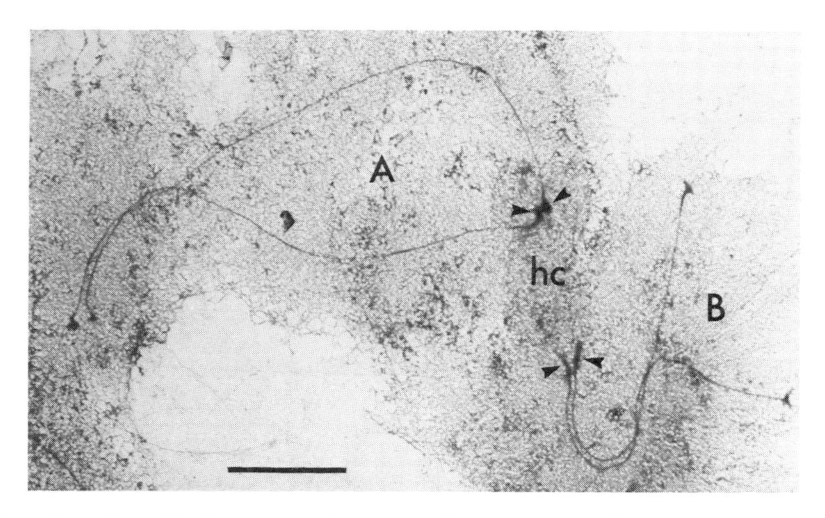

Fig. 8. Two diplotene autosomes from a spread mouse spermatocyte. Kinetochores (arrowheads) are virtually terminal and end in heterochromatic knobs (hc) which tend to fuse to form small chromocenters. In the bivalent at the left (A), the distal end remains synapsed, while on the right (B), the distal end has desynapsed. Although desynapsis in A undoubtedly started proximally, the kinetochore ends remain together, embedded in the terminal knobs. Desynapsis appears to have started at or near the bivalent ends. Magnification approximately 16,000 X. (Bar = 1 μm).

individualization, and the kinetochore is single. However, in the grasshopper, Melanoplus (26), late diplotene chiasma figures have been observed with attachment plaques and kinetochores, together with a suggestion of remnants of the axis (Fig. 9). Here the kinetochores are clearly double, suggesting that chromatid individuality is expressed only after the axis has broken down. In no instance has evidence been encountered to suggest that the SC is shed intact from the bivalent in mammals.

THE SC AND CROSSING OVER

There is apparently no exchange of SC lateral elements in crossing over; thus axes do not appear to participate directly in the chromatid reassortment that follows recombination. There is presently no direct evidence from which to conclude whether crossing over takes place in chromatin immediately associated with the SC or in that more peripheral to it. One observation, however, may be relevant to chiasma formation. "Recombination nodules" were observed by Carpenter (27) in serial section reconstructions of Drosophila oocytes to appear transiently associated with the SC at pachytene, in frequencies

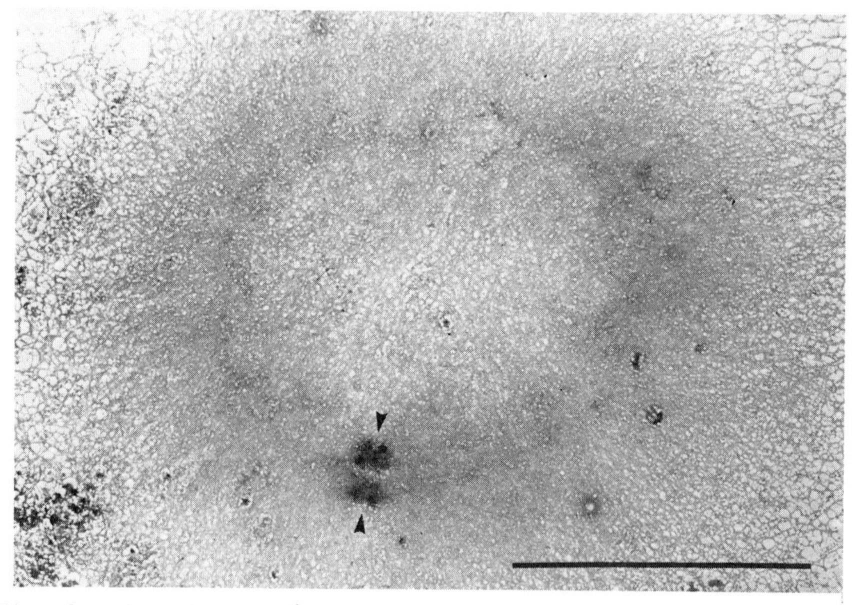

Fig. 9. Grasshopper (Melanoplus differentialis) spermatocyte: Counce-Meyer spread. Ring bivalent at late diplotene showing the bipartite structure of the two kinetochores (arrowheads), and remnants of SC material along the axes of the homologues. Magnification approximately 40,000 X. (Bar = 1 μm). From Solari and Counce (26) with permission of the Journal of Cell Science.

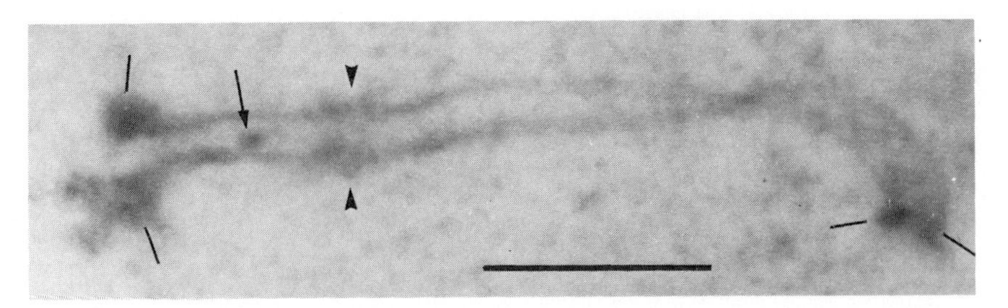

Fig. 10. Pachytene SC from the Syrian hamster: Counce-Meyer spread. Attachment plaques (-) and kinetochores (arrowheads) are indicated. The central element is indistinct in this preparation. A structure identified as a recombination nodule (arrow) lies on the SC about midway along the short arm. One or two similar bodies were found associated with each of 10 other SC's of the complement from which this example was taken. Magnification approximately 27,000 X. (Bar = 1μm).

and positions that suggest a parallel with crossing over. Comparable bodies have also been observed occasionally in spreads (Fig. 10; Moses, in preparation). They appear as ca. 90 nm dense granules centered on top of the SC, in the manner of a basketball on a ladder. There is seldom more than one per bivalent arm, and when one appears on an SC, most of the remainder of the SC complement show them as well. Although they are strongly implicated in crossing over, their exact role in the process is unclear. At diplotene, chiasma sites are marked by short residual segments of SC, according to serial section reconstructions by Solari (28). Such segments are also found in spread preparations (Fig. 11 ; Solari and Moses, in preparation), but again their significance to chiasma formation is not known. However, studies now in progress on species with localized chiasmata may help to clarify the relationship.

THE SEX CHROMOSOMES

The basic understanding of the mammalian sex chromosome comes largely from studies on sectioned material (29). Analysis of spread preparations shows (22) that the X and Y chromosomes do not follow the autosomal schedule of synapsis and desynapsis, and their unpaired axes undergo pronounced morphological changes during pachytene. In most mammals, the X and Y axes pair to form a length of SC that presumably marks regions of homology. Synapsis begins during late zygotene and early pachytene at one end where terminal plaques are closely

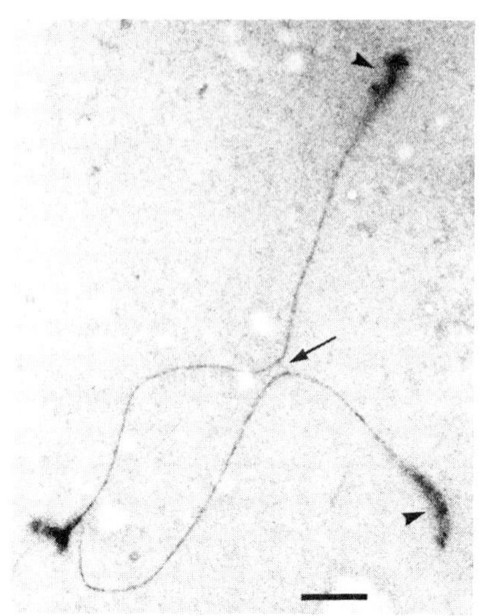

Fig. 11. Diplotene, mouse: Counce-Meyer spread. The ends of the axial elements carrying well-defined kinetochores (arrowheads) are widely separated, while the distal ends remain associated. A short length of SC (arrow) midway between the ends holds the homologues together in the region of the putative chiasma. The central element is not evident here, but it is in sections (28). Magnification approximately 7,700 X. (Bar = 1 μm). Micrograph by A. J. Solari.

associated, while those at the opposite end remain separated. The maximum SC length is attained early in pachytene. In the mouse, most of the distal length of the Y pairs with the distal portion of the X (Fig. 12); in the human, the entire short arm of the Y pairs with the distal portion of the short arm of the X (18); in the Chinese hamster the entire short arm of the Y pairs with part of what is tentatively identified as X long arm (21). The behavior of the XY-SC during most of pachytene varies with genus. In the Chinese hamster, the SC may remain essentially unchanged in length; in the Syrian hamster it undergoes partial reduction (Moses, unpublished); in the mouse it reduces early to a constant short length (29, and unpublished); in the human it reduces totally to an early association of ends (18). Desynapsis of the X and Y is generally complete by late pachytene-early diplotene, with only terminal or interstitial short lengths of SC remaining.

The presence of an SC in the paired region suggests homology, but is not alone sufficient to assure chiasma formation. However, from the weight of evidence that crossing over requires the presence of an SC (1, 2, 3, 4), the presence or absence of an SC during the time when recombination occurs may become limiting. Thus, in the human, where complete desynapsis occurs precociously during early-mid pachytene (18), recombination between X and Y may be prevented by the absence of an SC. The mouse may represent a similar situation; the long pairing region (Fig. 12) is reduced to a short distal SC before mid-pachytene.

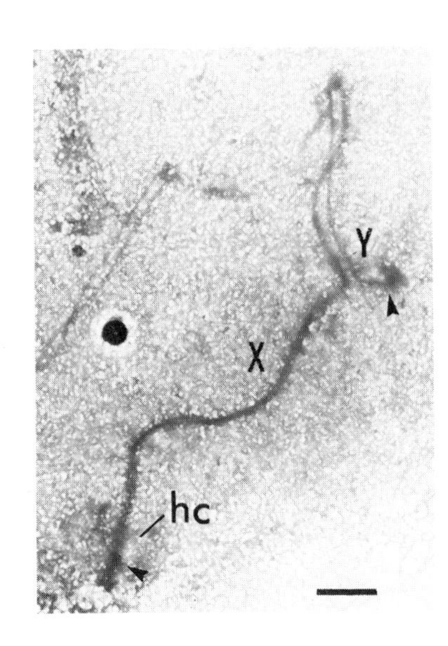

Fig. 12. Mouse XY pair: Counce-Meyer spread. At early pachytene most of the distal portion of the Y axis forms an SC with the distal portion of the X. By mid-pachytene desynapsis will have occurred reducing the SC at the distal end to about 1/5 of its present length. The unpaired axes are thicker than the lateral elements of the SC, but they have not developed pronounced differentiations. The kinetochore (arrowhead) of the Y is virtually terminal, while that of the X is sub-terminal, buried in the heterochromatic knob (hc). Magnification approximately 6,400 X. (Bar = 1 μm).

The unpaired arms of the X and Y usually remain separate until desynapsis is complete, at which point they sometimes approach each other and associate by their ends. During pachytene, the unpaired X and Y axes undergo conspicuous differentiations (29) that may be clearly followed in spread preparations. At zygotene, prior to XY synapsis, the axes resemble unpaired autosomal axes. After synapsis, the unpaired arms thicken and often appear double or triple. Although the axes undergo elaborations that vary in detail and extent from cell to cell, the overall pattern that they follow appears to be genus characteristic. Axes may develop hollow, fusiform thickenings, and increases may occur in mass of the axis and of material associated with it, in the form of coils, granules of various size, cloudy masses, etc. Many of the differentiations found are exemplified in the Chinese hamster (Fig. 13)(21). Following XY-SC reduction, the sex pair increases in mass, the axes become discontinuous, greatly elongated, and branched, leading to a complex network that is lost from view at the chromosomes condense.

QUANTITATIVE ANALYSIS OF SPREAD PREPARATIONS

It is apparent from the foregoing that the behavior of the autosomes and the sex chromosomes in meiotic prophase may be inferred from that of their axes. Thus, the axial elements at synapsis become the lateral elements of the SC, which in turn follows the behavior of the bivalent of which it is the axis (Fig. 5). These facts, augmented by the Counce-Meyer spreading technique, make possible a simply prepared, easily discernible representation of the bivalent, with clearly

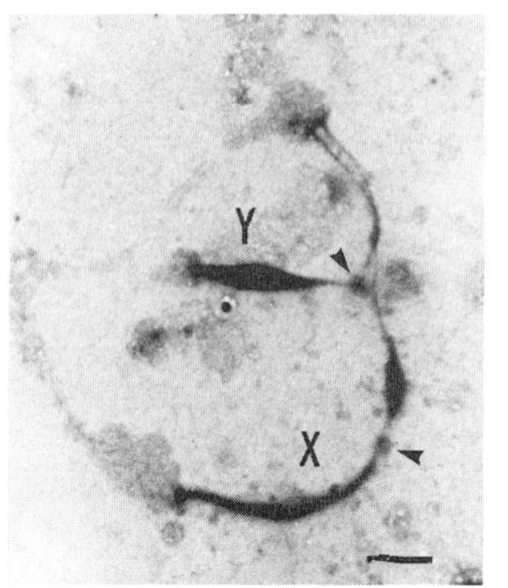

Fig. 13. Chinese hamster XY pair, mid pachytene. The unpaired axes have formed fusiform swellings which become hollow: one in the Y and two in the X. Kinetochores (arrowheads) are visible: the SC occupies an entire area of the Y and part of an area of the X. From measurements, the short arm of the Y appears to be paired with part of the long arm of the X. Magnification approximately 6,400 X. (Bar = 1 μm). From Moses (21), with permission of Chromosoma.

marked ends and a distinct kinetochore, that provides a high degree of resolution and detail. The potential of the SC for cytogenetic analysis has been suggested from the results of serial section reconstructions (25, 30, 31); it should now be more readily realizable if it can be demonstrated that the spreading approach is quantitatively as well as qualitatively reliable.

A quantitative analysis was made of measurements of SC lengths and arm ratios in over 50 spermatocyte nuclei from 7 Chinese hamster males (for details, see 22). Individual autosomal bivalents are identifiable by these characteristics, while X and Y are distinguishable by their presence in a sex body, and by the lengths and differentiations of their unpaired axes. The striking observation was made that despite sizeable absolute length variations, the relative length of any autosomal SC (i.e., its measured length ÷ the sum of all the SC lengths) is a constant (22). Thus, the distortions that might have been expected from the spreading procedure are in fact negligible. It also follows that the biological control of length in meiotic prophase is constant for all autosomes. The regularity of arm ratios also confirms both the lack of distortion and biological stability. As a consequence, relative length and arm ratio may be used together to characterize accurately each autosomal bivalent. An idiogram of the Chinese hamster autosomal complement is shown in Fig. 14 (dark bars). When the pachytene karyotype is compared with the mitotic metaphase karyotype (Fig. 14, light bars), the similarity is striking: the relative lengths and arm ratios are not significantly different at the 95% confidence level. The identity of SC and somatic arm ratios also proves that the differentiations on the SC identified as

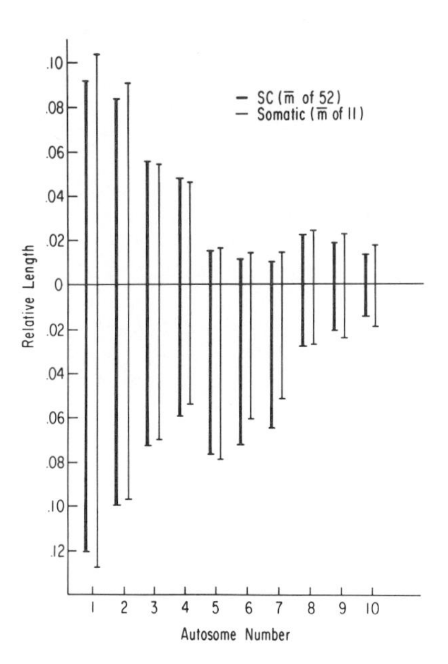

Fig. 14. Idiogram of Chinese hamster autosomal SC's, constructed from relative length measurements and arm ratios of 52 pachytene spermatocytes, compared with similar data from 11 sets of mitotic autosomes. From data in Moses et al (22).

kinetochores are in fact correctly positioned to represent that structure.

To determine whether the observed variations in absolute lengths of the autosomal SC's were random or were correlated with stage of pachytene, it was first necessary to establish a timing marker. In the Chinese hamster (21, 22), the progressive morphological differentiations of the X and Y axes are sufficiently distinct as to divide pachytene into 5 sequential periods, the length of each being proportional to the frequency at which it occurred. A correlation between SC length and stage of pachytene was in fact observed: a 30% decrease occurred between late zygotene and mid-pachytene, followed by a short, stable period, and then a 20% increase to late pachytene. The constancy of relative lengths during this period indicates that all autosomal SC's undergo the changes synchronously. This pattern of shortening and lengthening resembles that described in Drosophila from serial sections (31). The X and Y axes (but not the XY-SC) also undergo decreases in length, but these are not in proportion either to the autosomal SC's, or to each other. Thus, again, the sex chromosomes are, as they were in synapsis and desynapsis, out of phase with the rest of the complement.

The demonstration that autosomal relative lengths and

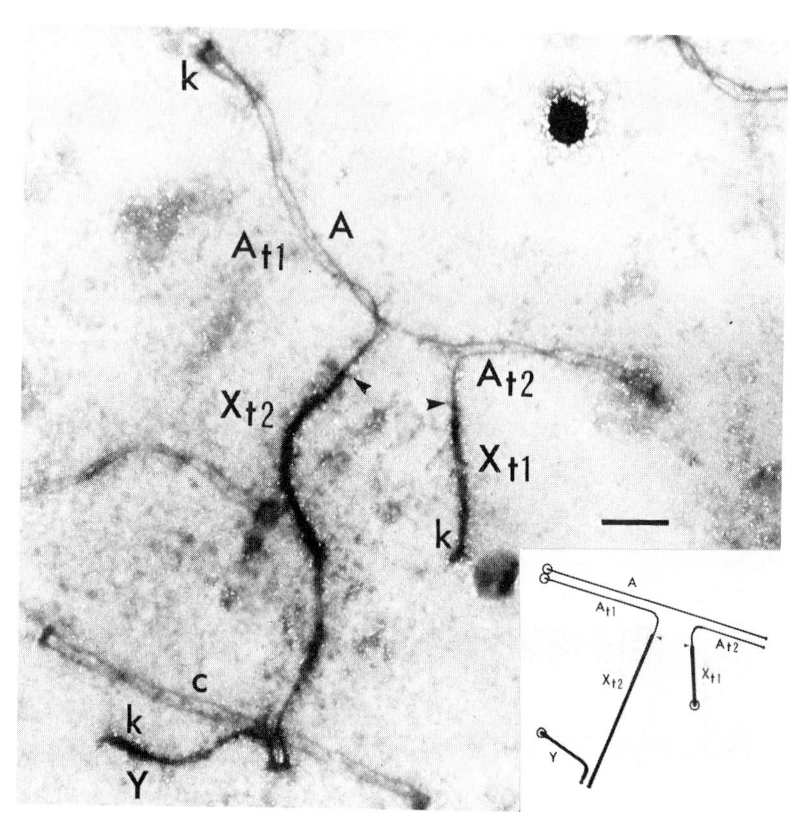

Fig. 15. Translocation quadrivalent from a male mouse heterozygous for an X-autosome translocation [T(X;7)2Rℓ]; compare with schematic in inset. A: non-translocated autosomal lateral element (axis); A_{t1} and A_{t2}: axes of translocated portions of autosome; X_{t1} and X_{t2}: translocated portions of X axis; _c_: indifferent autosomal SC; _k_: kinetochore; Y: Y axis; (-): distal attachment plaque; arrowheads: translocation breakpoints. In the long translocation product, $A_{t1}X_{t2}$, the X_{t2} portion consists of the distal 3/4 of the X. The X and Y are terminally synapsed via a short segment of SC. An autosomal SC (_c_) accidentally crosses the XY pair. Autosomal portions of the two translocated axes have either failed to pair with, or have stripped away from the non-translocated (A) axis for a short distance near the translocation breakpoints. Magnification approximately 9,000 X. (Bar = 1 μm). From Moses, Russell and Cacheiro (32) with permission of Science.

arm ratios alike are constant in the face of absolute length changes, and that there is a proportionality between autosomal

SC and mitotic metaphase length, validates the spreading method for SC karyotyping. The same constancy of relative length and of SC-mitotic proportionality has been observed in four other mammals (20; mouse, Moses in preparation; golden hamster, 17; lemur, 19; and man, 18) and SC karyotypes have been constructed for them.

CYTOGENETIC APPLICATIONS

One of the obvious applications of the technique to cytogenetic problems is in the analysis of chromosomal rearrangements. Pachytene is a difficult stage for such analysis with the light microscope, although it is potentially a most important one. A collaborative study of translocations in the mouse has been undertaken with Drs. Liane Russell and Nestor Cacheiro. Three have been investigated so far (32): two X-autosome [T(X;7)6Rℓ; T(X;7)2Rℓ] and one autosome-autosome translocation [T(10;18)12Rℓ]. In spread preparations of pachytene spermatocytes from heterozygous males of each of these stocks, the translocations were quickly and easily identified, and were diagrammatic in their clarity. An example is shown in Fig. 15 in which the breakpoints are sharply discernible because of differences in thickness between the translocated segments of the X axis and those of the autosomal axial segments to which they were translocated. It was calculated that 21% of the X axis and 60% of the autosome axis (#7) were proximal to the respective breakpoints. These figures compare well with 22% and 66% independently estimated from banded mitotic chromosomes.

The potential of this method for analyzing other chromosome irregularities is also under investigation. In this respect, Robertsonian translocations in lemur species have been studied (19; Moses, Karatsis and Hamilton, in preparation). A hybrid between _Lemur fulvus_ subspecies carried 5 metacentric chromosomes of one parent and 10 acrocentrics of the other. G-banding showed each of the acrocentrics to be homologous with an arm of a metacentric. In spread preparations, 5 heteromorphic trivalent SC's were found in each SC complement, representing the pairing of acrocentrics with metacentric arms, and confirming their homology. The three kinetochores were closely juxtaposed, with those of the acrocentrics in tandem along one axis and opposite to that of the metacentric. Such a disposition of kinetochores could play a role in disjunction of the trivalent and perpetuation of the rearrangement.

These two examples illustrate the opportunities for extending the method to other cytogenetic problems.

PROTEIN AXES OF MEIOTIC AND MITOTIC CHROMOSOMES

It is apparent that the axial element, both as chromosome axis and as the lateral element of the SC (which in turn is axial to the bivalent), is integral with chromosome structure at meiotic prophase. It is transient, appearing at synapsis and disappearing at or following desynapsis. The two most obvious functions attributable to it are to maintain a linear order of the chromosome, without which synaptic register could not be achieved, and to hold sister chromatids together as a unit structure. A means by which the latter may occur is suggested by the small amount of DNA that remains unreplicated after the pre-meiotic S period in lily (33), which could serve this purpose at the molecular level. Following the replication of this DNA at zygotene, the two chromatids should then be free to behave individually. However, their integration with a single protein axis would prevent this until the axis disperses following diplotene. This of course is just the time at which each chromosome is observed by light microscopy to become double.

With respect to the role of the axial element in maintaining the linear order of the chromosome, the relationship between autosomal SC and mitotic chromosome relative length raises an important question. A 1:1 relationship has been observed in every species so far examined (17, 18, 19, 20, 22, 30), indicating the generality of this phenomenon and underscoring the conclusion that similar length controls probably operate at both meiotic prophase and mitotic metaphase. While regulation of chromosome length may conceivably be determined by the folding and packing capacities of the chromatin alone, the demonstration that an axial structure at meiotic prophase follows chromosome length changes leads to the possibility that the axis, as a consequence of providing a framework along which the chromatin is arranged, may itself be a controlling factor in length determination. The critical period for such a function would be during the transition from a dispersed state of chromatin at interphase to a condensed one at metaphase. The obvious question that follows, then, is why no axis is seen in somatic chromosomes, where similar linearity and length controls are shown to operate.

A mitotic chromosome axis has, in fact, been inferred from observations on isolated and extracted mammalian chromosomes (34). However, axes have not been observed in thin sections or in spread preparations of somatic chromosomes (e.g., 35, 36, 37). On the other hand, to my knowledge, entire mitotic chromosomes have not been reconstructed by serial sectioning. Further, the spreading techniques generally

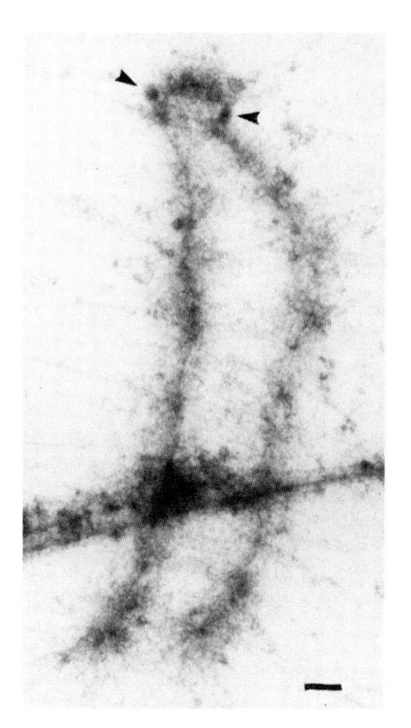

Fig. 16. Acrocentric mitotic chromosome from a HeLa cell spread and stained with the Counce-Meyer procedure and protected dried from Ficoll (39). The putative kinetochores are densely staining subterminal discs (arrowheads). The fuzzy appearances of the chromatin indicates that the chromosome is not significantly stretched. An axis extends in the chromatid on the left from the kinetochore nearly to the distal end. Fragments of the axis are visible in the chromatid on the right, which may be slightly stretched. Magnification approximately 4,300 X. (Bar = 1μm). From Moses, in preparation.

used may neither reveal nor preserve all chromosome structures, as attested to by the failure to observe kinetochores on either mitotic (36) or meiotic (10, 7, 8) chromosomes prior to the Counce-Meyer modifications (15, 37). Although the densely staining kinetochores of mitotic chromosomes are preserved, as shown _in vitro_ by their microtubule organizing capacity (38), axes are not found in mitotic chromosomes following this procedure. On the other hand, "protected drying" (39), a procedure that avoids the damaging surface tension stresses incurred during air drying, greatly improves the preservation of chromatin structure. Occasionally, in such preparations, mitotic chromosomes are found that have not been stretched during spreading and that show clear indications of a filamentous axis in each chromatid (Fig. 16) (40, Moses, in preparation). The axis appears to be continuous with the kinetochore; it stains strongly with ethanolic phosphotungstic acid as do meiotic chromosome axes, and disassembles at later stages, leaving an organization resembling that found at diakinesis (Fig. 9).

If the generality of such an axis in mitotic chromosomes is supported by more extensive observations, the presence of a similar axis in meiosis may be more an expression of a basic aspect of chromosome organization than a special characteristic of the meiotic process. Thus, synapsis, involving the uniting of homologous axes in an SC, via transverse filaments and the formation of a central element, would be the unique meiotic event. It should be emphasized that formation of an SC-like structure is not itself restricted to meiotic prophase chromosomes, as polycomplexes a) assemble apart from chromosomes, b) may persist in the cytoplasm, and c) may form in secondary spermatocytes and spermatids (1, 2), although they are not found outside germ line cells and their derivatives. It follows that SC formation as a concomitant of synapsis may then be regarded specifically as a meiotic phenomenon.

THE SC AND THE NUCLEAR ENVELOPE

In the mammals studied, meiotic events in which the chromosomal axes (and the SC) participate involve the nuclear envelope. Assembly of the axial elements and formation of the SC initiate at the nuclear envelope, forming a connection that is maintained until well after desynapsis. While such a relationship implies stability, it must also accommodate lability, in that chromosome ends move over the inner face of the envelope and thus alter their position with respect to each other, both before and after synapsis, as well as during pachytene (1, 41, 24).

The most conspicuous structure found associated with the attachment plaques are arrays of nuclear annuli, which often join the ends of SC's (Fig. 3). The connections provided by such bridges, some of which are so short as to constitute a virtual end-to-end junction, suggest maintenance of the kind of stability that could lead to tandem associations of specific ends. However, the annular associations have been shown to be completely random, and not to favor any particular chromosome combination (22). On the other hand, annuli, which are generally distributed over the nuclear surface, are also labile. They are known to congregate into closely associated clusters at meiotic prophase (42) and the arrays of annuli seen in spreads (5) and sections (24) probably reflect such aggregation. Movements of annuli somehow preserve the integrity of the membranous envelope with which they are intimately associated. The same is true for the attachment plaques of the SC, although the details of this membrane association are distinct from those of the annuli. Both annular and chromosome movement could be related. For example, it has been suggested (5) that if attachment plaques of the SC and annuli are intimately connected, movement of the latter during their aggregation could bring about movement of the chromosome ends.

Several lines of evidence point to annuli and chromosome axes having structural components in common. Most compelling is the observation on mosquito oocytes by Fiil and Moens (43) that annuli are generated in intimate conjunction with the lateral elements of polycomplexes; they associate into sheets together with fibrous material and eventually form annulate lamellae or become integrated with the nuclear envelope. The identity of SC lateral element and annular material was also inferred from similarities in staining with ruthenium red (44), and insofar as common staining properties with ethanolic phosphotungstic acid in Counce-Meyer spreads reflect similarities in chemical composition, annuli and chromosome axes have staining characteristics in common that suggest they may well be parts of the same structural system.

A proteinaceous inner lamina of the nuclear envelope, consisting of sheets of annuli connected by fibrous material, has been isolated and purified as a morphological entity from liver nuclei by Blobel and his associates (45). The preparations are characterized by three distinct proteins having molecular weights of 66, 68 and 69 x 10^3 daltons. This demonstration, together with the aforementioned morphological evidence, has stimulated a partially successful attempt to isolate inner lamina/annular complexes from pachytene spermatocytes in order to learn whether the lamina and SC behaved as a physical entity and could thus be separated as one (Walmsley and Moses, unpublished).

Nuclei were isolated by a modification of Aaronson and Blobel's procedure from Chinese hamster testicular suspensions, digested with DNase, and dispersed with Triton X-100. Intact, isolated SC's, mostly freed of chromatin, were observed. Many of them showed tufts of fibrillar material adhering to the ends (Fig. 17). Lateral elements, transverse filaments and central element, attachment plaques, and prominent kinetochores are all preserved; stubs radiating from the lateral elements probably represent DNase-resistant chromatin. The fact that the SC could remain intact after homogenization, centrifugation, etc., is promising. Of greatest interest is the observation that the tufts of fibrillar material continuous with the SC ends contain annuli, consistent with their being fragments of the fibrous lamina that are continuous with SC (Fig. 17). While these observations do not prove the unity of the SC and lamina, they are consistent with this idea.

Similar results to Aaronson and Blobel's have recently been obtained by Comings and Okada (46), who propose that the lamina is part of an extensive nuclear matrix (47) that is integrated with all chromosome structures, including the SC. Our limited results have no bearing on that proposition, as no morphological evidence for a matrix was obtained. Because our preparations were made from heterogeneous populations of

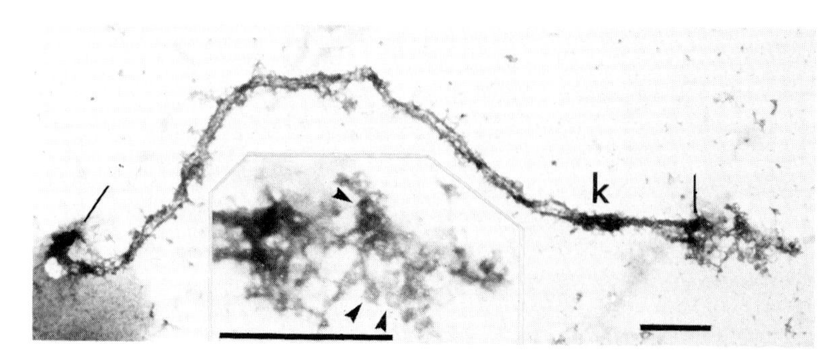

Fig. 17. Chinese hamster SC isolated from a testicular suspension by a procedure based upon that by which the pore-inner fibrous lamina complex has been isolated and characterized in liver nuclei (45). The preparation has been stained and dried by the Counce-Meyer procedure. As in intact nuclei, the lateral and central elements, the attachment plaques (-) and the kinetochore (k) are stained. Masses of fibrous material extending from the attachment plaques often contain annuli (arrowheads). Magnification approximately 9,000 X; inset, 22,000 X. (Bars = 1 μm). From Walmsley and Moses, in preparation.

testicular cells, the pellet containing the SC's was heavily populated with the residua of spermatozoa, spermatids, axonemes, manchettes, etc., and thus there was no hope of preparing a clean suspension of SC's for analysis. Nevertheless, the preliminary results suggest that SC's and annulus/fibrous lamina complexes can eventually be separated for analysis from enriched populations of spermatocytes (48). The questions of SC composition and relationships to the fibrous lamina cannot be settled without such preparations.

CONCLUSIONS

The morphological and functional continuity that embraces chromosome axis, kinetochore, attachment plaque, SC, and inner lamina/annular complex emerges as the common denominator of the observations discussed in this paper. In mammals, the inner face of the nuclear envelope appears to provide organizing centers from which assembly of the chromosome axis initiates. The material of these centers, possibly common to both axis and envelope, embodies syndetic properties by which homologous ends are brought into close proximity, thus setting the stage for initiation of the SC at the same place. In its formation, the SC literally creates the bivalent of which it is the axis. However, the extent to which SC formation requires the homology that it seems to keep in register, is not known.

The axial element, on the other hand, is seen as a transient backbone of the chromosome that functions in the transition from interphase to metaphase. It provides for genetic as well as chromosomal linear order. While the DNA molecule, as genophore, maintains gene sequences, populations of genes are kept aligned by the axis; the chromomeres of pachytene, and the bands of mitotic chromosomes known from light microscopy are probably the visible manifestations of such order. Accepting that an axis is common to all chromosomes, it is entirely possible that the linear, rod-like shape of the chromosome is set by the axis, and that for this reason, chromosomes are rods and not spheres. The axial protein(s) probably have DNA binding properties which serve to maintain order in the chromatin.

The kinetochore, as a modified locus on the axis, is probably determined by the DNA with which it is associated. It expresses singleness or doubleness of the underlying chromatids and ultimately serves its role in disjunction as an oriented microtubule organizing center. In a sense, through the axis, attachment plaque and inner lamina/annular complex, the kinetochore is continuous with the nuclear envelope. This is the surface on which the kinetochore appears to have had its evolutionary origins (49).

The nature of the recognition and pairing forces that are dependent on homology, and the precise role played by the SC in crossing over are central questions that remain unanswered. By contrast, it is notable that the SC now provides a useful handle for cytogenetic analysis at pachytene.

ACKNOWLEDGEMENTS

The research described was supported in parts by grants from the NSF (GB-40562) and NIH (GM-23047) to M. J. Moses, and 4-S01-RR-05404 and CA-14236 to Duke University. It is a pleasure to thank Todd Gambling and Garnett Slatton for their technical assistance, and Marlene Johnson for help in preparing this manuscript. I am particularly grateful for the collaboration at various times of Drs. S. J. Counce and A. J. Solari.

REFERENCES

1. Moses, M. J. (1968) Ann. Rev. Genet. 2, 363.
2. Moses, M. J. (1969) Genetics 61, 41.
3. Westergaard, M. and Wettstein, D. von (1972) Ann. Rev. Genet. 6, 71.
4. Gillies, C. B. (1975) Ann. Rev. Genet. 9, 91.
5. Moses, M. J. (1977a) Chromosoma (Berl.) 60, 99.
6. Moses, M. J., and Coleman, J. R. (1964) Proc. 23rd Symp. Soc. for the Study of Development and Growth; "Role of the Chromosomes in Development". (Ed., M. Locke), Academic Press, New York, p. 11.
7. Solari, A. J. (1972) Chromosoma (Berl.) 39, 237.
8. Solari, A. J., and Moses, M. J. (1973) J. Cell Biol. 56, 145.
9. Coleman, J. R., and Moses, M. J. (1964) J. Cell Biol. 23, 63.
10. Comings, D. E., and Okada, T. A. (1970) Chromosoma (Berl.) 30, 269.
11. Esponda, P., and Stockert, J. C. (1971) J. Ultrast. Res. 35, 411.
12. Chevaillier, P. (1974) J. Microscopie 19, 147.
13. Westergaard, M., and Wettstein, D. von (1970) Compt. Rend. Trav. Lab. Carlsb. 37, 239.
14. Kierszenbaum, A. L., and Tres. L. L. (1974) J. Cell Biol. 60, 39.
15. Counce, S. J., and Meyer, G. F. (1973) Chromosoma (Berl.) 44, 231.
16. Miller, O. L., and Beatty, B. R. (1969) Science 164. 955.
17. Moses, M. J., and Counce, S. J. (1974) in "Mechanisms in Recombination (R. F. Grell, ed.), New York, Plenum Publ. Corp., p. 385.
18. Moses, M. J., Counce, S. J., and Paulson, D. F. (1975) Science 187, 363.
19. Moses, M. J., Karatsis, P. A., and Hamilton, A. E. (1975) J. Cell Biol. 67, 297a.
20. Moses, M. J., and Counce, S. J. (1976) J. Cell Biol. 70, 131a.
21. Moses, M. J. (1977b) Chromosoma (Berl.) 60, 127.
22. Moses, M. J., Slatton, G., Gambling, T., and Starmer, C. F. (1977) Chromosoma (Berl.), 60, 345.
23. Moens, P. B. (1969) Chromosoma (Berl.) 28, 1.
24. Church, K. (1976) Chromosoma (Berl.) 58, 365.
25. Gillies, C. B. (1973) Chromosoma (Berl.) 43, 145.
26. Solari, A. J., and Counce, S. J. (1977) J. Cell Sci, in press.
27. Carpenter, A. T. C. (1975) Proc. Nat. Acad. Sci. (U.S.) 72, 3186.
28. Solari, A. J. (1970) Chromosoma (Berl.) 31, 217.

29. Solari, A. J. (1974) Intern. Rev. Cytol. 38, 273.
30. Moens, P. B. (1973) Cold Sp. Harb. Symp. Quant. Biol. 38, 99.
31. Carpenter, A. T. C. (1975) Chromosoma (Berl.) 51, 157.
32. Moses, M. J., Russell, L. B., and Cacheiro, N. L. (1977) Science, in press.
33. Stern, H., and Hotta, Y. (1973) Ann. Rev. Genet. 7, 37.
34. Stubblefield, E. (1973) Internat. Rev. Cytol. 35, 1.
35. Robbins, E., and Gonatas, N. K. (1964) J. Cell Biol. 21, 429.
36. Comings, D. E., and Okada, T. A. (1970) Cytogenetics 9, 436.
37. Moses, M. J., and Counce, S. J. (1974) J. Exptl. Zool. 189, 115.
38. Telzer, B. R., Moses, M. J., and Rosenbaum, J. (1975) Proc. Nat. Acad. Sci. (U. S.) 72, 4023.
39. Moses, M. J., and Solari, A. J. (1976) J. Ultrast. Res. 54, 109.
40. Moses, M. J., Counce, S. J., and Solari, A. J. (1974) J. Cell Biol. 63, 235a.
41. Rickards, G. K. (1975) Chromosoma (Berl.) 49, 407.
42. LaCour, L. F., and Wells, B. (1974) Phil. Trans. Roy. Soc. Lond. B. 268, 95.
43. Fiil, A., and Moens, P. (1973) Chromosoma (Berl.) 41, 37.
44. Engelhardt, P., and Pusa, K. (1972) Nature New Biol. 240, 163.
45. Aaronson, R. B., and Blobel, G. (1975) Proc. Nat. Acad. Sci. (U. S.) 72, 1007.
46. Comings, D. E., and Okada, T. A. (1976) Exptl. Cell Res. 103, 341.
47. Berezney, R., and Coffey, D. S. (1974) Biochem. Biophys. Res. Comm. 60, 1410.
48. Bellvé, A. R., Millette, C. S., Bhatnagar, Y. M., and O'Brien, D. A. (1977) J. Histochem. Cytochem. (Proc. 5th Internat. Symp. on Automated Cytology, 1976) in press.
49. Kubai, D. F. (1975) Intern. Rev. Cytol. 43, 167.

DIFFERENTIAL FLUORESCENT LABELLING OF CHROMOSOMES AND DNA WITH BASE PAIR SPECIFIC DNA BINDING ANTIBIOTICS

J. H. van de Sande[1], C. C. Lin[1,2], F. P. Johnston[3] and K. F. Jorgenson[4].

Divisions of Medical Biochemistry[1] and Pediatrics[2], Faculty of Medicine, and Department of Biology[3] and Department of Chemistry[4], The University of Calgary
Calgary, Alberta, Canada T2N 1N4

ABSTRACT. The anthracycline antibiotic daunomycin has been found to produce Q-like banding patterns on chromosomes. Solution interaction of this antibiotic with DNA of known base composition indicates that differential fluorescent quenching by DNA regions with specific base sequence arrangement can account for the observed bands on chromosomes. The carbohydrate moiety of daunomycin, daunosamine, stabilizes the interaction of this antibiotic with DNA or chromosomes, but it is not essential for the appearance of fluorescent bands.

The production of fluorescent bands on chromosomes by daunomycin or quinacrine is not a consequence of binding specificity of these antibiotics to DNA, but the variable quantum efficiency of these fluorochromes is a function of the base composition at the binding site. These results suggest that non-random nucleotide sequence arrangement along the chromosome is a primary determinant for the appearance of fluorescent bands.

Recently, the G-C specific DNA binding antibiotic olivomycin was found to produce characteristic reverse fluorescent banding patterns (R-bands) on human, bovine and mouse metaphase chromosomes. In solution, natural DNAs show an enhancement of olivomycin fluorescence related to their G-C content, which is an expression of the degree of antibiotic binding to DNA. In contrast, Hoechst 33258 fluorescence enhancement can be directly related to the A-T content of the interacting DNA. The solution findings and cytological observations suggest that fluorescent bands on chromosomes produced by olivomycin are the reverse of those produced by Hoechst 33258.

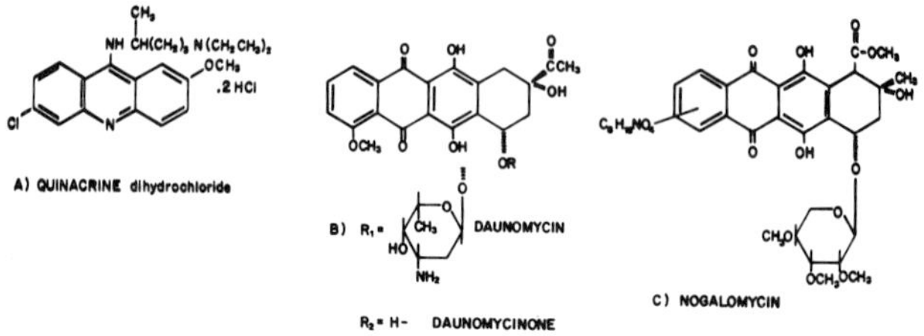

Fig. 1 Structures of quinacrine and the anthracycline antibiotics daunomycin, daunomycinone and nogalamycin.

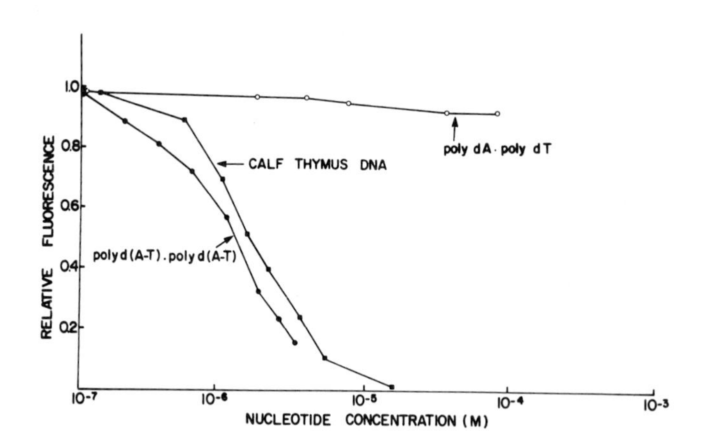

Fig. 2 Relative fluorescence of nogalamycin in the presence of varying concentrations of DNA. Nogalamycin (1×10^{-6} M) and polynucleotides interaction was carried out in 0.01 M Na phosphate buffer (pH 6.8). Excitation at 470 nm, emission at 545 nm.

INTRODUCTION

The differential staining of chromosome regions by the fluorescent dyes, such as quinacrine, permits the unambiguous identification of individual chromosomes and regions within a chromosome (1). The detailed mechanism by which fluorochromes produce chromosomes bands is still unclear. Fluorescent labels found to date show a great affinity for sequences rich in A-T base pairs and result in the production of Q-type bands (2-5). But the question of whether the production of fluorescent bands is due to variation in overall base composition or to the base sequence arrangement along the chromosome is still not resolved. The recent finding that daunomycin produces Q-type banding patterns on human metaphase chromosomes prompted a further investigation of the interaction of the anthracycline group of antibiotics with DNA and chromosomes (6,7). In addition, the chromomycin like antibiotics olivomycin, mithramycin and chromomycin A_3 were recently found to exhibit reverse banding patterns on human chromosomes (8). We present here also our findings on the interaction of olivomycin with natural DNAs. The relationship between the reverse bands on chromosomes produced by olivomycin and the Q-type bands on chromosomes produced by Hoechst 33258 has also been examined.

METHODS

Detailed methodology on the solution interactions between DNA and fluorochromes, and the preparation of banded chromosomes have been presented elsewhere (9, 10).

RESULTS

Interaction with Anthracyclines.

The antibiotics whose action on DNA and metaphase chromosomes preparations have been studied, are shown in Figure 1. The anthracycline antibiotics are all known to interact with DNA in solution and in doing so have been shown to exhibit altered fluorescence quantum yields (11,12). The anthracycline nogalamycin had been observed to prevent the *de novo* synthesis by *E. coli* polymerase I of the alternating copolymer poly d(A-T)·poly d(A-T), thus enabling the homopolymeric duplex poly dA·poly dT to be formed. Figure 2

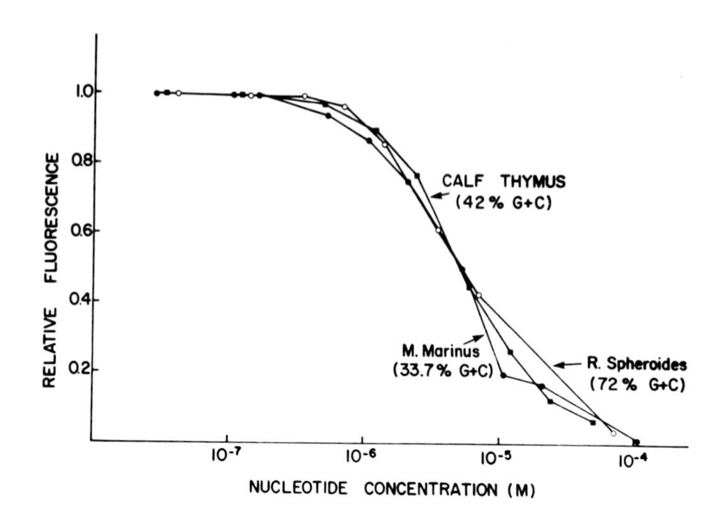

Fig. 3 Relative fluorescence of daunomycin in the presence of varying concentrations of DNA. Daunomycin (2 x 10^{6} M) and DNA interaction was carried out in 0.01 M Na phosphate (pH 6.8) buffer. Excitation at 485 nm, emission at 565 nm.

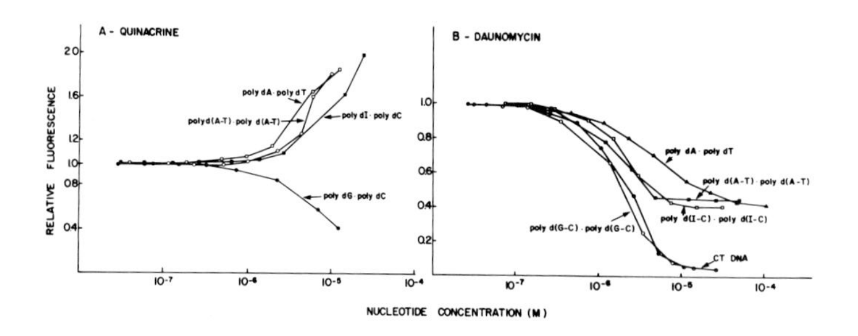

Fig. 4 Relative fluorescence of quinacrine and daunomycin in the presence of varying concentrations of synthetic polynucleotides.
A - Quinacrine (2.5 x 10^{6} M) and synthetic polynucleotides in 0.01 M Na phosphate buffer (pH 6.8). Excitation at 430 nm, emission at 495 nm.
B - Daunomycin (2.5 x 10^{6} M) and synthetic polynucleotides in 0.01 M Na phosphate buffer (pH 6.8). Excitation at 485 nm, emission at 565 nm.

shows that nogalamycin fluorescence is not affected by poly dA·poly dT, but is quenched to a similar extent by either calf thymus DNA or the repeating polymer poly d(A-T)·poly d(A-T). Nogalamycin does not produce fluorescent bands with human metaphase chromosome preparations.

Another member of the anthracyclines, daunomycin was recently found to produce fluorescent bands on human chromosomes which appeared to be similar to the banding patterns produced by quinacrine (10). The effect of the solution interaction of several natural DNAs with different base composition on the fluorescence emission of daunomycin was investigated and the results are shown in Figure 3. Variations in the base composition of the DNAs ranging from 33-72% G+C have little effect on the quenching of daunomycin fluorescence. At phosphate-dye ratios of 30 or more, the daunomycin fluorescence is completely quenched regardless of the base composition of the natural DNAs. A comparison between the effects of synthetic polynucleotides on the fluorescence of quinacrine or daunomycin is shown in Figure 4. The synthetic polymer poly d(G-C)·poly d(G-C) completely quenches the daunomycin fluorescence as was found for calf thymus DNA. The polymers poly d(A-T)·poly d(A-T) and poly dA·poly dT both quenched daunomycin fluorescence by only 50%, although higher concentrations of the latter DNA were required to reach this level of quenching. The binding of daunomycin to the synthetic A-T polynucleotides was studied in two ways. A stable complex between daunomycin and poly d(A-T)·poly d(A-T) could be isolated by gel filtration, but no stable complex could be detected between the drug and poly dA·poly dT. In addition, the effect of daunomycin on the thermal stability of the A-T polymers was determined. Table 1 shows that daunomycin has a greater effect on the thermal stability of poly d(A-T)·poly d(A-T) as compared to poly dA·poly dT.

The interaction between the chromophore of daunomycin, daunomycinone, and calf thymus DNA was investigated. It was found that daunomycinone fluorescence is quenched by DNA, but higher phosphate/dye ratio is required as compared to daunomycin. Daunomycinone also produces bands on chromosomes, but they are weaker compared to the bands produced by the parent compound daunomycin.

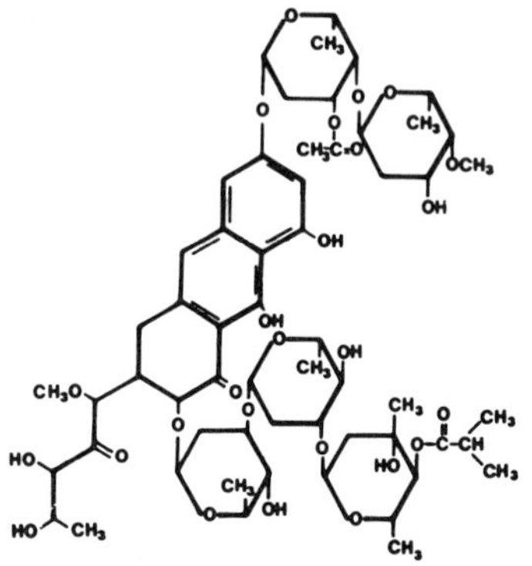

Fig. 5 Structure of olivomycin.

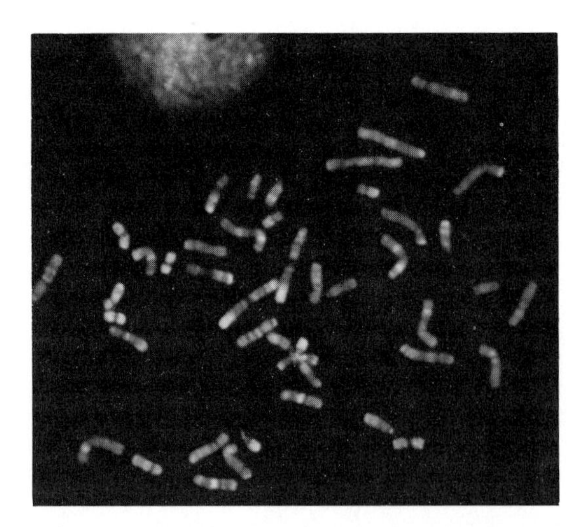

Fig. 6 A metaphase spread of a human lymphocyte from a female subject. The R banding pattern on chromosomes was produced by directly staining the slide of the chromosome preparation with olivomycin (1 mg/ml) in Sorrenson's buffer (pH 6.8).

TABLE 1

EFFECT OF DAUNOMYCIN ON THERMAL STABILITY OF POLYNUCLEOTIDES

Polynucleotide	Daunomycin / Phosphate	Tm(°C)
Poly dA·poly dT	0	58
	0.35	62
	0.70	64
Poly d(A-T)·poly d(A-T)	0	49
	0.35	61
	0.70	68

Interaction with Olivomycin.

Olivomycin contains a central chromophore to which five different carbohydrate moieties are attached (Figure 5). Recently, this antibiotic was found to produce characteristic reverse fluorescent bands (R-bands) on human chromosomes as is shown in Figure 6. The banding patterns along the chromosomes and the variable fluorescent regions on specific human chromosomes were in general the same as obtained by the acridine orange R banding procedure (13). This similarity has also been confirmed by using bovine and mouse chromosome preparations. The fluorescence intensity of olivomycin in the presence of several DNAs was measured as shown in Figure 7. Small spectral shifts are observed in the presence of DNA for both the excitation (405 to 440 nm) and emission (525 to 532 nm) spectra. The presence of Mg^{2+} is required for the interaction between olivomycin and native DNA. A linear relationship was observed between olivomycin fluorescence enhancement and the G-C content of the DNA. In contrast, the fluorochrome Hoechst 33258 exhibits an increase in fluorescence depending on the A-T content of the DNA. Figure 8 shows a comparison of the relative fluorescence intensities of the three dye's quinacrine, olivomycin and Hoechst 33258 in the presence of five DNAs of different base composition. An inverse relationship is observed between the fluorescence behavior of olivomycin and Hoechst 33258 in their interaction with

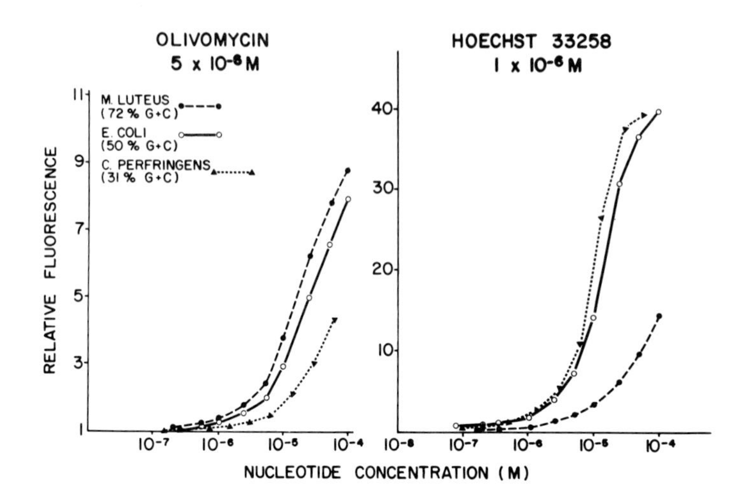

Fig. 7 Relative fluorescence of olivomycin and Hoechst 33258 in the presence of varying concentrations of DNA. The olivomycin (5 10^{-6} M) DNA interaction was carried out in 0.01 M Na phosphate (pH 6.8) containing 10^{-3} M $MgCl_2$ and 10^{-4} M EDTA. The Hoechst 33258 (1 x 10^{-6} M) DNA interaction was carried out in 0.01 M Na phosphate (pH 6.8) containing 0.15 M NaCl.

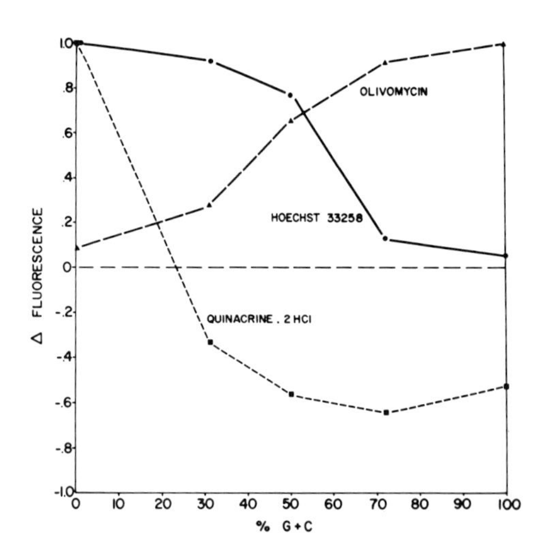

Fig. 8 Comparison of differences in fluorescence of quinacrine, olivomycin and Hoechst 33258 in the presence of DNA of different base composition. DNAs (1.78 x 10^{-5} M) used are poly d(A-T)·poly d(A-T), C. Perfringens DNA, E. Coli DNA, M. Lutens DNA and poly d(G-C)·poly d(G-C).

different DNAs. A similar observation is made at the cytological level as is shown in Table 2.

TABLE 2

THE EFFECT OF CHROMOSOMAL REGIONS ON THE FLUORESCENCE INTENSITIES OF QUINACRINE, OLIVOMYCIN AND HOECHST 33258

Chromosomal Region	Quinacrine	Olivomycin	Hoechst 33258
Human 1, 9, 16 Secondary constriction	-	+	-
Mouse centromere	-	+	-
Bovine centromere	-	-	+

- dull fluorescence
+bright fluorescence

Both the solution and cytological observations suggest that the bright bands produced by olivomycin are the reverse of the bright bands produced by Hoechst 33258.

DISCUSSION

An extreme in binding specificity between the two different A-T polymers is observed for the anthracycline drug nogalamycin (13). The nogalamycin fluorescence is not affected by poly dA·poly dT, in support of the finding that the antibiotic does not bind to this polymer. Excluding the poly dA·poly dT polymer, nogalamycin exhibits no differential fluorescence quenching with DNAs of different base composition. This agrees with the finding that no fluorescent bands are observed when human metaphase chromosome preparations are treated with this drug.

Close cytological examination of the chromosome bands produced by quinacrine or daunomycin failed to show any discernible difference in the location of bands along the chromosome. However, a marked difference was observed in the effect of DNAs on the quantum yield of quinacrine or daunomycin fluorescence in solution. Quinacrine has been found to be sensitive to changes in base composition of DNA, especially in the lower GC percent range (30-50%) (14). In contrast, the present findings on the interaction of daunomycin with natural DNAs indicate that the variation in mean base composition of the DNA has little effect on the extent of quenching. Synthetic polynucleotides containing exclusively A-T base pairs show considerable differences in their effect on quinacrine or daunomycin fluorescence intensities. Previously, it was shown that quinacrine fluorescence is enhanced by poly d(A-T)·poly d(A-T) or poly dA·poly dT (2,3). Daunomycin fluorescence, however, is quenched by these two polymers, but the extent of quenching levels off at approximately 50% of the initial daunomycin fluorescence. The small difference in daunomycin fluorescence quenching by the two A-T containing synthetic polynucleotides could be the result of a difference in affinity of the drug for these polynucleotides. Stable complex formation could be detected between daunomycin and poly d(A-T)·poly d(A-T), but no complex could be detected between daunomycin and poly dA·poly dT. In addition, the alternating polymer poly d(A-T)·poly d(A-T) is stabilized to a greater extent by daunomycin than is the homopolymeric duplex. The increase in Tm of poly dA·poly dT in the presence of daunomycin is an indication that binding of the drug to this polymer does take place. The fluorescence data indicate that similar nucleotide sequences (stretches of A-T base pairs) are responsible for the production of Q or D bands, but the mechanism is different. Quinacrine bands are visualized by fluorescence enhancement of the dye bound to these DNA regions, while daunomycin bands are the product of differential fluorescence quenching of the antibiotic bound to the same nucleotide sequences. Considerable evidence indicates that nucleotide sequence arrangements within eukaryotic DNA are non-random, and the quinacrine and daunomycin produced bands may reflect this heterogeneity of sequence arrangements.

The fluorescent banding patterns on chromosomes produced by olivomycin or Hoechst 33258 may be the direct consequence of another heterogeneity of the DNA within the eukaryote chromosome, namely base composition. The reverse bands produced by olivomycin treatment can be explained in terms of the requirements for G-C base pairs in the binding of this compound to DNA. Similarly, the faint Q-like bands observed on treatment of metaphase chromosomes with Hoechst 33258 have been attributed to the preferential binding of this dye to A-T rich DNA sequences (15). The fluorescence behavior of olivomycin and Hoechst 33258 upon interaction with DNAs, and the reversal of fluorescent bright bands on chromosomes produced by these two dyes, strongly supports our suggestion that olivomycin and Hoechst 33258 act in a complementary fashion, in which the former recognizes G-C rich areas on the chromosome and the latter is specific for A-T rich regions. It appears that although the mechanisms responsible for the revealing fluorescent banding vary with the different fluorochromes, they are all responding to the same underlying DNA sequence organization within a chromosome.

ACKNOWLEDGEMENTS

We like to thank Mrs. M. Heinz for typing this manuscript. This work was supported by Medical Research Council of Canada grants MA 4665 and MA 4885.

REFERENCES

1. Casperson, T., Farber, S., Foley, G.E., Kudynowski, J., Modest, E.J., Simonsson, E., Wagh, U., and Zech, L. (1968) *Exp. Cell Res.* 49, 219.
2. Weisblum, B. and de Haseth, P.L. (1972) *Proc. Nat. Acad. Sci. USA* 69, 629.
3. Pachmann, V. and Rigler, R. (1972) *Exp. Cell Res.* 72, 602.
4. Hilwig, I. and Gropp, A. (1972) *Exp. Cell Res.* 75, 122.
5. Disteche, C. and Bontemps, J. (1974) *Chromosoma (Berl.)* 47, 263.
6. Lin, C.C. and van de Sande, J.H. (1975) *Science* 190, 61.
7. Johnston, F.P., van de Sande, J.H. and Lin, C.C. (1976) *Can. J. Genet. Cyt.* 18, 565.
8. van de Sande, J.H., Lin, C.C. and Jorgenson, K.F. (1977) *Science* 195, 400.
9. Johnston, F.P., Jorgenson, K.F., Lin, C.C. and van de Sande, J.H. (1977) in press.
10. Lin, C.C., van de Sande, J.H., Smink, W.K. and Newton, D.R. (1975) *Can. J. Genet. Cyt.* 17, 81.
11. Kersten, W., Kersten, H. and Szybalski, W. (1966) *Biochemistry* 5, 236.
12. Calendi, E., Marco, A.D., Reggiani, M., Scarpinato, B. and Valentini, L. (1965) *Biochim. Biophys. Acta* 103, 25.

13. Olson, K., Luk, D. and Harvey, C.L. (1972) *Biochim. Biophys. Acta* 277, 269.
14. Comings, D.E. and Drets, M.E. (1976) *Chromosoma (Berl.)* 56, 199.
15. Latt, S.A. and Wohlleb, J.C. (1975) *Chromosoma (Berl.)* 52, 297.

THE ORGANIZATION OF H1 HISTONE ON CHROMOSOMES

Jim Gaubatz, Ross Hardison, Joyce Murphy,
Mary Ellen Eichner and Roger Chalkley

Department of Biochemistry, University of Iowa
Iowa City, Iowa 52242

ABSTRACT. This paper addresses itself to the organization of H1 on the chromosome. Crosslinking studies indicate in contrast to other histones which have a dimer as a key intermediate, that H1 can be crosslinked to form poly H1. This argues that the H1 molecule must in some way span the outside of the nucleosome structure. This notion is supported by studies of the extension of chromatin in increasing concentrations of urea. H1 appears to stabilize against extension unless it is previously cleaved with trypsin. These observations are interpreted in terms of a detailed model for chromatin structure.

The eucaryotic chromosome is thought to consist of groups of the non-H1 histones situated every 180-200 base pairs along the DNA strand (1,2,3). The non-H1 histones interact with 140 base pairs of DNA (4), developing a compact subunit known as a nucleosome (5). The nucleosomes are connected by some 40-60 base pairs of linker (or spacer) DNA (4,6,7). It has been generally assumed that the H1 histone interacts with the spacer DNA (8,9,10), thus permitting the overall structure to become even more compact.

An initial impression from these results is that a single H1 histone occupies the internucleosomal spacer region with relatively little overlap across the nucleosome itself. However, as an outcome of our studies on chromatin structure, we realized that the H1 histone could span the nucleosome, interacting with a part of the spacer DNA on each side. Thus two H1 molecules would share a given internucleosomal region and might in fact come into close proximity one with another. This last point is particularly telling since at early times in the reaction of various crosslinking reagents with chromatin one observes the formation of relatively homogeneous poly H1 (11, 12,13,14).

Our basic strategy has been to disrupt the chromatin structure in some way and to assay for the role that H1 is playing in this process. One way to disrupt chromatin structure is by the selective use of nucleases. A second method employs urea to extend the chromosomal material. Still a third method utilizes trypsin to dissect H1 molecules. We have found that the degree of extension in urea depends not only upon the presence of H1, but also upon its structural integrity.

The formation of poly-H1 at early stages in the crosslinking of chromatin has been described for both short range (12) and longer range crosslinking reagents (11,13,14). A direct demonstration of this effect is presented in Figure 1.

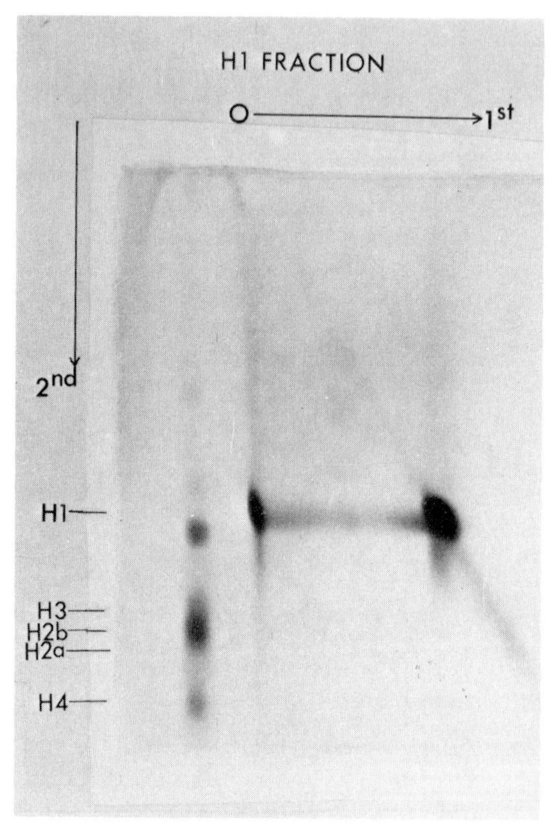

Fig. 1. Poly H1 in chromatin. Chromatin (15) was reacted with 1 mg/ml MMB, pH = 7.4, 4° for 4 hr and dialyzed against 6 M urea, 0.5 mM cacodylate, pH = 6.0, 4° overnight. This chromatin was crosslinked by oxidation with 35 mM H_2O_2, 20 min, 21° followed by treatment with 10 mM iodoacetamide 30 min, 21°, in the dark to block any free sulfphydrl groups and dialysis against 0.5 mM cacodylate, pH = 6.0 to remove excess reagents. Histones were extracted with 0.4 N H_2SO_4 from MMB-crosslinked chromatin and subjected to chemical fractionation (16). The H1 fraction was electrophoresed in a 14 cm long, 3 mm diameter acid-urea gel (17 ; 3.25 M urea) for 16 hr at 85 V (horizontal in this figure). This gel was soaked in 2 M 2-mercaptoethanol, 4 M urea, 0.09 M HOAc for 2 hr, 21° to reduce the disulfide crosslinks between histones, shrunk in 95% ethanol for 30 min, 21°, placed atop a pre-electrophoresed 14 x 16.5 x 0.3 cm acid-urea slab gel (3.25 M urea) and electrophoresed in the second dimension (vertical) at 180 v for 8.5 hr. A control whole histone sample imbedded in polyacrylamide was placed to the left of the first dimension gel and electrophoresed simultaneously. The gel was stained in 0.1% amido black and destained electrophoretically. "O" denotes the origin, which is the top of the first dimension gel. Similar results can be obtained from compact chromatin in 1 mM $MgCl_2$ by omitting the dialysis steps and extracting H1 and associated polymers with 5% perchloric acid immediately after oxidation and amidomethylation.

In this figure we see the results of an experiment in which calf thymus chromatin was crosslinked with methy-4-mercapto-butyrimidate (MMB) up to the stage at which about 15% of histone monomers are crosslinked. The histones (both crosslinked and unreacted) were subsequently isolated and the H1 fraction separated by standard means (16). The H1 fraction was then run in an acid-urea gel prior to reversal of crosslinking by mercaptoethanol. The reversed material was electrophoresed in a second dimension (downward in the figure) along with a control of whole histone (left hand side of figure). Much of the higher molecular weight material which migrated slowly in the first dimension now moves as monomer histone and it can be seen that H1 appears to be the only component of these polymers. At later stages in the crosslinking the yield of poly-H1 decreases as it becomes crosslinked to other histone fractions (14). Furthermore, these basic observations appear to hold for both extended chromatin and for compact nuclear material (data not shown). This result is unlikely to be a simple artifact of histone migration as free H1 forms polymers only very slowly (18) and crosslinking of monomers does not yield poly H1 (19).

The role of H1 histone and its contribution to the overall structrue of chromatin was analyzed by studying the viscosity of chromatin from which histones were selectively removed. The viscosity of compact chromatin is relatively insensitive to salt over the range 0-0.5 M as shown in Figure 2. Coinciding with the removal of H1 histone from compact chromatin at 0.6 M NaCl there is a fivefold increase in viscosity of the chromosomal material. Subsequently as the other four histones are removed over the ionic strength range 0.6-2.0 M NaCl there is a steady increase in viscosity amounting to an additional 2.5 fold increment. If the H1 histone is cleaved by trypsin treatment, the fragments remain associated with the chromatin and there is no increase in viscosity as a function of time of trypsin treatment (data not shown).

The simplest interpretaion of these observations is that the H1 histone is neutralizing much of the internucleosomal charge on the DNA. After removal of H1 histone, the internucleosomal DNA becomes much more extended though compact nucleosomal regions are still preserved to a considerable degree. As more histones are dissociated the DNA becomes more and more extended as the subunit components are lost. Of course these observations simply provide evidence that H1 histone interacts with the internucleosomal DNA. There is no indication whether a given H1 histone is located only between two nucleosomes or if it spans a nucleosome thus sharing spacer DNA with an adjacent H1 molecule.

We have attempted to probe this problem further by comparing the urea-induced extension of chromatin with or without H1.

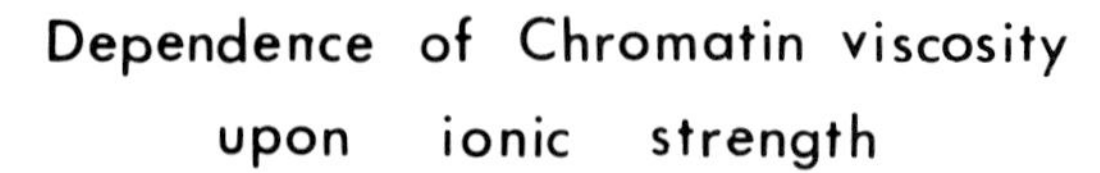

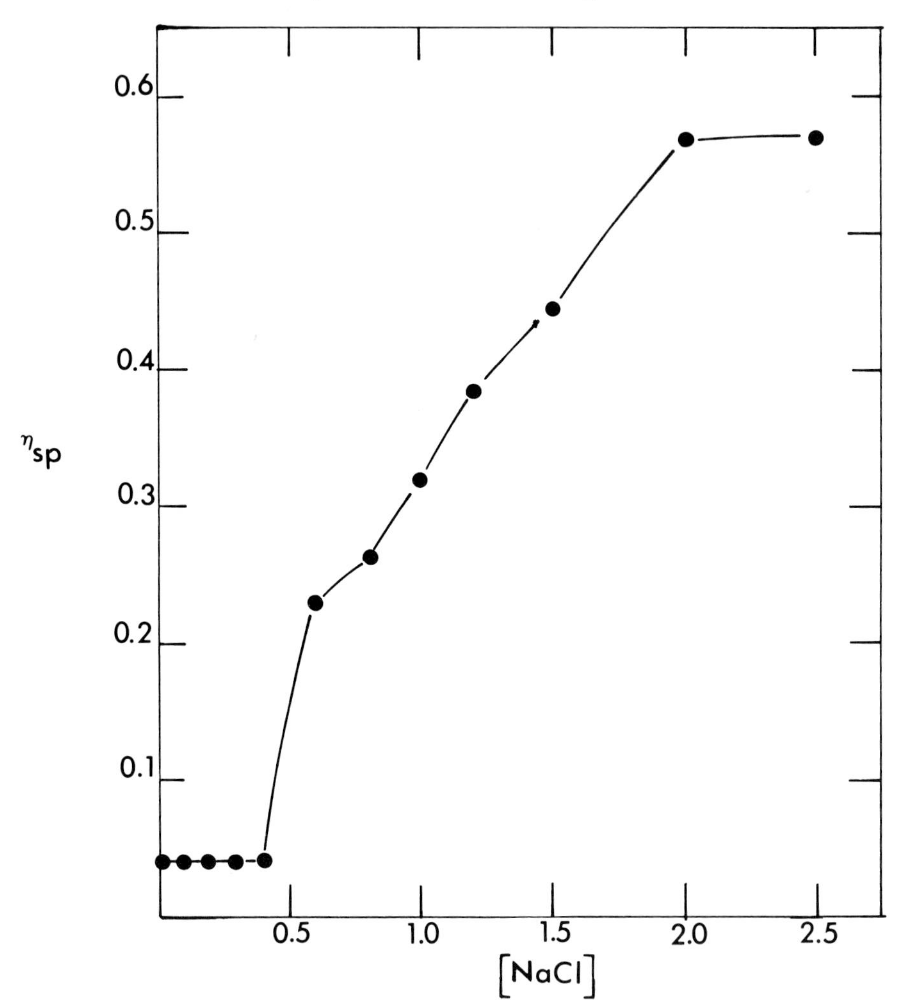

Fig. 2. The specific viscosity of chromatin as a function of ionic strength.

Chromatin lacking H1 was prepared by standard means (20) utilizing exposure to 0.6 M NaCl followed by centrifugation. Controls showed that H1 was indeed efficiently removed and also that the material was free of endogenous proteolysis.

The ability of increasing concentrations of urea to extend the chromosomal material is shown in Figure 3. All viscosity measurements were carried out at 0°. The extension of

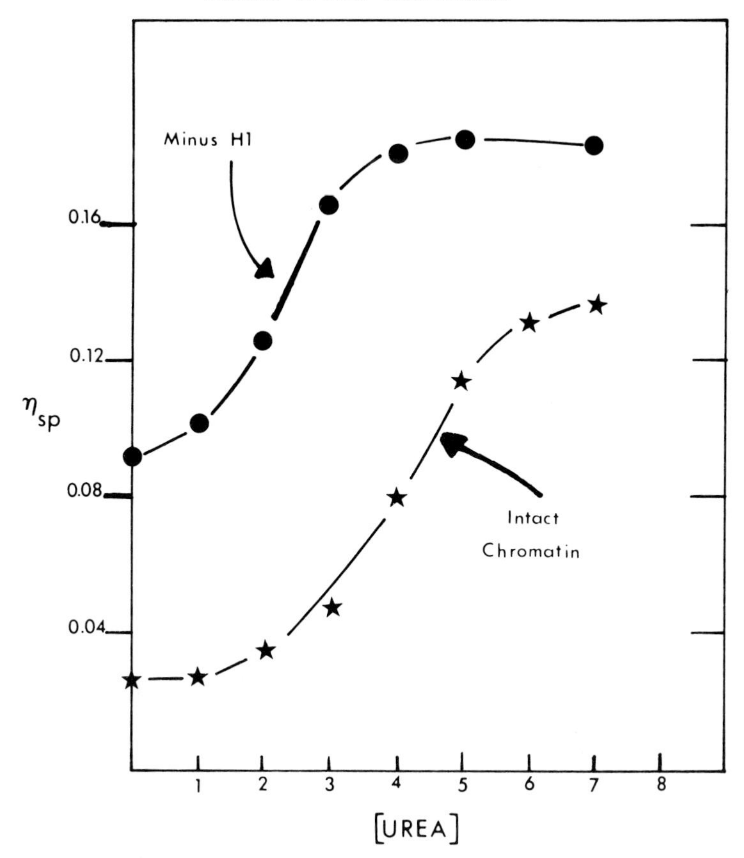

Fig. 3. The viscosity of chromatin material as a function of urea concentration. Chromatin lacking H1 was prepared as described in the text. Urea was freshly dissolved each day to avoid ammonium isocyanate accumulation. All measurements were recorded in an Ostwald viscometer in 0.01 M Tris pH 7.4 at 0° in the presence of ice.

intact chromatin occurs over a fairly large range of urea concentrations. One sees an initial increase in 2 M urea, and the transition is not complete until a concentration of 7 M is obtained. The concentration of urea which produces 50% extension is 4.4 M. The extension profile of chromatin lacking H1 is radically different. Extension begins in 1 M urea and is complete in 4 M urea. The concentration of urea producing 50% extention is 2.7 M. Presumably in the case of chromatin lacking H1 we are seeing the effect of disrupting intranucleosomal binding forces, i.e., breaking histone-histone interactions. One may surmise that in intact chromatin the intranucleosomal binding forces are equally susceptible to urea but that the presence of H1 inhibits extension at lower urea concentrations. However, full extension can occur at high urea concentrations. In urea, fully extended chromatin lacking H1 is more viscous than fully extended intact chromatin, presumably due to the presence of 20% protein free DNA, the increased negative charge generating a more rigid molecule.

We conclude that the presence of H1 histone along with the other histones leads to a cooperative stabilization of chromatin against urea-induced viscosity increases. This interpretation would appear to fit better with the notion that the H1 histone spans the nuclesome rather than interacting with the internucleosomal spacer DNA.

In order to test the idea that H1 stabilizes the nucleosome against extension at low urea concentrations but not in 7 M urea, we have sought to selectively destroy the H1 molecule by gentle trypsin treatment. This results in cleavage of the H1 molecule without allowing dissociation from the chromatin. The results of such an approach are shown in Figure 4. If chromatin is digested in trypsin and analyzed for viscosity changes in 3 M urea as a function of time of digestion, we see that as the H1 histone is digested a twofold increase in extension is observed. The increase in viscosity correlates very well with the cleavage of the H1 molecule as is shown in Figure 4a, so that when H1 has been degraded the increase in viscosity shows a distinct plateau. In contrast there is no change in the response to 6 M urea of chromatin even when it has been incubated with trypsin for up to 180 mins as shown in Figure 4b. The viscosity of chromatin in the absence of urea exhibits no change over the time periods of digestion used in these studies.

If the alteration in viscosity after trypsin treatment is a reflection of a critical role played by the H1 histone in chromatin structure, one would expect that tryptic digestion of chromatin lacking H1 would not lead to a change in its response to urea treatment. That this is indeed the case is shown in Figure 5 in which chromatin lacking H1 histone was incubated with trypsin for 180 min and assayed at various times for its extension in 2 M urea. Chromatin minus H1 exhibits no increase on viscosity in urea throughout the entire time of the incubation.

We have systematically assayed for the effect of prior trypsin treatment upon the ability of urea to extend the chromatin structure. In each case the chromatin was digested with trypsin and the viscosity determined in the presence of a given concentration of urea at various time intervals. In Figure 6a we show a summary of all the data obtained in this manner.

Dependence of η_{sp} in 3M Urea on time of trypsin digestion in 0.01 M Tris pH 7.4

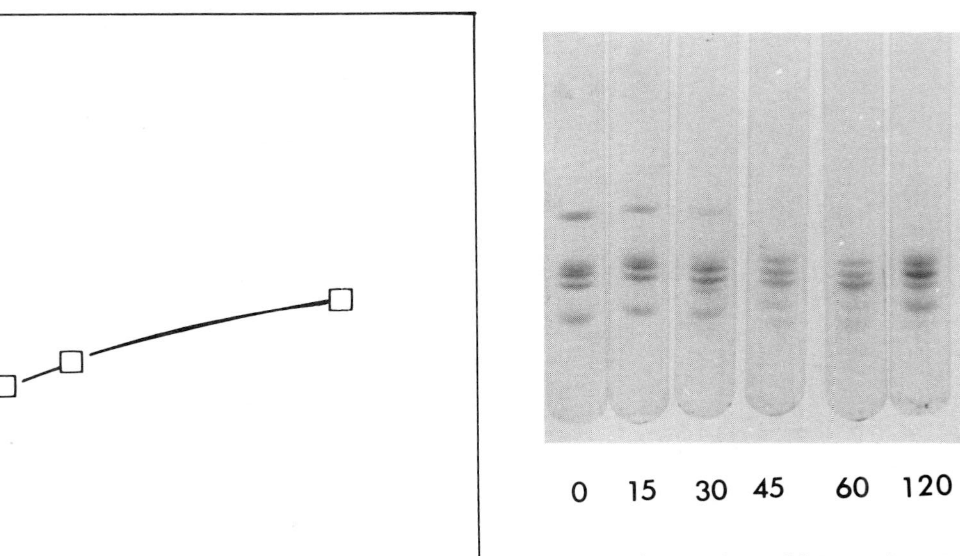

Fig. 4a. The effect of trypsin treatment of chromatin upon its ability to extend in 3 M urea.

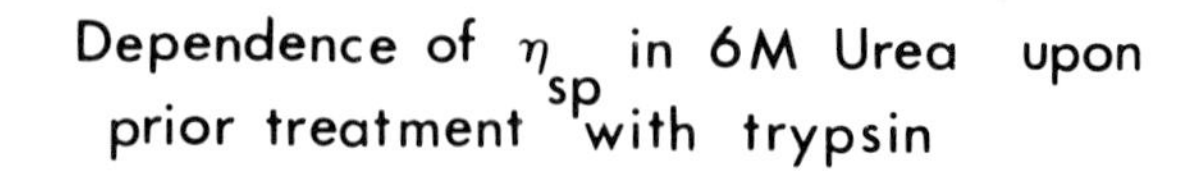

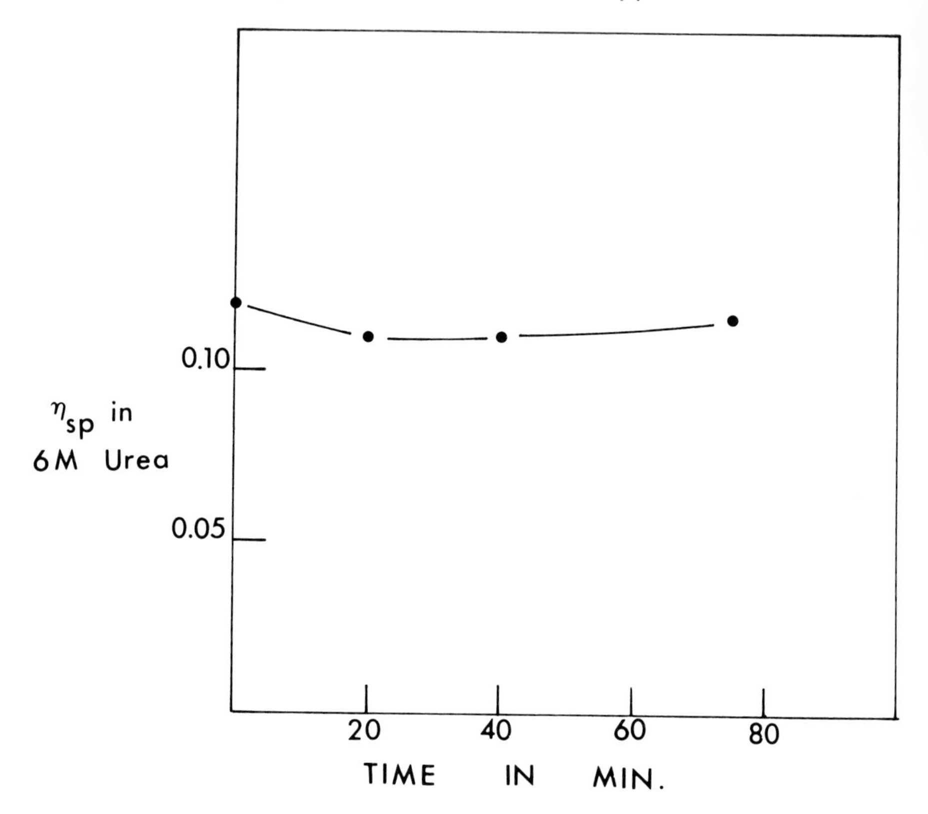

Fig. 4b. The effect of trypsin treatment of chromatin upon its ability to extend in 6 M urea. Chromatin was suspended in 0.01 M tris pH 7.4 and trypsin (0.5%) added. The incubation was performed on ice. Samples were withdrawn at the indicated times and adjusted to a final urea concentration of either 3 M or 6 M as shown.

The viscosity at various urea concentrations before trypsin treatment provides a "baseline" for any subsequent changes, and the plateau value viscosity observed after extended trypsin treatment is indicated by the arrows in Figure 6a. It is clear that after trypsin treatment the response of the chromatin to urea is changed in as much as the material extends at lower urea concentrations. In fact, the extension profile of trypsin treated chromatin now closely resembles that of untrypsinized chromatin lacking H1 as shown in Figure 6b.

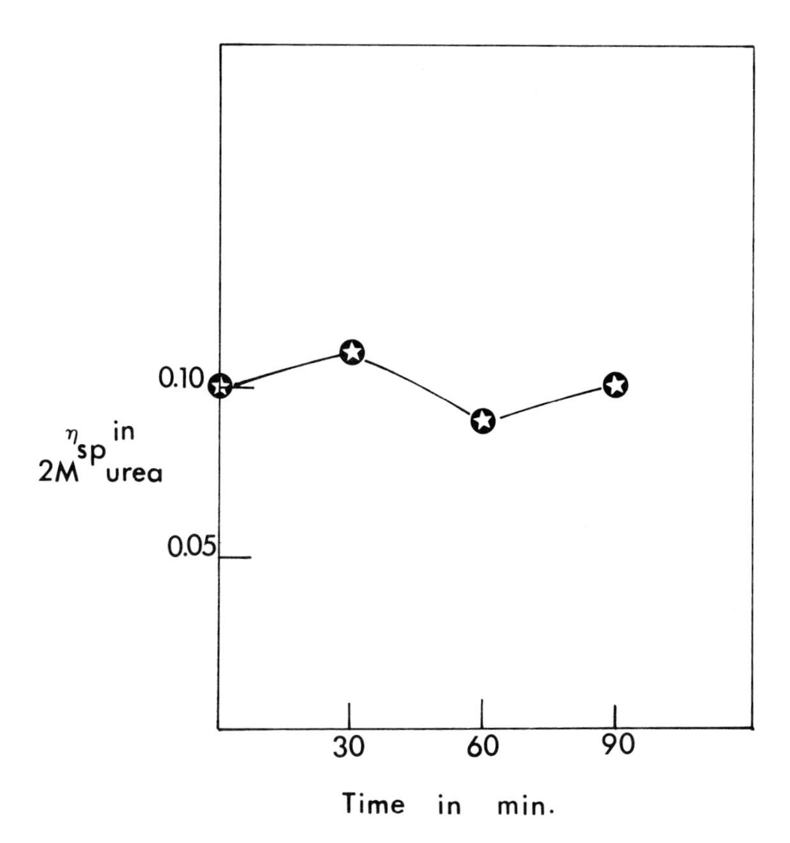

Fig. 5. The effect of trypsin treatment of chromatin lacking H1 upon the ability to extend in 2 M urea. Trypsin treatment was identical to that shown in the legend to Figure 4.

The extension profile of chromatin lacking H1 is unaffected by a prior trypsin treatment in contrast to the effect seen with whole chromatin.

Thus simply by cleaving H1 molecules, one can shift the urea extension profile from that of intact chromatin to that resembling chromatin lacking H1. Evidently the stabilization of intact chromatin against urea extension at intermediate urea concentrations requires that H1 not only be present, but that

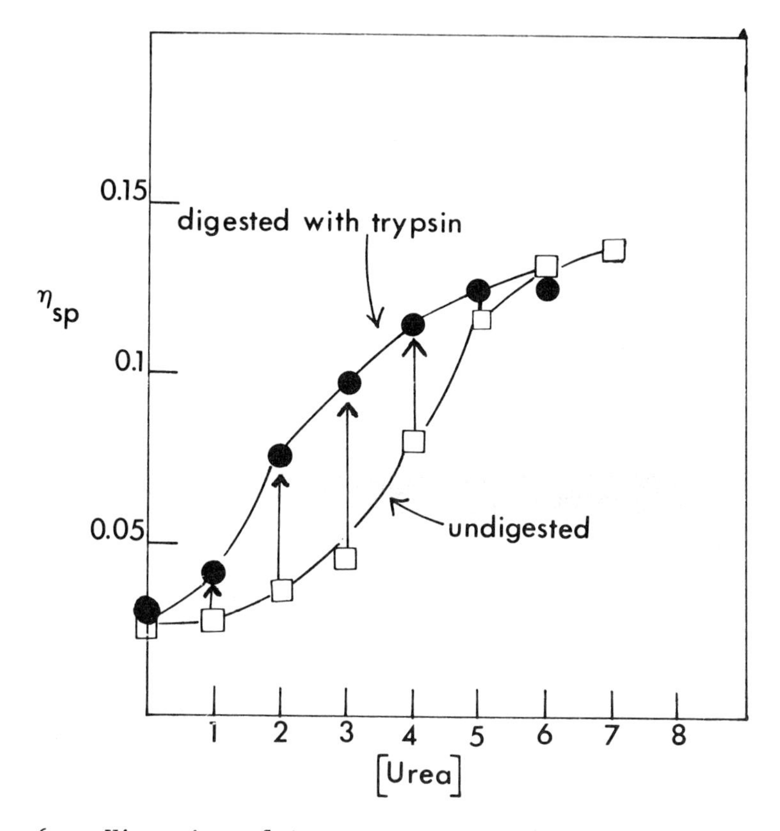

Fig. 6a. Viscosity of intact chromatin (-□-□-) and of trypsin-treated chromatin (---●---●---) as a function of urea concentration.

it be intact as well.

A simple model accommodating these observations can be developed as follows. We argue that H1 spans the nucleosome, binding to half of the spacer DNA on either side, i.e., to a total of 40-60 base pairs. The nucleosome is stabilized by hydrophobic (and possibly hydrogen) bonds and also by the presence of the H1 molecule (presumbalby due primarily to an electrostatic neutralization of DNA charge). We may surmise that the primary destabilizing force in chromatin will come from electrostatic repulsion of what is still a highly negatively charged complex.

A possible model for chromatin structure is shown in

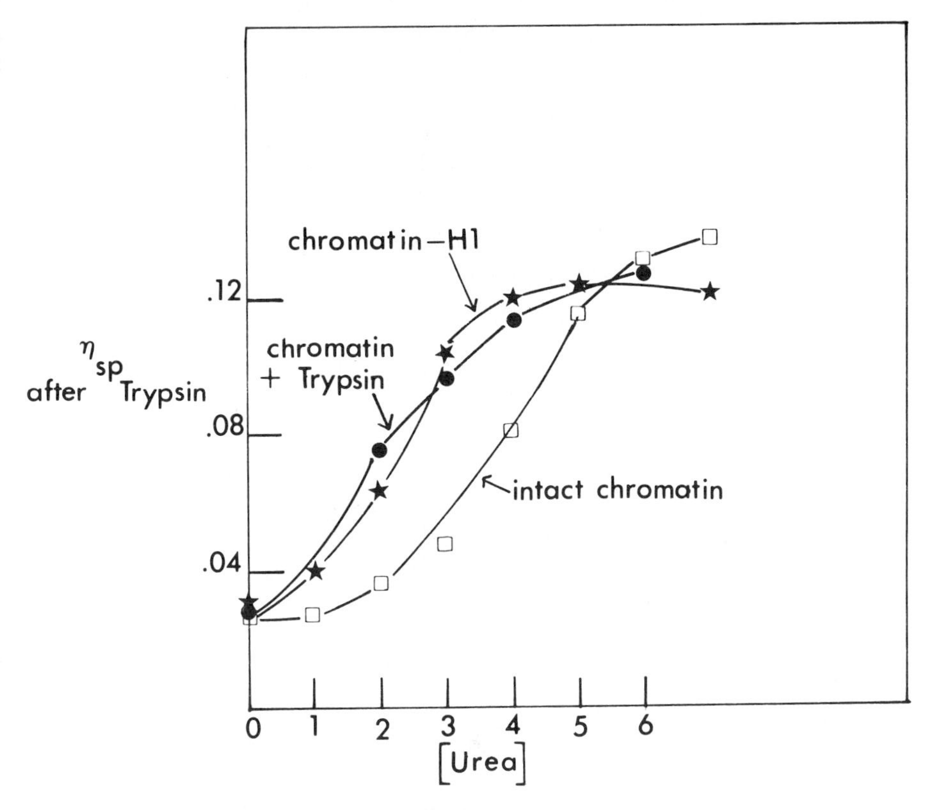

Fig. 6b. Superimposed upon the data of figure 6a we have the viscosity of chromatin lacking H1 as a function of urea concentration (-★-★-).

Figure 7. The internucleosomal spacer DNA is defined by postulating polarity requirements of the four non-H1 histones (21). In this instance the organization of the DNA is superhelical (even the spacer DNA) and the H1 histone is envisaged as interacting with spacer DNA on both sides of the nucleosome region. Two possible schemes of interaction are possible, either (a) an end to end interaction of adjacent H1 molecules or (b) an overlapping interaction. It should be possible to distinguish between these two possibilities by nucleolytic

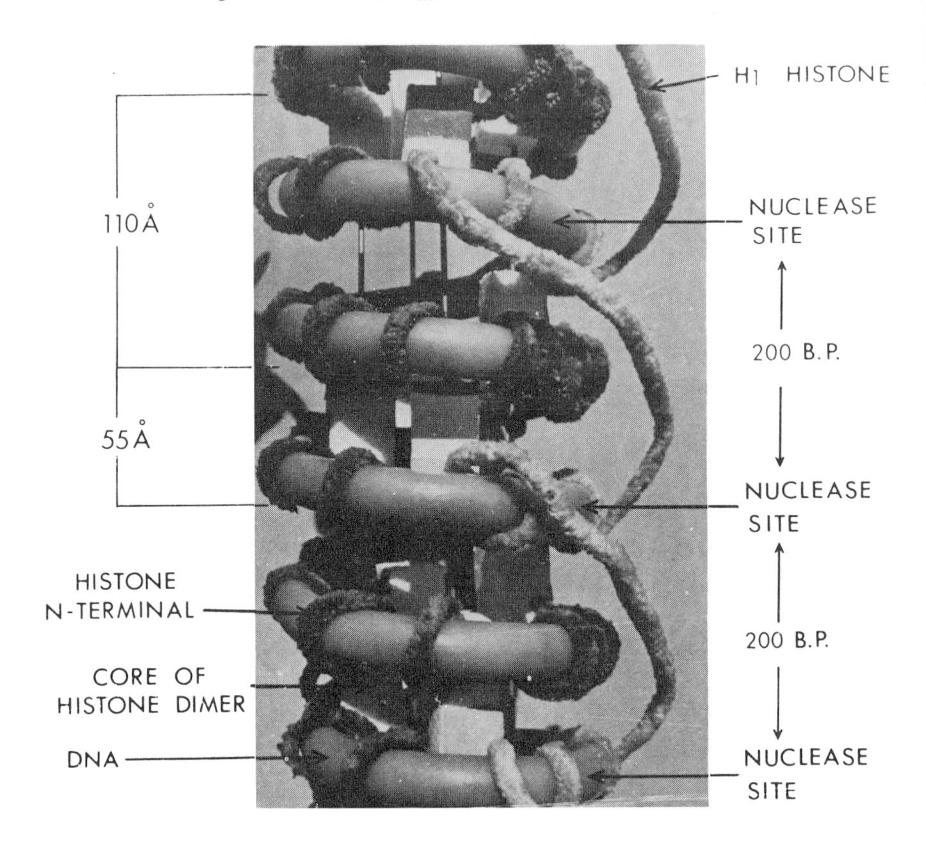

Fig. 7. Scale model of chromatin studies based on symmetry predictions for histone dimer interactions. DNA (solid tubular structure) is coiled so that there are 1-3 turns per 140 b.p. The non-H1 histones (dark tails and white cubes) interact with the DNA so that the histones are uniformly organized along the extended DNA molecule, but the tails (N-terminus ends are polarized according to symmetry postulates (21). This leads to the formation of discrete units at 4 sets of histone dimer which stabilize the nucleosome. Their center of mass arises every 110 A°. As a result of the polarity if the N-termini ends of the non-H1 histone regions of the DNA are bare of histone every 200 b.p. This represents the nuclease site. H1 (light colored tubes on right hand side) is envisaged as spanning the nucleosome, binding to the spacer DNA region.

cleavage with micrococcal nuclease because model (a) should lead to simple cleavage of chromatin, whereas model (b) should lead to nucleosomal units crosslinked by H1 even though the connecting DNA strand has been severed.

Nuclei were digested with micrococcal nuclease until about

19% of the DNA was rendered acid soluble. After inhibiting the nuclease activity, the reaction mixture was applied to a sucrose density gradient. Centrifugation gave the profile shown in Figure 8a. Material in the region D is low molecular weight nucleotide material that is soluble in acid. Region C is the monosome fraction containing DNA, 145 b.p. in size (see lower panel to Figure 8a). Region B is the oligomer region, an area of vital interest to us, since we wondered if it reflected the size of the DNA or if it simply consisted of small fragments of DNA held together by H1 crosslinks. Accordingly we removed the protein from the material in this region of the gradient and analyzed for DNA sizes electrophoretically (bottom panel of Figure 8a). It appears that in general, the size of the chromatin oligomers is a reflection of DNA size though there is clearly a trace of lower molecular weight DNA in the higher molecular weight regions of the sucrose gradient. This is most critically seen for the pellet the DNA of which covers a range of sizes including the 145 base pair fragments. However, the proportion of material in the pellet is very small in this instance (it can increase at later stages of digestion).

Since H1 is removed in 0.6 M NaCl, we also analyzed the nuclease product from this same digestion on a sucrose gradient containing 0.6 M NaCl. The results are shown in Figure 8b. The size distribution is now somewhat different with about 20% more material in the more slowly sedimenting region of the gradient. The sizes of DNA at any given point in the gradient are more sharply representative of the size of the chromatin fragment. Thus, we conclude that most of the H1 is not overlapping in the internucleosomal region. However, in order to account for the observations reported in this experiment the possibility remains that either about 20% of the H1 might be overlapping or that a small fraction may have rearranged during the course of the partial digestion. Similar data have been obtained by Altenburger and his colleagues (22).

We have tested for the possibility of rearrangement by analyzing the binding of H1 to various size components of the chromatin during the time courses of nuclease digestion. Nuclei were digested to varying degrees with nuclease and fractionated on sucrose gradients in a manner analogous to that shown in Figure 8. The gradients were separated into three general size groups, the monosome region, the oligomer and the pellet fractions. After separation the three types of material were extracted with 0.4 N H_2SO_4 and the relative amount of H1 determined. The amount of H1 relative to that present in total chromatin is lower in the monosome fraction and elevated in the pellet (Figure 9a). The amount of H1 present on the intermediate size material is comparable to that present in total chromatin.

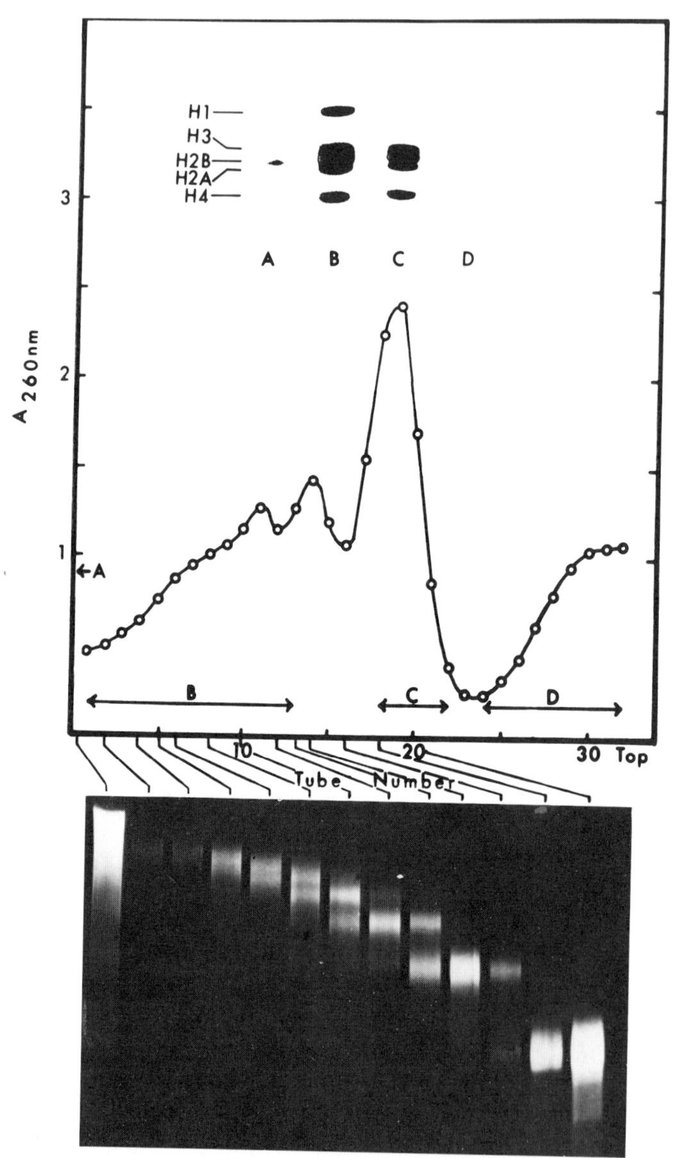

Fig. 8a. Sucrose gradient fractionation of chromatin fragments from Micrococcal nuclease-digested, calf thymus nuclei (19% acid-soluble DNA). The digest (35 $A_{260\ nm}$ units) was sedimented through 30-ml, 7-30% linear sucrose gradients containing 1 mM EDTA, 5 mM cacodylate-HCl, pH 6.5 and no salt (Fig. 8-a) or 0.6 M NaCl (Fig. 8-b). Sedimentation was with a SW 27 rotor at 25,000 rpm and 4°C for 18 hrs. The histones and DNA species from the various fractions are shown above and below the absorbance profile respectively.

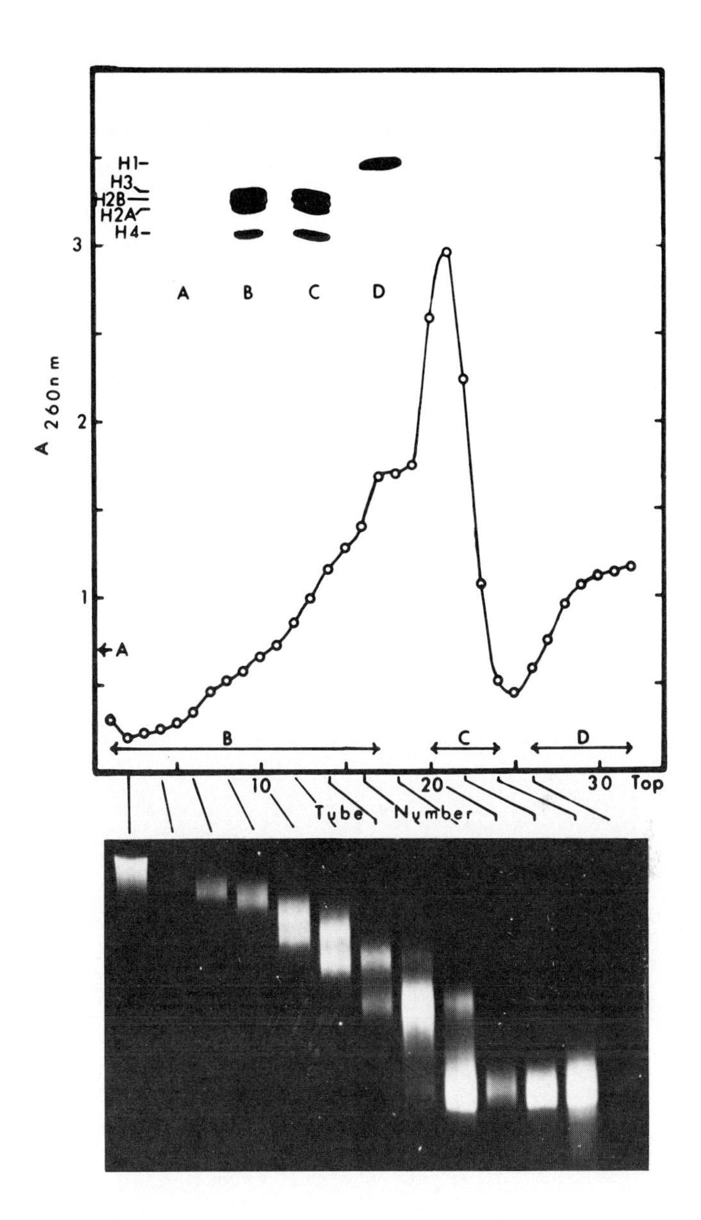

Fig. 8b. Sucrose gradient containing 0.6 M NaCl.

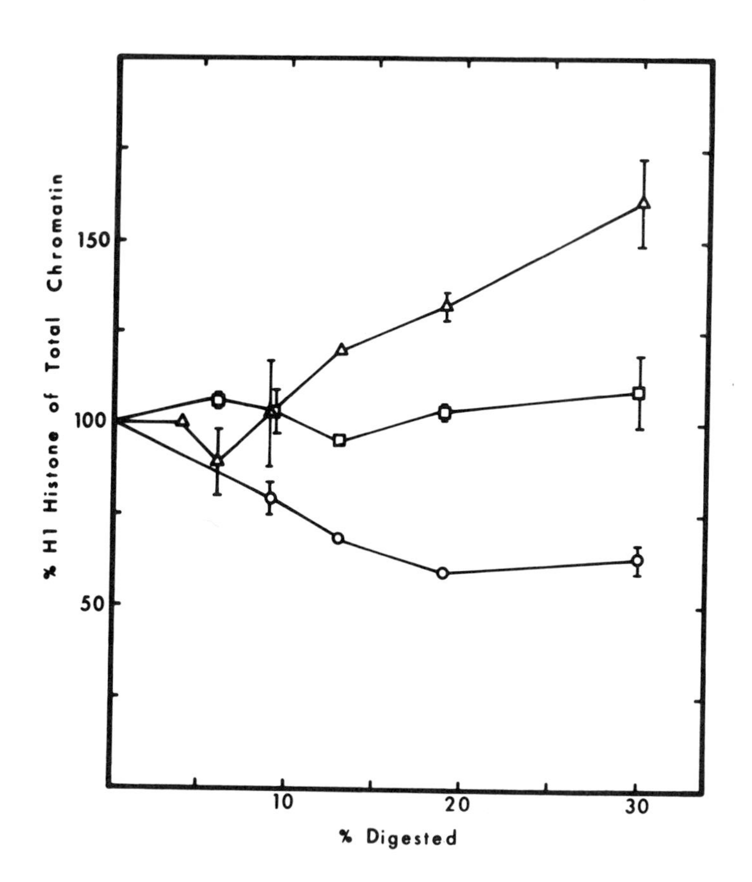

Fig. 9b. The relative amount of H1 histone found with isolated nucleosomes at low ionic strength (Figure 9a) or 0.3 M NaCl (Figure 9b). The amount of H1 histone from mononucleosomes (○), oligonucleosomes-dimer to hexamer (□), and pelleted chromosomal material (△) compared to the amount of H1 from total nuclease-chromatin was determined as a function of the percent-digestion. The components were separated as described in the legend to Figure 8. Brackets indicate the standard error found for 3 or more determinations.

The experiment was repeated in an identical manner except that the gradients now contained 0.3 M NaCl so that any weakly bound H1 might be removed. The results of this analysis are shown in Figure 9b. The pelleted material now shows the presence of a normal complement of H1, whereas both the monomer and the oligomer have lost much of their H1 histone at this ionic strength. The simplest interpretation appears to be that as the spacer DNA is digested the H1 histone collapses on the nucleosome though now binding more weakly and therefore becomes extractable in 0.3 M NaCl. Some of this weakly bound histone may migrate to interact with other more intact nucleoprotein particles possibly causing them to precipitate and to appear in

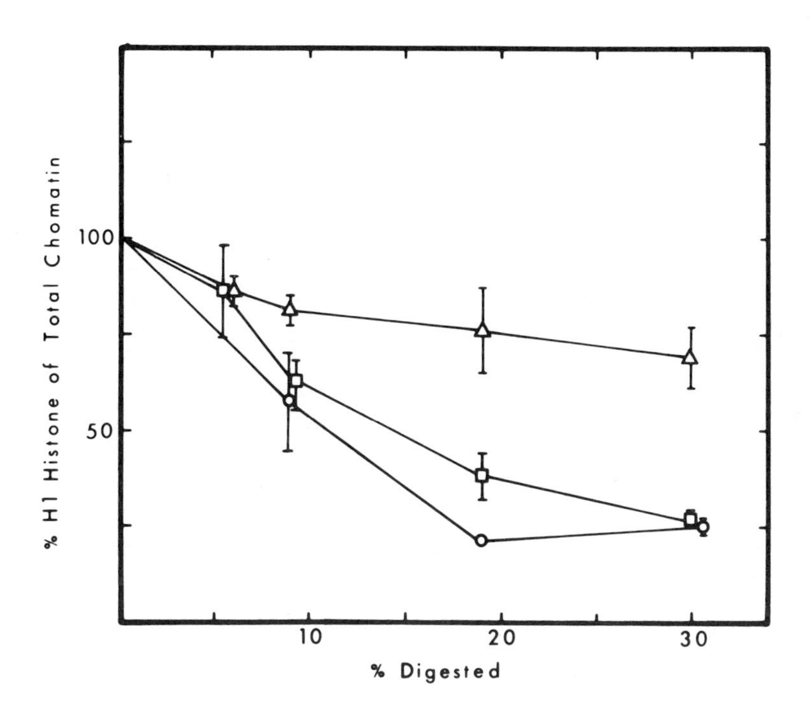

Fig. 9b. See legend to Figure 9a for details.

the pellet. The surplus H1 histone in the pellet can be washed off in 0.3 M NaCl.

The weaker binding of H1 to the oligomers is also not surprising since the bulk of oligomer material is made up of dimers and trimers. In this case the overall binding of H1 would be expected to be lower since the amount of DNA available for H1 binding has been seriously reduced (though not to the same degree as the monosomes) as a result of the nucleolytic activity. Thus the H1-induced crosslinking of nucleosomes observed as a minor component in the gradient at low ionic strength is quite possibly obtained as nuclease-induced artifact.

We conclude that H1 spans the nucleosome, but does not span over greater distances (such as across the gyres of a solenoid, otherwise more extensive crosslinking of oligomers by H1 would have been detected). Furthermore, the orientation of each H1 molecule to adjacent H1 molecules is such that overlap does not occur to any significant degree.

ACKNOWLEDGEMENTS

This work was supported by grants from the USPHA #CA 10871, CA-17224 and CA 09119.

REFERENCES

1. Kornberg, R. (1974). Science 184, 868-871.
2. Hewish, D. R. and Burgoyne, L. A. (1973). Biochem. Biophys. Res. Comm. 52, 504-510.
3. Noll, M. (1974), Nature 251, 249-251.
4. Shaw, B., Herman, T., Kovacic, R., Baudeau, G. and Van Holde , K. E. (1976). Proc. Nat. Acad. Sci. USA 73, 505-509.
5. Oudet, P., Gross-Bellard, M., and Chambon, P. (1975). Cell 4, 281-300.
6. Lohr, D., Corden, J., Tatchell, K., Kovacic, R. and Van Holde, K. E. (1977). Proc. Nat. Acad. Sci USA 74, 79-83.
7. Noll, M. and Kornberg, R. (1977). J. Mol. Biol. 109, 393-404.
8. Noll, M. (1976) Cell 8, 349-355.
9. Morris, N. R. (1976). Cell 8, 357-363.
10. Morris, N. R, (1976). Cell 9, 627-632.
11. Olins, D. and Wright, E. (1973). J. Cell Biol. 59, 304-317.
12. Bonner, W. and Pollard, H. (1975). Biochem. Biophys. Res. Comm. 64, 282-288.
13. Thomas J. O. and Kornberg, R. (1975) FEBS Letters 58, 353-358.
14. Hardison, R., Eichner, M. E., and Chalkley, R. (1975).

Nucleic Acids. Res. 2, 1751-1770.
15. Panyim, S., Bilek D. and Chalkley, R. (1971). J. Biol. Chem. 246, 4206-4216.
16. Oliver, D., Sommer K., Panyim, S., Spiker, S., and Chalkley, R. (1972) Biochem. J. 129, 349-353.
17. Panyim, S. and Chalkley, R. (1969). Arch. Biochem. Biophys. 130, 337-345.
18. Hyde, J. E. and Walker, I. O. (1975) FEBS 50, 150-154.
19. Thomas , J. O. and Kornberg, R. (1975). Proc. Nat. Acad. Sci. USA 72, 2626-2630.
20. Bartley, J. and Chalkley, R. (1972). J. Biol. Chem. 247, 3647-3655.
21. Jackson , V., Hoffmann, P., Hardison, R., Murphy, J., Eichner, M. E. and Chalkley, R. (1977) in *Molecular Biology of the Mammalian Genetic Apparatus—Its Application to Gene Expression, Carcinogenesis and Aging*, ed. T'so, P. (Elsevier North-Holland Biomedical Press, Amsterdam) pp. 281-299.
22. Altenburger , W., Horz, W. and Zachau, H. G. (1976) Nature 264, 517-522.

ORGANIZATION AND FUNCTION OF THE RIBOSOMAL GENES IN *PHYSARUM POLYCEPHALUM*

Vincent G. Allfrey, Edward M. Johnson, Irene Y.-C. Sun and Virginia C. Littau

The Rockefeller University
New York, New York 10021

and

Harry R. Matthews and E. Morton Bradbury

Portsmouth Polytechnic Institute
Portsmouth PO1 2DZ, Hants. UK

ABSTRACT. The reiterated ribosomal cistrons of *Physarum polycephalum* can be isolated as protein-associated components of highly-purified nucleoli, and as elements of a dense DNA satellite separable from the main-band DNA by isopycnic density gradient centrifugation. The structure of the satellite (rDNA), as revealed by restriction endonuclease digestion, is palindrome-like — in which two ribosomal genes of opposite polarity are separated by a long spacer sequence. Electron microscopy of dispersed nucleoli reveals transcription complexes at either end of the spacer.

The organization of the ribosomal genes in nucleoli involves their interaction with sets of structural and regulatory proteins. Nuclease digestions have been employed to compare the subunit organization of ribosomal genes and main-band DNA in chromatin and to evaluate the role of histones in nucleolar chromatin as compared to bulk chromatin structure. Evidence for the presence of ribosomal DNA sequences in particles containing nucleosome–sized lengths of DNA is presented.

Nucleolar non-histone proteins have been extracted from purified nucleoli and fractionated by affinity chromatography on *Physarum* DNA covalently attached to Sephadex G-25. DNA-binding fractions have been compared with regard to their affinities for ribosomal DNA and main-band DNA sequences. Evidence for selective binding of some nucleolar proteins to rDNA is presented.

Conditions have been developed for isotopic labelling of ribosomal RNA in intact nuclei, using α-amanitin to suppress the synthesis of HnRNAs, and identifying the newly-synthesized transcripts by hybridization to the ribosomal DNA satellite and to its restriction fragments.

INTRODUCTION

The mechanisms of transcriptional control in eukaryotic cells involve the participation of structural and regulatory proteins which interact with DNA to affect its conformation and influence its template function. The primary structural controls exerted by the histones (1,2) and modulated by post-synthetic modifications such as histone acetylation (3) and phosphorylation (4,5) are supplemented by more specific interactions between DNA and non-histone chromosomal proteins which control both the rate of transcription (e.g.(6-8) and the nature of the transcript (e.g (9-11)). As yet, the molecular mechanisms which determine the selectivity of transcription in a given cell type are not understood, but — based on previous experience with prokaryotic systems — they are likely to involve DNA-sequence recognition by specific subsets of DNA-binding proteins.

Cell nuclei contain large numbers of non-histone proteins associated with their chromatin; for example, the total number of electrophoretically-distinct proteins in HeLa cell nuclei is estimated at over 470 (12). Many of these proteins are structural components of the nucleus; others are involved in DNA and RNA synthesis and processing, in histone modification, in the "packaging" of nascent RNA chains, and in the assembly of ribosomal subunits in the nucleolus. The problem is to identify which of the remaining nuclear proteins act to promote RNA chain initiation at specific sites and which influence rates of chain elongation or termination. On the premise that DNA-sequence recognition is a prime prerequisite for such functions, we have concentrated on DNA-affinity methods for the fractionation of nuclear non-histone proteins. The experiments to be described deal with a highly-favorable system for the detection of proteins interacting with a specific DNA sequence — that encoding for the ribosomal precursor RNA in the myxomycete *Physarum polycephalum*. The availability of the reiterated ribosomal genes as components of a highly-purified

DNA satellite, the likelihood that proteins controlling ribosomal RNA synthesis will be present in thousands of copies per diploid genome, and the ability to test for the fidelity of ribosomal gene transcription in isolated nuclei and nucleoli by hybridization of the transcript to the rDNA satellite, all contribute to the feasibility of analysis of transcriptional control in this system.

In addition, the system lends itself to experiments designed to investigate the structural differences between "active" and "inactive" genes in *Physarum*, using limited nuclease digestions as a probe of chromatin organization and protein localization.

ISOLATION OF THE RIBOSOMAL DNA SATELLITE

The genes for ribosomal RNA in *Physarum*, as in *Xenopus* (13, 14) and other eukaryotes, are localized in the nucleolus (15), where they occur in multiple copies (16). A diploid *Physarum* nucleus contains about 1 pg of DNA (17) of which 0.4% is hybridizable to the 19S and 26S ribosomal RNA sequences of the organism (18-20). This corresponds to about 200-400 copies per nucleus of the sequences coding for 19S and 26S ribosomal RNA. The number of copies varies depending upon the stage in the growth cycle; except for a brief pause in early S-phase (16), rDNA synthesis occurs at all stages of the cycle, and about 50% of the ribosomal cistrons are synthesized during the G_2-phase, after the synthesis of bulk DNA has ceased (16, 21, 22). The reiteration of the ribosomal genes in *Physarum* (while not as phenomenal as that seen in the oogonia of Xenopus, which may contain 30 pg of of rDNA at the end of pachytene (23, 24)) is extensive, and combined with the ease of growing the organism and preparing nucleoli on a large scale (25), it permits the isolation of rDNA in milligram amounts.

In the experiments to be described, the nucleoli were isolated from microplasmodia of the a x i strain (chromosome number = 22) of *Physarum polycephalum* as the first step in the purification of the rDNA satellite. The diploid nuclei of the microplasmodia (the growth form of *Physarum* in shaking-culture (26)) divide synchronously with a doubling-time of about 9 hours. Procedures have been developed for the isolation of nuclei at successive stages during the cell cycle

(17,25) and for the subsequent purification of the nucleoli (25). The method involves homogenization of the microplasmodia in 0.25 M sucrose, 0.1 M Tris-HCl (pH 7.2), 0.1% Triton X-100, 1 mM $CaCl_2$, recovery of the nuclei by low-speed centrifugation, sonication to disrupt the nuclei, and purification of the nucleoli by zonal centrifugation in a 5-30% sucrose gradient (25). Figure 1 shows a typical nucleolar preparation as viewed by transmission electron microscopy. The absence of visible contamination has also been confirmed by scanning EM (30). Such preparations include the rDNA satellite in a highly-enriched state, comprising from 30 to over 70% of the DNA present.

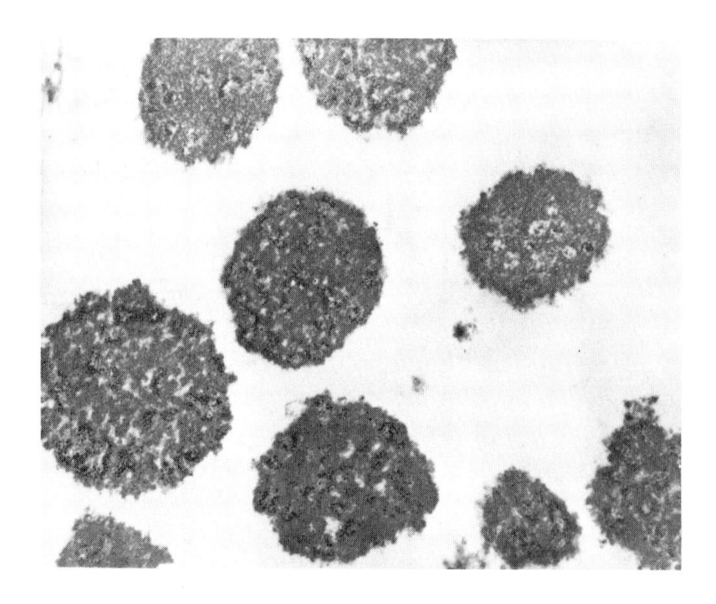

Fig.1. Transmission electron micrograph of nucleoli of *Physarum polycephalum* isolated as described (25). Note the absence of significant contamination by chromatin fragments.

The rDNA was extracted by lysis of the isolated nucleoli in sodium dodecylsulfate and treated with pronase (or proteinase K) and ribonuclease, prior to purification by CsCl gradient centrifugation. The rDNA of Physarum, like that of Xenopus (27) and other organisms (28,29), has a higher buoyant density than does the main-band DNA; the Physarum rDNA satellite has a density of 1.713 g/ml, as compared with 1.703 g/ml for the main-band (16,21,25,31). The purification of the rDNA is illustrated by the sedimentation analyses summarized in Figure 2.

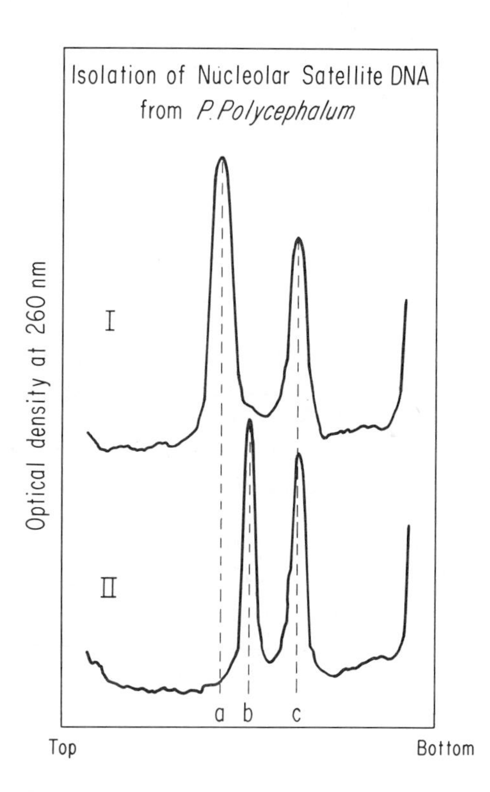

Fig. 2. Isopycnic centrifugation of total Physarum DNA (I) and of the isolated rDNA satellite (II) in neutral CsCl. The total Physarum DNA has a prominent main-band of buoyant density = 1.703 g/ml (peak a) ; the nucleolar satellite has a density = 1.713 g/ml (peak b); the position of the density standard (M. lysodeikticus DNA, ρ = 1.731) is given by peak c.

The upper panel (I) shows the sedimentation profile of total Physarum DNA with a prominent main-band at ρ = 1.703 g/ml and a small shoulder at the satellite density ρ = 1.713 g/ml. The lower panel (II) shows the corresponding analysis of the rDNA purified from isolated nucleoli.

GENETIC MAPPING OF THE rDNA SATELLITE

The sequences coding for Physarum ribosomal RNAs are localized on linear duplex DNA molecules of median length 17.8 μm (32) and molecular weight 39 x 10^6 (33). The structure has been analyzed by restriction nuclease digestion followed by sizing of the fragments (32, 33) and hybridization to ribosomal RNAs (33). Restriction endonucleases Eco R1 and Hind III each cut the linear rDNA molecule to produce one large and two small fragments which are represented twice in the intact rDNA chain, one at each end. The sites of cleavage by each enzyme have been determined (33). The resulting restriction fragments were separated by electrophoresis in 1% agarose gels, denatured, and transferred to Millipore membrane filter strips by the Southern procedure (34). The strips were cut in half down the middle, and each half was hybridized to either 19S ^{32}P-labelled RNA or 26S ^{32}P-labelled RNA (33). The results show that the DNA sequence complementary to 26S RNA includes both Eco R1 sites and one Hind III site ; the 19S complementary sequence includes the other Hind III site. A map of Physarum rDNA based on the available data indicating the localization of the 19S and 26S ribosomal sequences is shown in Figure 3. The rDNA molecule contains two genes for preribosomal RNA arranged at opposite ends of a long spacer in inverted polarity (33). This palindrome-like arrangement is similar to that observed in the rDNA of Tetrahymena pyriformis (35, 36). Such inverted gene polarity in the rDNA of Physarum indicates that both DNA strands of the linear duplex can code for the ribosomal RNA sequences, an important factor in considering probes for the fidelity of rDNA transcription in in vitro systems. It is not yet clear whether the long spacer region includes transcribed sequences, although electron microscopy indicates that much of it is transcriptionally inert.

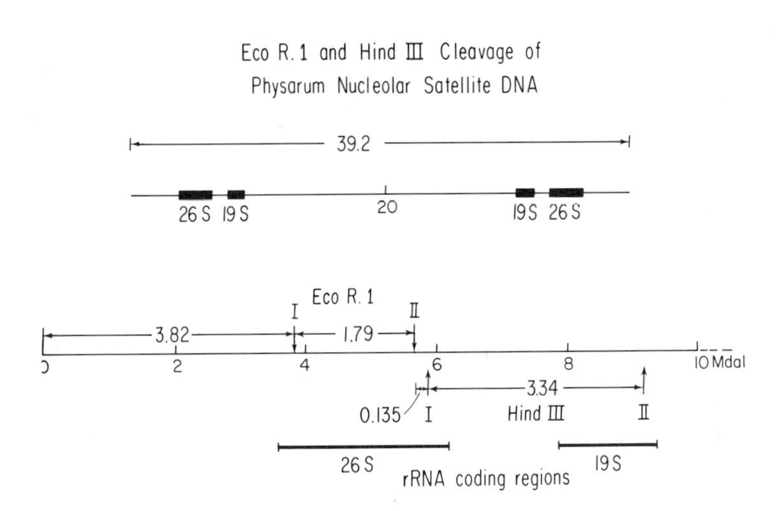

Fig. 3. Genetic map of the rDNA satellite of Physarum polycephalum, as revealed by restriction nuclease digestion and hybridization of the fragments to radioactive 19S and 26S ribosomal RNAs (33). Note the palindrome-like arrangement of the ribosomal genes at either end of the long spacer.

COMPARATIVE STRUCTURAL ORGANIZATION OF RIBOSOMAL GENES AND PHYSARUM CHROMATIN

An important basic problem is whether the ribosomal genes in rDNA are associated with histones to form nucleosomes or whether such actively-transcribing genes occur in a different configuration. Electron microscopic visualization of ribosomal transcription complexes in Oncopeltus have indicated that the nascent ribosomal RNP fibrils are attached to an unbeaded chromatin strand (37). Similarly, the transcribing matrix of ribosomal genes in Notopthalmus does not appear to contain the 70-100 A spherical particles seen in the non-transcribing spacer regions (38). The transcribed chromatin often appears thicker than a double-strand of DNA, suggesting the presence of associated proteins (37, 38). However, measurement of the DNA-packing ratio in the ribosomal genes (e.g., 1.2 μm of B structure DNA per μm of chromatin (37)) indicates that the DNA exists in an extended configuration, unlike that beaded chromatin strands. Spacer regions of Drosophila rDNA contain nucleosome-like particles (38, 63) although the protein composition of the particles remains to be determined (70).

Electron microscopy of chromatin spreads of Physarum reveals the usual beads-on-a-string organization (39). Figure 4 shows a typical chromatin spread prepared by the Miller and Bakken procedure (40).

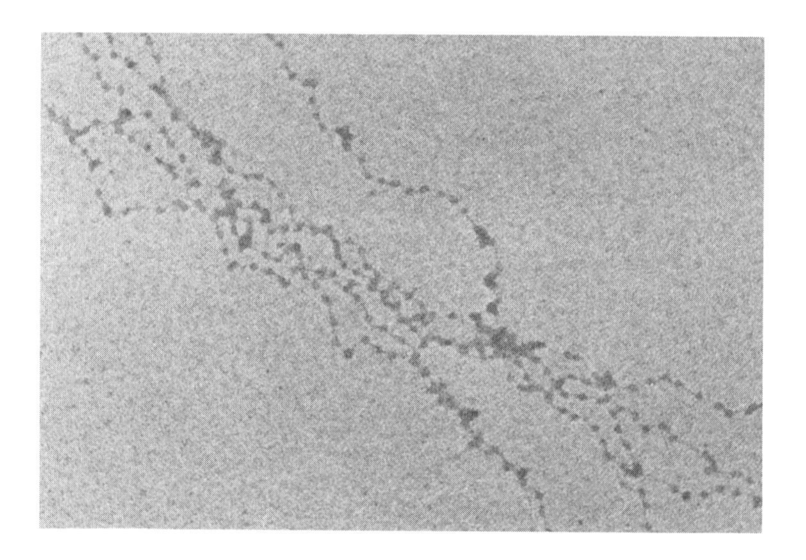

Fig. 4. Electron micrograph of Physarum chromatin after fixation with formalin and staining with phosphotungstic acid.

The size of the nucleosome "beads" averages 8.5 ± 1.4 nm in diameter. Interbead distance is variable, ranging from no visible separation up to 13 nm separation, but generally averages 7.5 ± 2.8 nm. A total length of 16 nm between bead centers is an approximate median value for the Physarum nucleosome repeat length. The measured lengths of the interconnecting DNA strands (range 4.7 - 10.3 nm) correspond to 14-30 nucleotide-pairs of DNA in the B-helical conformation.

The subunit structure of *Physarum* chromatin has been analyzed by staphylococcal nuclease digestion and electrophoretic separation of the resulting DNA fragments (39). Lengths of the DNA segments were determined by comparison with DNA size-standards obtained from polyoma A-2 DNA treated with restriction nuclease Hpa II, and λ dv-1 phage DNA digested with restriction nuclease Bsu. After limited digestion, the *Physarum* DNA segments migrate in a pattern characteristic of DNA lengths which are multiples of a basic repeating unit. The average DNA repeat length decreases as digestion time increases; the longest repeat length, observed after 1 minute of mild digestion, is about 190 base-pairs; the average repeat length observed in a 10 minute digest is 174 base-pairs. After extensive nuclease digestion, *Physarum* chromatin DNA is converted to a subunit monomer of average length about 159 base-pairs (39). Recent experiments indicate that this 159 bp nuclease-resistant segment is an intermediate in digestion to a monomeric subunit containing about 140 DNA base-pairs. The 159 bp fragment of *Physarum* is more prominent than the equivalent fragment from mammalian chromatin which is digested more readily to the 140 bp length typical of DNA in the nucleosome "core particles". Another difference between *Physarum* and mammalian chromatins is the shorter repeat length of the myxomycete chromatin. Figure 5 compares the fragment sizes of *Physarum* chromatin with those of rat liver chromatin after partial digestion with staphylococcal nuclease. The rat nucleosome DNA repeat length, calculated by averaging differences between adjacent multimer band sizes is 201 $\pm$ 5 base-pairs, as compared with 173 $\pm$ 10 base-pairs for *Physarum*. The lengths of the nuclease-accessible nucleosome-interconnecting DNA segments have been calculated as varying from 13-31 base-pairs (39). This estimate is in good agreement with the lengths of the interconnecting DNA strands (14-30 base-pairs) seen in electron micrographs (although spreading conditions may influence the latter).

The nucleosome monomers and oligomers produced during limited digestions with staphylococcal nuclease can be separated by centrifugation on sucrose density gradients. When such fractionations are carried out at moderate ionic strengths (0.3 M NaCl), the mononucleosome peak appears as a doublet with a lighter component (peak A) containing DNA of the same size as the heavier monomer peak.

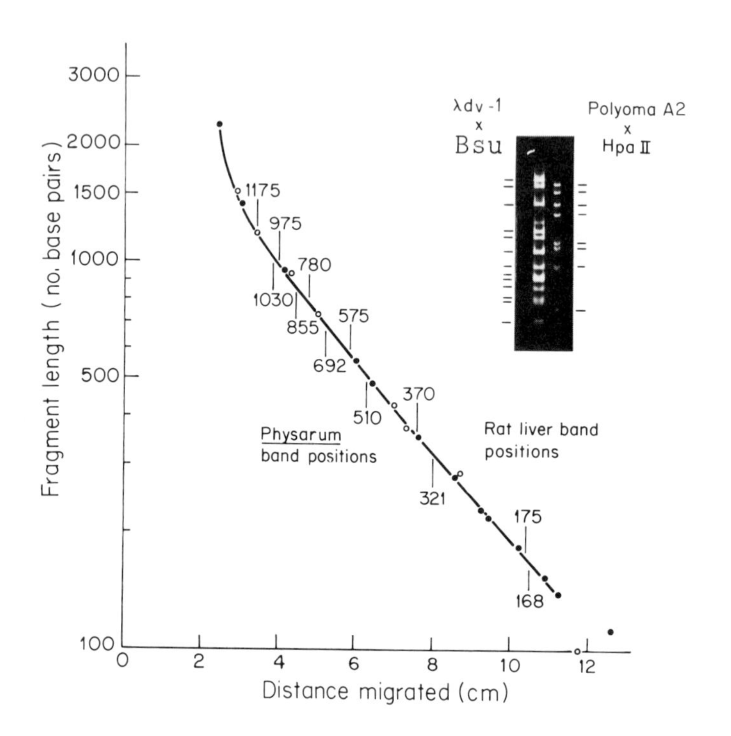

Fig. 5. Size distribution of DNA fragments released from Physarum and rat liver nuclei during limited digestions with staphylococcal nuclease. The fragments were separated by electrophoresis on 3.5% polyacrylamide gels and their sizes were determined by comparison with polynucleotide markers of known length obtained by restriction nuclease digestion of polyoma A-2 and λ dv-1 DNAs. Open circles show the position of the polyoma A-2 fragments while closed circles show the λ dv-1 standards. Note the close agreement between standards over the range 140-1200 bp. Lines drawn to the top of the curve show the positions of the rat chromatin fragments; lines drawn to the bottom of the curve show the positions of the Physarum chromatin bands.

Hybridization of ^{32}P-labelled 19S RNA and 26S RNA to the various nucleosome monomers and oligomers shows that the peak A monomers contain a higher proportion of rDNA sequences than do the more rapidly-sedimenting mononucleosomes (Figure 6).

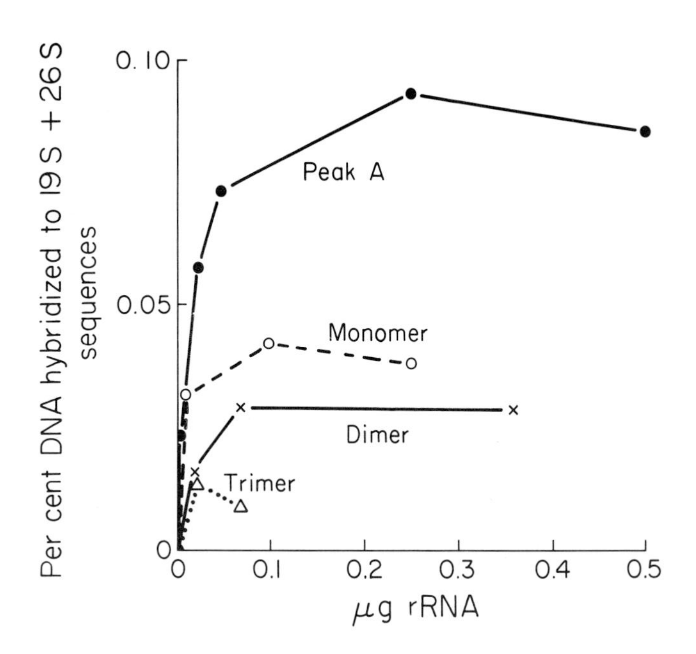

Fig. 6. Hybridization of radioactive 19S and 26S ribosomal RNAs to Physarum nucleosomal fractions generated by limited digestion with staphylococcal nuclease and separated by sucrose density-gradient centrifugation. Two monomeric forms of the nucleosomes are separable; peak A has a greater content of sequences hybridizable to the ribosomal RNAs.

Electrophoretic analyses of the proteins in the two monomeric nucleosome peaks indicate differences in their protein compositions (41). These results are comparable in some respects to those of Bakayev et al. (42) who observed that staphylococcal nuclease digestion of mouse Ehrlich ascites tumor chromatin yielded deoxyribonucleoprotein particles which sediment more slowly than the subunits.

The occurrence of DNA sequences hybribizable to ribosomal RNAs in the light (peak A) deoxyribonucleoprotein particles of Physarum suggests that the ribosomal genes may exist in a characteristic conformation. Higher oligomers are deficient in ribosomal sequences relative to the smaller monomer particles. We have visualized the transcription complex by electron microscopy and have confirmed the extended and apparently non-beaded state of the ribosomal genes. The results of the nuclease digestion experiments clearly indicate that at least a certain proportion of the ribosomal genes in Physarum are released as if they were organized into nucleosomes. It is not known whether these protected segments are derived from the transcription complexes. Their protection from random nuclease attack is in accord with earlier observations on the ribosomal genes of Xenopus (43, 44) and Tetrahymena (45). We believe that these results are compatible with the view that there are fundamental differences in the nature (or post-synthetic modification) of histones and other proteins associated with DNA in the transcription complex. These changes would permit extension of the DNA template, but the DNA-associated proteins would continue to confer protection against random nuclease attack.

IDENTIFICATION OF NUCLEOLAR PROTEINS WHICH PREFERENTIALLY COMBINE WITH rDNA

On the premise that proteins involved in the control of ribosomal RNA synthesis will be localized in the nucleolus and interact with rDNA sequences, we have begun to study the DNA-binding properties of Physarum nucleolar proteins. The proteins were extracted from purified nucleoli (25) under non-denaturing conditions, using procedures similar to those employed for the purification of Physarum RNA polymerases (46-48). The proteins in a 0.5 M $(NH_4)_2SO_4$ nucleolar extract were fractionated by affinity chromatography on Physarum DNA covalently-attached to Sephadex G-25 (49, 50). The affinity chromatography of nuclear proteins under these conditions has been shown to be highly reproducible. The columns fractionate nuclear proteins into subsets of characteristic DNA-binding and electrophoretic properties. The method, involving successive increments in the salt concentration of the eluting buffer to displace different classes of proteins, has been shown to preserve both enzyme activity (51) and hormone-receptor function (52) of the DNA-binding fractions. Comparisons of protein elution profiles from parallel columns of low C_0t, intermediate C_0t, and high C_0t DNA show that the various non-histone nuclear proteins can be characterized according to their affinities for repeated and unique sequences (49,50,53).

When Physarum nucleolar proteins were fractionated on columns of Physarum DNA-Sephadex G-25, it was found that more than 70% of the proteins in the nucleolar extract had little or no DNA affinity. The DNA-binding proteins were subsequently eluted in a stepwise salt gradient. Protein fractions eluted at different ionic strengths were compared with respect to their capacity to preferentially bind the rDNA satellite, as compared to main-band DNA. Binding was measured by retention of ^{125}I-labelled rDNA or ^{125}I-labelled main-band DNA on nitrocellulose filters (51,54-56), both types of DNA being sheared to equal lengths. The results summarized in Figure 7 show that protein fraction IV (a fraction eluting from the DNA column at high ionic strength (1 M KCl)) has a higher affinity for rDNA than for main-band DNA.

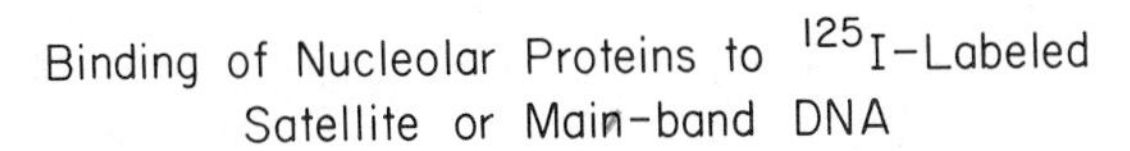

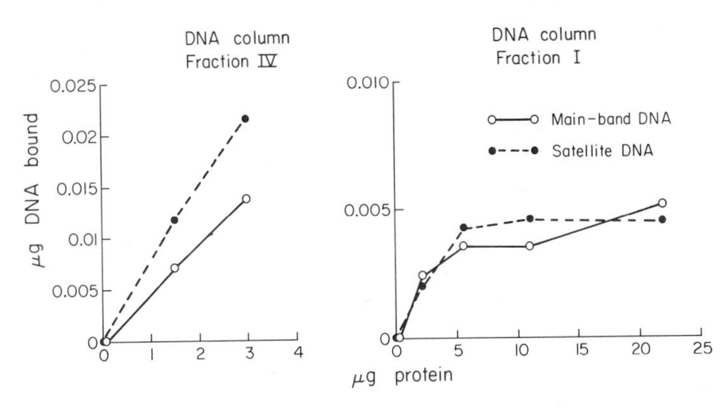

Fig. 7. Differential DNA-binding by nucleolar protein fractions eluted at different ionic strengths from a column of Physarum DNA-Sephadex. The binding of the protein fractions to ^{125}I-labelled rDNA or main-band DNA is compared. Note that fraction IV has preferential affinity for the rDNA satellite; fraction I does not.

Such preferential binding to the rDNA satellite is not seen in the nucleolar protein fraction (I) which elutes from the DNA column at low ionic strength (0.05 M KCl). This indication that nucleolar extracts contain proteins which preferentially combine with ribosomal DNA sequences was further tested by competition-binding studies in which the same nucleolar protein fraction (IV) was allowed to interact with ^{125}I-rDNA in the presence of increasing amounts of either unlabelled rDNA or unlabelled main-band DNA. It was found that excess rDNA can completely displace the ^{125}I-rDNA in the complex; excess main-band DNA could significantly displace the ^{125}I-rDNA but appears to be less effective. More detailed experiments to investigate the basis of the difference are now in progress.

RIBOSOMAL GENE TRANSCRIPTION IN ISOLATED PHYSARUM NUCLEI

The eventual aim of these studies is to identify which of the rDNA-binding proteins influence the transcription of ribosomal RNA. Based on previous evidence for the lack of fidelity of transcription of free *Xenopus* ribosomal DNA in *in vitro* systems (57), as compared with the strand and sequence-selective readout of ribosomal genes in isolated *Xenopus* kidney nuclei (58), we have focused on ribosomal RNA synthesis in isolated *Physarum* nuclei.

Conditions have been found that permit extended incorporations of radioactive nucleotides into *Physarum* nuclear RNA. In agreement with earlier observations (59), we find that the nuclei have a high basal level of α-amanitin insensitive RNA synthesis. We have established that ribosomal RNA synthesis in the isolated Physarum nucleus is resistant to α-amanitin, by hybridization of the newly-synthesized radioactive RNA to the purified rDNA satellite. The results shown in Figure 8 compare the hybridization properties of newly-synthesized RNA in the presence and absence of α-amanitin. It is clear that transcription of the ribosomal genes is not inhibited under these conditions. Similar resistance to α-amanitin has been noted for transcription of rDNA in *Xenopus* nuclei (58).

Studies in progress indicate that that the isolated nuclei of *Physarum* synthesize a high-molecular-weight precursor of the 19S and 26S ribosomal RNAs (60). This is in accord with earlier observations on intact cells (61, 62) and with the EM morphology of the transcription complex.

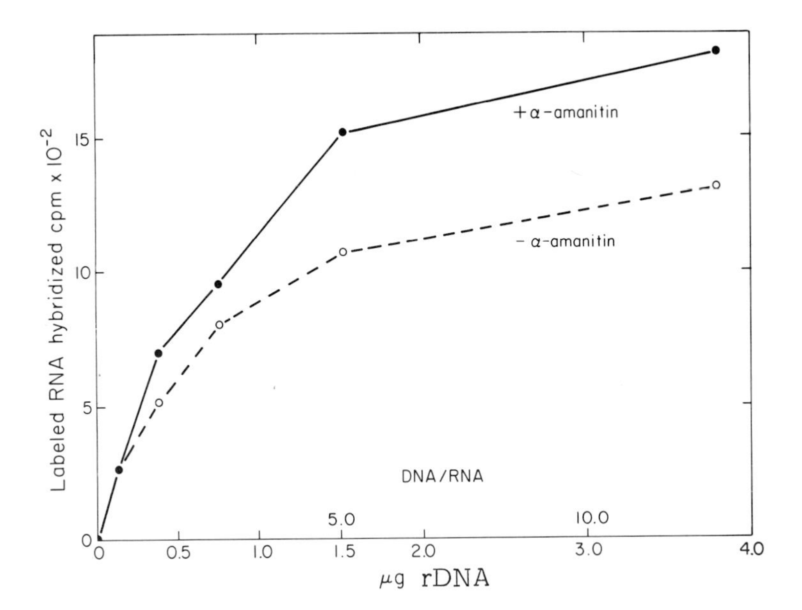

Fig. 8. Hybridization analysis of the RNA synthesized in suspensions of isolated Physarum nuclei. The RNA, labeled with ^{3}H-GTP in the presence or absence of α-amanitin, was hybridized to increasing amounts of the rDNA satellite as indicated in the abscissa. The presence of newly-synthesized ribosomal RNA sequences is clearly indicated.

In view of the many indications that ribosomal RNA synthesis is subject to physiological controls (e.g. (63-68)) and is not simply limited by the number of rDNA copies or the pool size of RNA polymerase I (57, 63, 69), we believe that control of transcription of Physarum ribosomal DNA is regulated by its association with nucleolar proteins which control the readout of the appropriate sequences and restrict transcription from the non-transcribed spacer regions.

The availability of milligram amounts of the purified rDNA satellite, the existence of nucleolar proteins which bind rDNA preferentially, and the ability of isolated nuclei to synthesize ribosomal RNAs in the presence of α-amanitin all contribute to a growing potential for a more incisive analysis of the mechanisms controlling the transcription and processing of preribosomal RNA.

ACKNOWLEDGEMENTS

This research was supported in part by grants from the American Cancer Society (NP-228G), the USPHS (NIH Grant GM 17383) and the National Science Foundation (PCM76 - 19926). Collaboration between workers at the Rockefeller University and Portsmouth Polytechnic was supported by a grant from NATO.

REFERENCES

1. Allfrey, V.G., Bautz, E.K.F., McCarthy, B.J., Schimke, R.T. and Tissieres, A. (eds.) "Organization and Expression of Chromosomes", Life Sciences Report 4, Dahlem Konferenzen, Berlin. 1977.
2. Elgin, S.C.R. and Weintraub, H. (1975). Annu. Rev. Biochem. 44, 725.
3. Allfrey, V.G. (1977) In "Chromatin and Chromosome Structure" (H.J. Li and R. Eckhardt, eds), pp. 167-191, Academic Press, New York.
4. Johnson, E.M. and Allfrey, V.G. (1977) In "Biochemical Actions of Hormones" Vol. 5 (G. Litwack, ed.) in press.
5. Langan, T.A. and Hohmann, P. (1975) In "Chromosomal Proteins and their Role in the Regulation of Gene Expression" (G.S. Stein and L.J. Kleinsmith, eds.), pp. 113-125, Academic Press, New York.
6. Teng, C.S., Teng, C.T. and Allfrey, V.G. (1971) J. Biol. Chem. 246, 3597.
7. Kamiyama, M. and Wang, T.Y. (1971) Biochim. Biophys. Acta 228, 563.
8. Shea, M. and Kleinsmith, L.J. (1973) Biochem. Biophys. Res. Commun. 50, 473.
9. Chiu, J.F. and Hnilica, L.S. (1977) In "Chromatin and Chromosome Structure" (H.J. Li and R. Eckhardt, eds.), pp. 193-254, Academic Press, New York.
10. Tsai, S.Y., Harris, S.E., Tsai, M.J. and O'Malley, B.W. (1976) J. Biol. Chem. 251, 4713.

11. Gilmour, R.S. and Paul, J. (1975) In " Chromosomal Proteins and their Role in the Regulation of Gene Expression" (G.S.Stein and L.J.Kleinsmith, eds.), pp. 19-33, Academic Press, New York.
12. Peterson, J.L. and McConkey, E.H. (1976) J.Biol.Chem. 251, 548.
13. Brown, D.D. and Gurdon, J.M. (1964) Proc.Nat.Acad.Sci. USA 51, 139.
14. Wallace, H. and Birnstiel, M.L. (1966) Biochim.Biophys. Acta 114, 296.
15. Ryser, U., Fakan, S. and Braun, R. (1973) Exptl.Cell Res. 78, 89.
16. Zellweger, A., Ryser, U. and Braun, R. (1972) J.Mol.Biol. 64, 681.
17. Mohberg, J. and Rusch, H.P. (1971) Exptl.Cell Res. 66, 305.
18. Ryser, U. and Braun, R. (1974) Biochim.Biophys.Acta 361, 33.
19. Hall, L., Turnock, G. and Cox, B.J. (1975) Eur.J.Biochem. 51, 459.
20. Bohmert, H.J., Schiller, B., Bohme, B. and Sauer, G.W. (1975) Eur.J.Biochem. 57, 361.
21. Newlon, C.S., Sonenshein, G.E. and Holt, C.E. (1973) Biochemistry 12, 2338.
22. Guttes, E. and Guttes, S. (1969) J.Cell Biol. 43, 229.
23. Bird, A.P. and Birnstiel, M.L. (1971) Chromosoma 35, 300.
24. Watson-Coggins, L. and Gall, J.G. (1972) J.Cell Biol. 52, 569.
25. Bradbury, E.M., Matthews, H.R., McNaughton, J. and Mølgaard, H.V. (1973) Biochim.Biophys.Acta 335, 19.
26. Daniel, J.W. and Baldwin, H.H. (1964) In " Methods in Cell Physiology" (D.M.Prescott, ed.), pp 9-41, Academic Press, New York.
27. Birnstiel, M.L., Spiers, J., Purdom, J., Jones, K. and Loening, U.E. (1968) Nature 219, 454.
28. Birnstiel, M.L., Chipchase, M. and Spiers, J. (1971) Prog. Nucleic Acid Res.Mol.Biol. 11, 351.
29. Sinclair, J.H. and Brown, D.D. (1971) Biochemistry 10, 2761.
30. deHarven, E. and Johnson, E.M. (1977) unpublished experiments.

31. Holt, C.E. and Gurney, E.G. (1969) J.Cell Biol. 40, 484.
32. Vogt, V.M. and Braun, R. (1976) J.Mol.Biol. 106, 567.
33. Molgaard, H.V., Matthews, H.R. and Bradbury, E.M. (1976) Eur.J.Biochem. 68, 541.
34. Southern, E.M. (1975) J.Mol.Biol. 98, 503.
35. Karrer, K.M. and Gall, J.G. (1976) J.Mol.Biol. 104, 421.
36. Engberg, J., Andersson, P., Leick, V. and Collins, J. (1976) J.Mol.Biol. 104, 455.
37. Foe, V.E., Wilkinson, L.E. and Laird, C.D. (1976) Cell 9, 131.
38. Woodcock, C.L.F., Frado, L.L.Y., Hatch, C.L. and Ricciardiello, L. (1976) Chromosoma 58, 33.
39. Johnson, E.M., Littau, V.C., Allfrey, V.G., Bradbury, E.M. and Matthews, H.R. (1976). Nucleic Acids Res. 3, 3313.
40. Miller, O.L.Jr. and Bakken, A.H. (1972) Acta Endocrin. Suppl. 168, 155.
41. Johnson, E.M., Matthews, H.R. and Allfrey, V.G. (1977) manuscript in preparation.
42. Bakayev, V.V., Melnickov, A.A., Osicka, V.D. and Varshavsky, A.J. (1975) Nucleic Acids Res. 2, 1401.
43. Reeves, R. and Jones, A. (1976) Nature 260, 495.
44. Reeves, R. (1976) Science 194, 529.
45. Mathis, D.J. and Gorovsky, M.A. (1976). Biochemistry 15, 750.
46. Burgess, A.B. and Burgess, R.R. (1974). Proc.Nat.Acad. Sci.USA 71, 1174.
47. Gornicki, S.Z., Vuturo, S.B., West, T.V. and Weaver, R.F. (1974) J.Biol.Chem. 249, 1792.
48. Weaver, R.F. (1976) Arch.Biochem.Biophys. 172, 470.
49. Allfrey, V.G., Inoue, A., Karn, J., Johnson, E.M. and Vidali, G. (1974) Cold Spring Harbor Symp.Quant.Biol. 38, 785.
50. Allfrey, V.G., Inoue, A. and Johnson, E.M. (1975) In "Chromosomal Proteins and their Role in the Regulation of Gene Expression" (G.S.Stein and L.J.Kleinsmith, eds.) pp. 265-300, Academic Press, New York.
51. Johnson, E.M., Hadden, J.W., Inoue, A. and Allfrey, V.G. (1975) Biochemistry 14, 3873.
52. Inoue, A., Silva, E., Oppenheimer, J. and Allfrey, V.G. (1977) manuscript in preparation.
53. Allfrey, V.G., Inoue, A., Johnson, E.M., Good, R.A. and Hadden, J.W. (1975) CIBA Foundation Symp. 28, 199.

54. Johnson, E.M., Inoue, A., Crouse, L.J., Allfrey, V.G. and Hadden, J.W. (1975) Biochem.Biophys.Res.Commun. 65, 714.
55. Bourgeois, S. (1972) Acta Endocrinol.Suppl. 168, 178.
56. Riggs, A.D., Suzuki, H. and Bourgeois, S. (1970) J.Mol. Biol. 48, 67.
57. Roeder, R.G., Reeder, R.H. and Brown, D.D. (1970) Cold Spring Harbor Symp.Quant Biol. 35, 727.
58. Reeder, R.H. and Roeder, R.G. (1972) J.Mol.Biol. 67, 433.
59. Grant, W.D. (1972) Eur.J.Biochem. 29, 94.
60. Sun, I.Y.C., Johnson, E.M., Matthews, H.R. and Allfrey, V.G. (1977), unpublished experiments.
61. Jacobson, D.N. and Holt, C.E. (1969) J.Cell Biol. 43, 57a.
62. Melera, P.W., and Rusch, H.P. (1973) Exptl.Cell Res. 82, 197.
63. McKnight, S.L. and Miller, O.L., Jr. (1976) Cell 8, 305.
64. Brown, D.D. and Littna, E. (1964) J.Mol.Biol. 8, 669.
65. Emerson, C.P. and Humphreys, T.D. (1971) Science 171, 898.
66. Hildebrandt, A. and Sauer, H.W. (1977) Biochem.Biophys. Res.Commun. 74, 466.
67. Bastien, C. (1977) Biochem.Biophys.Res.Commun. 74, 1109.
68. Hall, L. and Turnock, G. (1976) Eur.J.Biochem. 62, 471
69. Roeder, R.G. and Rutter, W.J. (1970) Biochemistry 9, 2543.
70. Zentgraf, H., Scheer, V., Franke, W. and Trendelenberg, M. F. (1976) J.Cell Biol. 70, 390 a.

THE SIGNIFICANCE OF STRUCTURAL VARIATIONS OF LYSINE-RICH HISTONES

R. David Cole, Myrtle W. Hsiang, George M. Lawson, Rodney O'Neal and Shirley L. Welch

Department of Biochemistry, University of California, Berkeley, California 94720

ABSTRACT. There are multiple species of H1 histones in most tissues. Their role in chromatin structure in not understood but it is clear they introduce compositional diversity into chromatin at the nucleosomal level or immediately above. This report presents evidence suggesting that the compositional diversity introduced by H1 may result in functionally significant irregularities of chromatin conformation.

Changes in the quantitative recipe of H1 histone subfractions correlate with differentiation (phenotype of tissue), embryonic development, spermatogenesis, mitotic index, and hormonal induction.

In addition to other evidence that histone H5 might be functionally related to H1, studies on amino acid sequence suggest about 60% homology. A similar homology probably exists between conventional H1 histone and special H1 histones found in the gonads of two echinoderms, sea cucumber and sea urchin. These possible homologies are of functional significance because of the extreme condensation of the specialized chromatins of avian erythrocytes and maturing sperm cells.

The various H1 histones of a particular organism show diversity in their interactions with DNA as observed by circular dichroism, viscosity, and binding affinities (filter binding assays and DNA-cellulose chromatography). Subfractions of H1 histone can be selectively extracted from chromatin while the latter is still in the nucleus.

INTRODUCTION

With the development of the idea that chromatin contains a string of subunits called nu bodies or nucleosomes (1-3), a consensus has developed for the roles of four (H2A, H2B, H3, H4) of the major classes of histones. Although there is more to learn about the uniformity of nucleosome distribution and composition, and the meaning of the post-translational modification of their histone cores, the first level of function of the core histones is understood to be the condensation of DNA into a very specific conformation. The role of the fifth major class of histone, H1, has not yet been clarified, even to the primitive level at which we understand the others.

A role of H1 histone in higher orders of condensation has seemed likely since the early work of Littau, et al (4). The frequently observed swelling of chromatin when H1 histone is selectively extracted by acid (5) or salt (6) has been interpreted (7) as a loss of cross links formed by H1 histone. More recently Finch and Klug (8) have reported that H1 histone plays a role in stabilizing solenoids, the next level of condensation above the nucleosome, according to their postulate. While recent suggestions (9, 10) that H1 histone determines the spacing between nucleosomes do not directly address themselves to chromatin condensation by H1, they can easily be interpreted in that way. The condensation aspect is attractive because two or three more levels of condensation above the nucleosome level must be accomplished to fit chromosomal DNA into the nucleus. It is generally agreed that H1 histone is on the surface of chromatin (11), and the imposition of such a structural element on the surface of the string of nucleosomes would seem to be an obvious way to develop the next level of folding.

A diversity of amino acid sequences was demonstrated in calf thymus H1 histone a decade ago (12) by use of peptide maps and other arguments. The diversity of primary structure has been observed in many other tissues and cell types since. The natural stoichiometry (2) in many systems allows about one copy of H1 per nucleosome, but this allowance must be shared by several subfractions of H1 that differ in primary structure. Therefore it is clear that H1 histone must impart a compositional diversity to chromatin.

The question addressed in the present report is whether the compositional diversity introduced into chromatin by H1 histone develops a diversity of physical conformations as H1 presumably leads to a higher level of folding. Beyond that lies the question whether that structural diversity is involved in regulation of any of the several functions of chromatin.

CORRELATION BETWEEN H1 RECIPES AND PHYSIOLOGICAL PROCESSES

Support for the notion that the diversity H1 histone is expressed as functional diversity, comes from a variety of observations. Pronounced differences (Fig. 1) were observed in the relative amounts of the various H1 histones in a number of tissues (13-15). This dependence of H1 histone recipes with (tissue) phenotype can be taken as a correlation of H1 diversity with differentiation. A correlation was also observed with embryonic development in the sea urchin, where

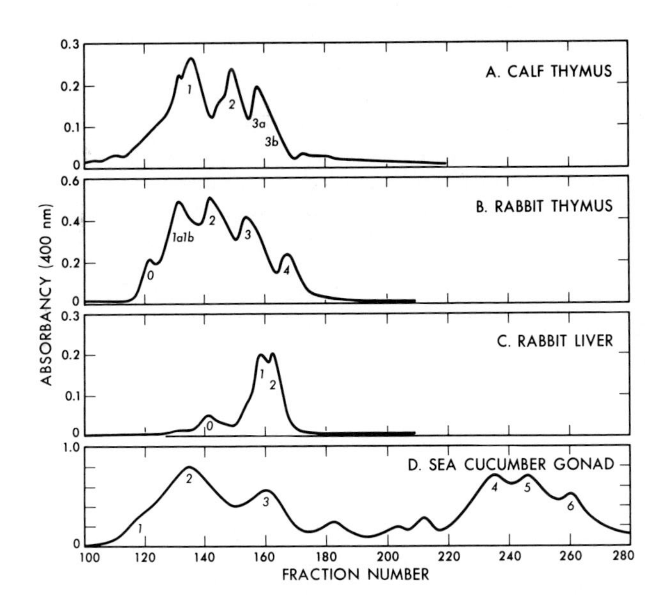

Fig. 1. Chromatography on Amberlite-IRC 50 of H1 histone from various tissues. The columns for A, B and C were 2.3 x 30 cm, eluted with 1700 ml gradients of guanidinium chloride from 8.5 to 17% (13). A 2 x 15 cm column was used for sample D, with a 500 ml gradient (26). Protein concentration was measured by turbidity.

the gastrula and morula stages of development clearly differed in the nature of their H1 histone (16). A significant change in the pattern of H1 histones was seen in rat testis during spermatogenesis (17). A fourth correlation between H1 histone recipes and physiological function was seen in the hormonal induction of milk proteins in organ culture (18). In this system, lactogenic hormones were found to alter the recipe of H1 histones as new chromatin was being formed about a day before milk products were synthesized. Yet another correlation of H1 histone recipe and a function of chromatin was established by Panyim and Chalkley (19) who observed a particular H1 species, called H1°, in nondividing cells; Marks (20) later extended this work to establish a quantitative inverse proportionality between the amount of this H1 species and mitotic index.

If histone H5 might be considered an extreme form of H1, then the apparent role of H5 in the extreme condensation of mature avian erythrocyte chromatin could be considered a functional correlation with H1 (and H5) recipe. The arguments for considering H5 as a particular species of H1 include a unique solubility in perchloric and trichloroacetic acids, which they share. More importantly, H5 seems to occur at the expense of H1 (21), and H1 displaced H5 during the reactivation of erythrocyte chromatin in heterokaryons (22). Recently, amino acid sequences have been partially worked out for chicken H5 (23, 24) and for one of the chicken H1 subfractions (25). The amino terminal halves of the two histones seem to be about 60% homologous and it is most likely that the carboxyl terminal halves will be even more homologous since the composition of that part of the histone is dominated by a few amino acids (arginine, lysine, alanine and proline).

A homology between a special family of H1 histones and the more usual H1 was observed in sea cucumber gonads (26). Peptide maps suggested about 65% homology by that rough measure. Possibly related to this is the H1 histone from the sperm of sea urchin, another echinoderm. The partial amino acid sequence worked out by Strickland, et al (27) for this H1 suggests it is about 60% homologous to both rabbit H1 and to chicken H5. These special H1 species ought to be considered in the light of the unusual physiology of sperm and ripe gonads.

The correlation of H1 histone recipes with differentiation, embryonic development, hormonal induction and mitotic index, seem to take on added significance in the light of the organization of histone genes. With the possible exception of the gene for H5 (28), the histone genes appear to be arranged tandemly, and reiterated to form a single block (29, 30). With such an arrangement it is difficult to imagine why different recipes of H1 histone are developed in various tissues, hormonal states, etc., unless the different H1 histones are being selected for different functions.

SELECTIVITY IN THE INTERACTION OF VARIOUS H1 HISTONES WITH DNA

It can be demonstrated that the various H1 histones differ in the manner in which they interact with DNA, non-histone protein, and chromatin. First consider the interaction of Hl histones and DNA. Following the procedures of Fasman et al (31) complexes can be formed between T7 DNA and individual Hl histones. Four different Hl histones from rabbit thymus were found to differ profoundly in the extent to which they distorted the circular dichroic spectrum of the DNA (Fig. 2).

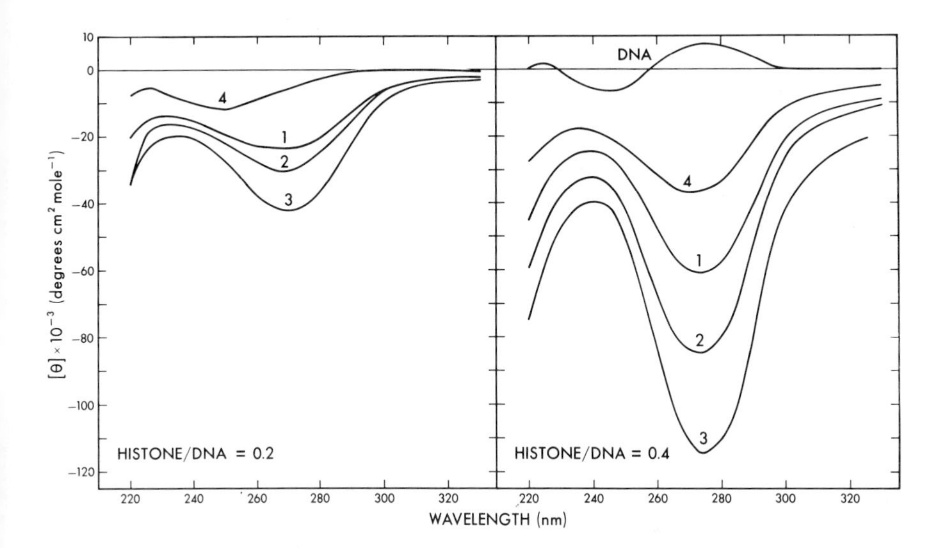

Fig. 2. Circular dichroism of complexes between T7 DNA and different rabbit thymus Hl histones. Subfractions of rabbit thymus lysine-rich histone were obtained by the method of Bustin and Cole (13). Complexes between T7 DNA and individual Hl histones were made by step gradient dialysis following the procedure of Fasman, et al (31) except all solutions were buffered with 0.015 M citrate pH 7. Curves 1-4 correspond to complexes made with T7 DNA and Hl histones RTL 1-4 respectively. The curve labelled DNA is the spectrum of deproteinized T7 DNA. All CD spectra were recorded at 23°C, 1 cm pathlength. The concentration of DNA in all complexes was 100 µg/ml. a: the weight ratio of histone: DNA is 0.2. b: the weight ratio of histone: DNA is 0.4.

The precise structures that are present in these complexes are not understood, but it is clear that each H1 histone handles the DNA differently. Preliminary data suggest that the hydrodynamic viscosities of the complexes differ, and filter binding assays indicate a difference among the H1 subspecies in their affinities for DNA. Support for the latter conclusion was obtained by passing a mixture of H1 histones down a DNA-cellulose column; the resolution of the several H1 histones reveals differences in their binding properties.

Smerdon and Isenberg (32) have reported that two non-histone proteins form complexes with H1 histone. Moreover, the HMG proteins are selective among the H1 subfractions with regard to this interaction. Confirming this report is the work of Yu and Spring (33), who passed the HMG proteins through columns to which individual H1 histones were bound covalently. Once again, selective interactions were observed.

SELECTIVE EXTRACTION OF INDIVIDUAL H1 HISTONE SUBFRACTIONS FROM NUCLEI

To show that the selective interactions observed in binary complexes (DNA and histone or histone and non-histone protein) might apply to the more complicated interaction of H1 histones and native whole chromatin, we have shown that one H1 species of HeLa cells can be differentially extracted from intact nuclei. If aliquots of nuclei are carefully washed in buffers of different pH the results shown in Fig. 3 can be obtained. At pH 3.8, no significant amount of H1 is extracted, and at pH 3.0 all the H1 is displaced, but at intermediate pH increments in the series it is observed that the H1 subfraction of higher electrophoretic mobility is displaced in clear preference to the slower H1 component.

None of the evidence that H1 histones differ in their interactions with DNA, with non-histone proteins, and with chromatin, proves that they produce irregularities in higher levels of chromatin folding. Nevertheless, the evidence on interactions, coupled with the correlations with physiological functions, make it seem attractive to postulate that as H1 histone develops a higher level of folding of the string of nucleosomes, it produces functionally significant irregularities in the conformation.

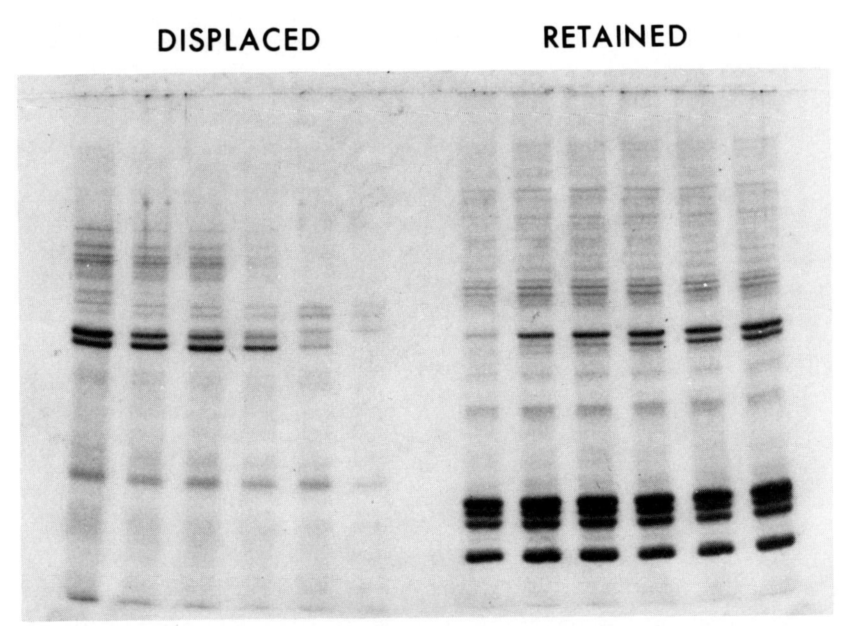

Fig. 3. Selective displacement of H1 subfractions from intact HeLa cell nuclei by modification of the procedure of Mirsky and Silverman (34). Nuclei were extracted twice for 15 min. each by stirring gently at 4°C. in citric acid sodium phosphate buffers ranging in pH from 3.0 to 3.8, and in total molarity from 60 mM to 68 mM (35), and also made 0.25 M in sucrose, 25 mM in KCl, 1 mM in $CaCl_2$ and 1 mM in $MgCl_2$. Combined supernatants were dialyzed, lyophilized, and electrophoresed on SDS slab gels according to Laemmli (36). Histone (protein) retained within the nuclei was solubilized in 0.4 N H_2SO_4, dialyzed, lyophilized, and electrophoresed. Lanes 1-6, proteins displaced from nuclei; 7-12, proteins retained within nuclei.

ACKNOWLEDGEMENTS

These studies were supported by NSF Grant GB 38658, USPHS grant GMS 20338 and Contract N01-CB-43866 from the NCI, as well as the University of California Agricultural Experimental Station.

REFERENCES

1. Olins, A. L., and Olins, D. E. (1974) *Science* 183, 1330.
2. Kornbeg, R. D. (1974) *Science* 184, 868.
3. Van Holde, K. E., Sahasrabuddhe, C. G. and Shaw, B. R. (1974) *Nucleic Acid Res.* 1, 1599.
4. Littau, V. C., Burdick, C. J., Allfrey, V. G., Mirsky, A. E. (1965) *Proc. Nat'l. Acad. Sci. U.S.A.* 54, 1204.
5. Murray, K., (1966) *J. Mol. Biol.* 15, 409.
6. Wilhelm, X., and Champagne, M. (1969) *Eur. J. Biochem.* 10, 102.
7. Bradbury, E. M., Carpenter, B. G., and Rattle, H. W. E. (1973) *Nature* 241, 123.
8. Finch, J. T. and Klug, A. (1976) *Proc. Nat'l. Acad. Sci. U.S.A.* 73, 1897.
9. Compton, J. L., Bellard, M. and Chambon, P. (1976) *Proc. Nat'l. Acad. Sci. U.S.A.* 73, 4382.
10. Morris, N. R. (1976) *Cell* 9, 627.
11. Baldwin, J. P., Boseley, P. G., Bradbury, E. M., and Ibel, K. (1975) *Nature* 253, 245.
12. Kinkade, J. M. and Cole, R. D. (1966) *J. Biol. Chem.* 241, 5798.
13. Bustin, M. and Cole, R. D. (1968) *J. Biol. Chem.* 243, 4500.
14. Kinkade, J. M. (1969) *J. Biol. Chem.* 244, 3375.
15. Nelson, R. D., and Yunis, J. J. (1969) *Expt. Cell Res.* 57, 311.
16. Ruderman, J. V., Baglioni, C. and Gross, P. R. (1974) *Nature* 247, 36.
17. Kistler, W. S. and Geroch, M. E. (1975) *Biochem. Biophys Commun.* 63, 378.
18. Hohmann, P. and Cole, R. D. (1971) *J. Mol. Biol.* 58, 533.
19. Panyim, S. and Chalkley, R. (1969) *Biochem. Biophys. Res. Commun.* 37, 1042.
20. Marks, D. B., Kanefsky, T., Keller, B. J., and Marks, A. D. (1975) *Cancer Res.* 35, 886.
21. Seligy, V. L. and Neelin, J. M. (1970) *Biochim. Biophys. Acta* 213, 380.
22. Appels, R., Bolund, L., and Ringertz, N. R. (1974) *J. Mol. Biol.* 87, 339.
23. Garel, A., Mazen, A., Champagne, M., Sautiere, P., Kmiecik, D., Loy, O., and Biserte, G. (1975) *FEBS Lett.* 50, 195.
24. Sautiere, P., Kmiecik, D., Loy, O., Briand, G., Biserte, G., Garel, A. and Champagne, M. (1975) *FEBS Lett* 50, 200.
25. Spring, T. G. and Cole, R. D. unpublished observation.
26. Phelan, J. J., Subirana, J. A. and Cole, R. D. (1972) *Eur. J. Biochem.* 31, 63.

27. Strickland, W. N., Schaller, N., Strickland, M. and VonHolt, C. (1976) *FEBS Lett* 66, 322.
28. Scott, A. C. and Wells, J. R. E. (1976) *Nature* 259, 635.
29. Kedes, L. H. and Birnstiel, M. (1971) *Nature, New Biol.* 230, 165.
30. Kedes, L. H. (1976) *Cell* 8, 321.
31. Fasman, G. D., Schaffhausen, B., Goldsmith, L. and Adler, A. (1970) *Biochem.* 9, 2814.
32. Smerdon, M. J. and Isenberg, I. (1976) *Biochem.* 15, 4242.
33. Yu, S. H. and Spring, T. G. (1977) *Biochim. Biophys. Acta* in press.
34. Mirsky, A. E. and Silverman, B. (1972) *Proc. Nat. Acad. Sci.* 69, 2115.
35. Gomori, G. (1955) *Methods in Enzymology* 1, 138.
36. Laemmli, U. K. (1970) *Nature* 227, 680.

MICROCELL-MEDIATED CHROMOSOME TRANSFER

R. E. K. Fournier and F. H. Ruddle

Department of Biology, Kline Biology Tower, Yale University, New Haven, Connecticut 06520

ABSTRACT. Microcell-mediated chromosome transfer is a technique by which single or limited numbers of intact chromosomes can be transferred from one mammalian cell to another. This somatic cell genetic approach consists of the following steps: 1) micronucleation of the donor cells, 2) enucleation of micronucleate cells to yield free microcells, 3) purification of isolated microcells, 4) fusion of microcells with intact recipients, 5) recovery of viable microcell hybrids. In this report we describe in detail the use of this new genetic tool.

INTRODUCTION. There are presently three somatic cell genetic manipulations which can be used to transfer genetic information from one mammalian cell to another. These operationally distinct modes of genetic exchange result in the transfer of different amounts of genetic material from donor to recipient.

The first genetic procedure is somatic cell hybridization. By fusing two intact cells it is possible to combine whole genomes in a single hybrid cell. In interspecific hybridizations, unilateral chromosome elimination generally occurs resulting in the production of proliferating hybrid cells containing only a subset of the genetic material of one of the parents. Such cell lines are thus useful genetic tools, and have led to the chromosomal assignment of over 200 gene loci in man and other mammalian species (1).

A more recently developed mode of genetic exchange is chromosome-mediated gene transfer. This technique, which was first convincingly demonstrated by McBride and Ozer (2) in 1973, utilizes isolated metaphase chromosomes as vectors for the transfer of small pieces of genetic material. By selecting for the expression of a given gene locus and analyzing the resulting clones for co-transfer of tightly linked markers, it has been possible to estimate the size of the transferred genetic material ("transgenome"). Such estimates suggest a transgenome size on the order of 5×10^6 nucleotide base pairs (3).

The newest somatic cell genetic strategy is microcell-mediated chromosome transfer (4). In this procedure subnuclear particles isolated from donor cells are fused with intact recipients. This results in the transfer of single or limited numbers of intact chromosomes from donor to recipient. The transferred chromosomes are stably maintained and expressed in the resulting microcell hybrids.

In this report we describe the production of proliferating microcell hybrids and present relevant observations concerning each step in the application of this new somatic cell genetic strategy.

RESULTS. The production of proliferating microcell hybrids is a multi-step process which is illustrated diagrammatically in Fig. 1. This procedure was developed from the original observations of Ege and Ringertz (5), who pointed out the potential usefulness of subnuclear particles in hybridization experiments. The essential features of each step of the procedure can be outlined as follows. (1) Micronucleation of the donor cells results in the distribution of donor chromosomes into discreet subnuclear packets. (2) Enucleation of micronucleate cells produces subnuclear particles (microcells) appropriate for fusion. (3) Purification of isolated microcells removes contaminating whole nuclei (minicells) and intact donors. (4) Fusion of purified microcells with intact recipients can be efficiently accomplished with the use of inactivated Sendai virus. (5) Recovery of viable microcell hybrids depends critically on the sate of the donors at the time of enucleation. Each of these steps is discussed separately and in detail below.

(1) Micronucleation of the Donor Cells.

When rodent cells are subject to prolonged mitotic arrest, the cells eventually escape the metaphase block and enter a G1-like configuration. Under these conditions the chromosomes decondense and the nuclear membrane reforms. This occurs somewhat aberrantly, however, and leads to the production of cells containing numerous micronuclei, each containing a limited number of chromosomes. Such micronucleate cells (Fig.2) remain morphologically intact for days or weeks, depending upon the particular cell line.

It has been possible to successfully micronucleate all rodent lines tested to date. A partial listing of lines and effective micronucleation conditions are presented in Table I. Colcemid is the mitotic arrest agent of choice, and effective levels, which are cell line dependent, range from 0.01-0.1 μg/ml. Micronucleation is a highly concentration-dependent process, and the empirically-determined optimal

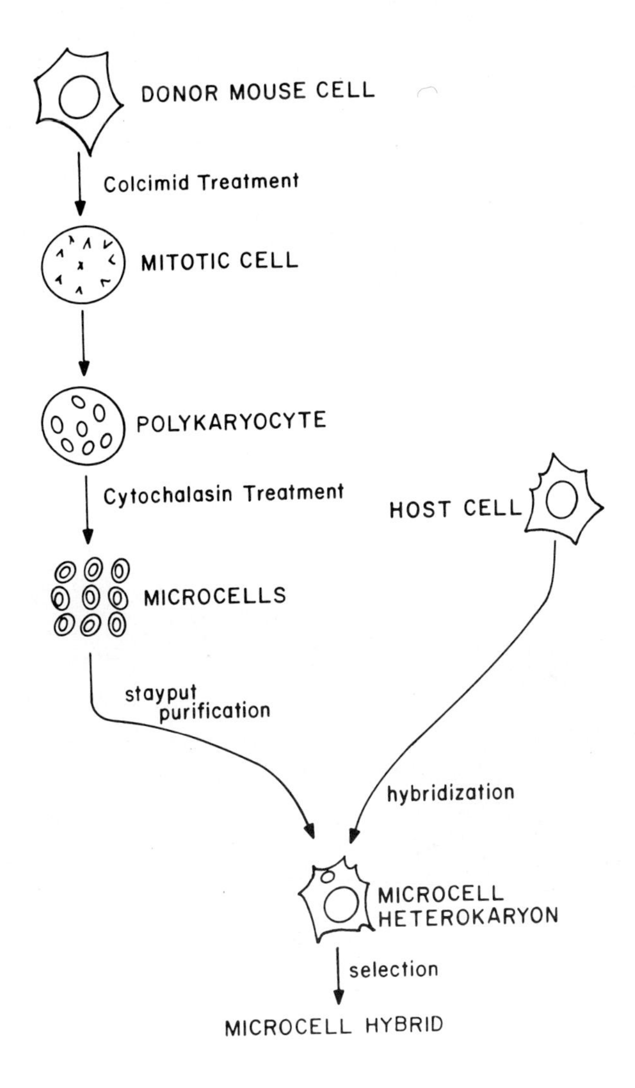

Fig. 1. Diagrammatic illustration of the preparation of microcell hybrids. For explanation, see text.

concentration may differ only 2-fold from concentrations which are ineffective. Higher concentrations generally produce strong cytotoxic effects.

For cells susceptible to cytolysis by colcemid, nitrous oxide (N_2O) at 5 atm is a useful alternate blocking agent. This technique has been used extensively for micronucleation of Chinese hamster cells.

Micronucleation of human cells is much more difficult. Although populations of micronucleate human cells can be produced under certain conditions, such cells are not viable and will not attach to glass or plastic. Since enucleation of micronucleate cells requires firm attachment of the cells to a substratum (see below), these cells are useless for our purposes. In order to prepare human microcells ("mini-segregants"), a differently conceived approach developed by Johnson and co-workers (6,7) can be employed.

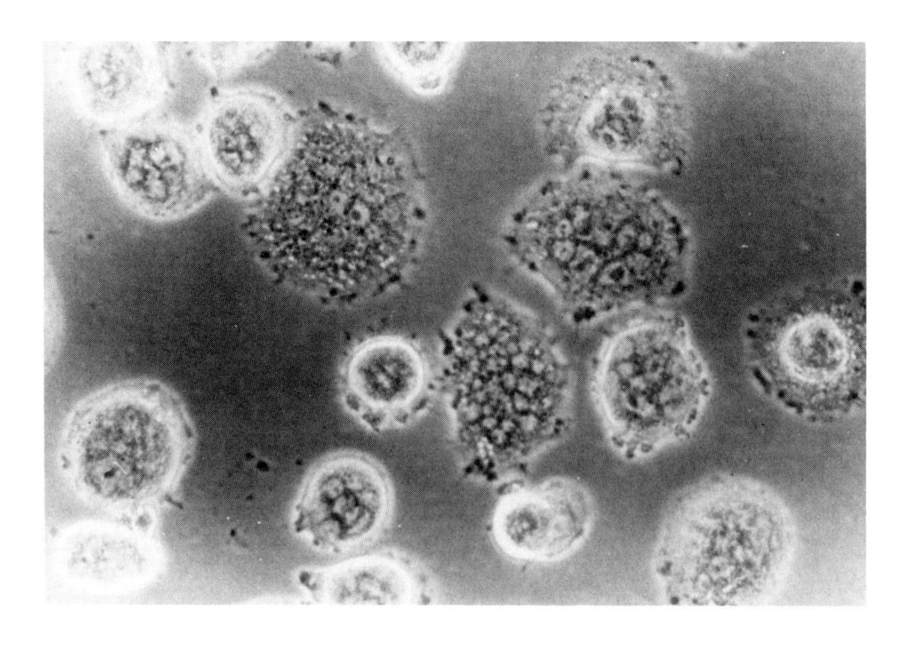

Fig. 2. Morphology of micronucleate mouse A9 cells. Exponentially growing cultures were exposed to 0.1 μg/ml colcemid for 48h. Phase contrast photomicrograph of living material x 400.

TABLE I. Effective Micronucleation Conditions for Various Rodent Cells

cell line	description	mitotic arrest agent	effective duration	yield of micronucleate cells
A9	mouse L-cell derivative $HPRT^-$ $APRT^-$	colcemid (0.1μg/ml)	48h	>90%
B82	mouse L-cell derivative TK^-	colcemid (0.05μg/ml)	48h	>90%
$CT11C_I$	Stable gene transfer line derived from A9. Expresses human HPRT	colcemid (0.1μg/ml)	48h	>90%
MEF	primary mouse embryo fibroblasts	colcemid (0.05μg/ml)	36h	60-70%
E36	Chinese hamster $HPRT^-$	N_2O (5 atm)	24h	80-90%
ECm-2	E36 x $CT11C_I$ microcell hybrid	N_2O (5 atm)	24h	80-90%
RG6A-tgA	rat glioma $HPRT^-$	colcemid (0.08μg/ml)	48h	70-80%

(2) Enucleation of Micronucleate Cells

Once micronucleation of the donor cells has been satisfactorily achieved, the micronuclei are removed from the cells using cytochalasin B (CB). The method of choice is to centrifuge plastic discs to which the cells are attached (8,9,10). A general outline of the micronucleation and enucleation steps of this procedure follows.

Plastic discs are cut from tissue culture plates using a heated 15/16 cork borer. The discs are then marked since only the inner surface has been treated for cell attachment. Sterilization of the discs is accomplished by rinsing in 95% ETOH and exposing to a germicidal UV lamp. Eight sterile discs are transferred to a 100mm tissue culture plate, and the donor cell suspension seeded into the plate. Cells are plated at a density such that 1-2 doublings are required to attain confluence. The cells are allowed to attach to the substratum for 12h or more, and the mitotic arrest agent added.

After micronucleation the cells are enucleated essentially as described by Ege and Ringertz (5). The discs are placed cell-side-down in 50 ml round bottom polycarbonate tubes containing 6ml complete medium and spun at 12,000 x g for 10 min. at 34°. This step removes loosely attached cells which would otherwise contaminate the microcell preparation. The discs are then centrifuged in the presence of 10 μg/ml CB at 39,000 x g for 20 min. at 34°. During this spin, individual micronuclei are drawn out of the cells on long cytoplasmic threads. These threads eventually break yielding free microcells consisting of a single (usually) micronucleus surrounded by a thin rim of cytoplasm and an intact plasma membrane. Such structures can be efficiently fused with recipients, which results in the insertion of an intact micronucleus into the cells.

The enucleation step described above results in >95% enucleation of the cells on the discs. The discs contain most of the cytoplasmic material and are discarded. The microcell pellets are resuspended and pooled. At this point, the crude preparation typically contains 60-70% microcells of various sizes,30-40% small cytoplasmic vesicles, and 0-10% intact cells. The level of contamination by whole cells depends, in general, on the cell line. For L-cells the value is usually 5-7%. For primary fibroblasts, which are much more firmly attached, whole cells comprise >0.1% of the pellet material.

Wigler and Weinstein (11) have developed a method for enucleating mononucleate cells in suspension using discontinuous Ficoll gradients containing CB. This technique has limited utility for enucleating micronucleate cells since only partial enucleation results, and yields are low.

(3) Purification of the Crude Microcell Preparation.

Purification of the crude microcell preparation by unit gravity sedimentation (STAPUT gradient) accomplishes two objectives. First, by using only the smallest microcells it is possible to transfer only a very limited number of chromosomes from donor to recipient. Second, such a purification by separating according to size eliminates intact donor cells from the preparation. It is therefore possible to use one-sided-selection in microcell hybridization experiments since intact donors are not present in the fusion mixture.

The apparatus routinely used for 1g sedimentations is a simple device consisting of a gradient mixing chamber (LKB), a peristaltic pump (Cole-Parmer), and a bubble trap and gradient chamber (5 and 50 ml disposable syringes respectively) connected to the line by 3-way valves (B-D). The crude microcell preparation, suspended in 2 ml 0.5% bovine serum albumin (BSA) in PBS, is added to the gradient chamber, and a 1-3% linear BSA gradient (in either PBS or serum-free medium) pumped into the chamber from below. For single hybridization experiments, gradients with a total volume of 50 ml are allowed to stand for 3.5 h at room temperature. For fusion, the top 15-20 ml of the gradient (carefully excluding the sample zone) are collected. Using material obtained from 16 discs, 4-8 x 10^6 purified microcells can be recovered from this fraction.

Such a gradient fractionation efficiently separates the initial heterogeneous mixture into various size classes. Virtually all whole cells in the gradient are confined to the bottom 15-20 ml. A typical STAPUT gradient fractionation has been published elsewhere (4). Such a purification scheme allows microcells of different sizes to be selected for fusion, and thus it is possible to gain some control over the number of chromosomes transferred per hybridization event.

(4) Fusion of Purified Microcells with Intact Recipients.

Purified microcells can be efficiently fused with intact recipients using β-propiolactone-inactivated Sendai virus according to standard somatic cell genetic procedures. Using the protocol outlined below it is possible to routinely obtain heterokaryon frequencies on the order of 1%. This value compares favorably with that of cell x cell fusions performed under similar conditions. The final titer of virus used for fusion should be 500-750 HAU/ml since extensive lysis of microcells occurs in the presence of high titers of Sendai virus.

For fusion, the purified microcell preparation is suspended in 0.5 ml serum-free medium. This suspension is added to a washed, chilled, near-confluent monolayer of recipient cells in a 25 cm^2 flask. 0.5 ml inactivated Sendai virus suspension in serum-free medium is then added, and the flask allowed to stand for 30 min at 4°. After a 60-90 min incubation at 37°, the flask is rinsed with complete medium to remove virus and non-adhering microcells. The fusion mixture is then incubated 18-36 h at 37° before being distributed into twenty 25 cm^2 flasks.

(5) Recovery of Proliferating Microcell Hybrids.

Proliferating microcell hybrids have been generated by fusing murine microcells from various sources with intact mouse, Chinese hamster, and human recipients (4). We have also successfully fused Chinese hamster microcells with rat cells, and both HAT (12) and ouabain (13) selective systems have been employed in these experiments. The yield of microcell hybrids is necessarily lower than that observed in similar cell x cell fusions using the same parents since only a fraction of the microcell heterokaryons contain the donor chromosome carrying the complementing gene. When this fact is taken into account, intraspecific microcell hybridizations yield viable hybrids at frequencies comparable to similar cell x cell fusions. Interspecific microcell hybridizations tend to be less efficient and yield 3-10 fold fewer hybrid clones than most intraspecific fusions.

We have found (4) that the conditions used for micronucleation of the donor cells are critical factors influencing the recovery of viable microcell hybrids. In general, exposing the donors to effective concentrations of the mitotic arrest agent for more than two cell generations severely diminishes the yield of microcell hybrids. For example, microcells isolated from A9 populations (doubling time about 20 h) exposed to colcemid for 48 h are more than ten times as efficient at yielding microcell hybrids than similar preparations isolated from A9 exposed to colcemid for 72 h (4). This result was observed in spite of the fact that both microcell preparations were morphologically indistinguishable and yielded virtually identical heterokaryon frequencies upon fusion. This is a general observation and has been found for various intra- and interspecific microcell hybridizations. It is therefore necessary to carefully control micronucleation conditions of the donor material in all microcell hybridization experiments.

DISCUSSION. Microcell-mediated chromosome transfer is a technique which allows single or limited numbers of intact chromosomes to be transferred from one mammalian cell to another. This new somatic cell genetic tool has several important features which are worthy of note.

First, this procedure allows the direction of chromosome segregation to be experimentally determined. Thus, it is now possible to construct human cell lines containing limited numbers of mouse (or other rodent) chromosomes, a situation in contrast to the usual pattern of chromosome elimination in man x mouse hybrid cells. Such hybrid lines will be particularly useful for mapping murine genes, especially in view of the large body of information concerning the discrimination of homologous murine and human phenotypes. These kinds of hybrid cells are presently in use for a number of gene mapping studies (C. Kozak, R.E.K. Fournier, and F.H. Ruddle, manuscript in preparation).

A second important feature of microcell hybrids is their karyotypic simplicity. Using traditional cell x cell hybrids, lengthy periods of cultivation , re-analysis, and subcloning are often necessary before such lines can serve as useful genetic tools. This laborious task is largely circumvented with the use of microcell hybrids. We have taken advantage of this fact in recent studies of cell lines generated by chromosome-mediated gene transfer. Using stably-transformed lines (heteroploid mouse cells expressing human HPRT) as donors in chromosome transfer experiments, it has been possible to provide strong genetic evidence for the insertion of the human transgenome into host murine chromosomes (R.E.K.Fournier and F.H.Ruddle, manuscript in preparation). This analysis was tremendously simplified by using microcell rather than whole cell hybrids.

A third important consideration is that microcell hybrids containing only a limited amount of genetic material of the donor cells tend to strongly resemble the recipient cell line. It is therefore possible to generate hybrid lines with particular properties (e.g. ability to grow in suspension) by appropriately choosing the recipient. This has potentially important implications regarding the study of genetic factors regulating epigenotype expression, and we believe that more meaningful experiments in this area can now be performed using microcell hybrids.

Finally, microcell hybridization allows more control over the input of foreign genetic material in the hybrid cells. This can be accomplished with respect to numbers of transferred chromosomes by using appropriately-sized microcells for fusion. The input of particular chromosomes can be realized using appropriate selective conditions for isolation of the hybrid clones. It is obvious that as more selectable markers either become available for or are introduced into different chromosomes, our control over which specific chromosomes are present in the hybrids will become even more precise.

ACKNOWLEDGEMENTS

We gratefully acknowledge the technical assistance of Ms. C. Colmenares, Ms. J. Lawrence, and Ms. E. Nichols. This manuscript was skillfully prepared by Ms. M. Reger. R.E.K.F. is a Leukemia Society of American Fellow. These studies were also supported by a grant from The National Institutes of Health (GM 09966).

REFERENCES

1. Ruddle, F.H. and Creagan, R.P. (1975) Ann. Rev. Genet. 9, 407.
2. McBride, W.O. and Ozer, H.L. (1973) Proc. Nat. Acad. Sci. USA 70, 1258.
3. Willecke, K., Lange, R., Kruger, A., and Reger, T. (1976) Proc. Nat. Acad. Sci. USA 73, 1274.
4. Fournier, R.E.K. and Ruddle, F.H. (1977) Proc. Nat. Acad. Sci. USA 74, 319.
5. Ege, T. and Ringertz, N. R. (1974) Exp. Cell Res. 87, 378.
6. Johnson, R.T., Mullinger, A.M., and Skaer, R.J. (1975) Proc. R. Soc. Lond. 189, 591.
7. Schor, S.L., Johnson, R.T., and Mullinger, A.M. (1975) J. Cell Sci. 19, 281.
8. Prescott, D.M., Myerson, D., and Wallace, J. (1972) Exp. Cell Res. 71, 480.
9 Wright, W.E., and Hayflick, L. (1972) Exp. Cell Res. 74, 187.
10. Poste, G., and Reeve, P. (1972) Exp. Cell Res. 73, 287.
11. Wigler, M. H. and Weinstein, I.B. (1975) Bioch. Biophys. Res. Comm. 63, 669.
12. Littlefield, J.W. (1964) Science 145, 709.
13. Kucherlapati, R.S., Baker, R.M., and Ruddle, F. H. (1975) Cytogenet. Cell Genet. 14, 362.

REGIONAL MAPPING OF GENE LOCI ON HUMAN CHROMOSOMES 1 AND 6 BY INTERSPECIFIC HYBRIDIZATION OF CELLS WITH A t(1;6)(p3200;p2100) TRANSLOCATION AND BY CORRELATION WITH LINKAGE DATA.

Uta Francke, Donna L. George and Michele A. Pellegrino*

Department of Pediatrics, University of California San Diego, School of Medicine, La Jolla, CA 92093

*Department of Molecular Immunology, Scripps Clinic and Research Foundation, La Jolla, CA 92037

ABSTRACT. Human fibroblasts containing a balanced reciprocal translocation between the short arms of chromosomes 1 and 6 - designated t(1;6)(p3200;p2100) - were fused with an established Chinese hamster cell line. Hybrid clones segregating human chromosomes were studied for the presence of the translocation chromosomes 1^T and 6^T and their normal homologs 1 and 6. Six clones that had retained 1^T, five clones with 1 and 6^T, three clones with 6 and 6^T and one clone with 1,6 and 6^T as control, were analyzed for expression of human genes known to be located on the short arm of chromosome 1 and on chromosome 6. The results indicate that known linkage groups had been disrupted by the breakpoints in both chromosomes leading to the t(1;6) translocation.

On chromosome 6, HLA maps distal and PGM-3 and ME-1 proximal to the breakpoint in 6p2100. On chromosome 1, PGD, ENO-1 and UMPK map distal and PGM-1 proximal to the breakpoint in 1p3200.

Linkage map distances (in centimorgans), previously established by family studies, were correlated with detailed cytological maps, based on G-banded early metaphase chromosomes. This resulted in more precise regional localization of genes on chromosomes 6 and 1p.

INTRODUCTION

In recent years, human gene mapping has made rapid progress due to advances in somatic cell genetic and chromosome identification methods. The study of rodent-human somatic cell hybrids segregating human chromosomes has permitted the assignment of more than 100 gene loci which are expressed in cultured cells to specific chromosomes (1,2). In order to determine the linear order and subregional localization of genes within chromosomes, human cells carrying chromosomal rearrangements have been used for cell fusion experiments (3). In particular, the use of reciprocal translocations provides information on regional gene mapping on more than one chromosome. In the experiment described here, human cells carrying a reciprocal exchange between near equal-sized pieces of the short arms of a chromosome 1 and a chromosome 6 (TH-5 cell strain) were hybridized with a Chinese hamster cell line (called 380-6). The resulting cell hybrid lines were informative for regional mapping of genes previously assigned to chromosomes 1 and 6.

METHODS

Cell fusion was carried out in suspension in the presence of inactivated Sendai virus (4). Isolation and cloning of the TH-5/380-6 hybrid lines has been described (5). Chromosome studies were carried out at the same passage as cell harvests for HLA and isozyme studies. At least 25 metaphases from each hybrid clone were analyzed after trypsin-Giemsa or Quinacrine mustard banding.

Hybrid clones were tested for expression of the human forms of the following gene products:

On chromosome 1 (short arm):	Method
ENO-1 (E.C.4.2.1.11) Enolase-1 (6,7)	C (8)
PGD (E.C.1.1.1.44) 6-Phosphogluconate dehydrogenase (9-11)	C (12)
UMPK (E.C.2.7.4*) Uridine monophosphate kinase (13,14)	C (15)
PGM-1 (E.C.2.7.5.1) Phosphoglucomutase-1 (9-11)	S (16)
On chromosome 6:	
HLA-A Human leukocyte antigens A-locus (17,18)	M (5,19)
PGM-3 (E.C.2.7.5.1) Phosphoglucomutase-3 (20,21)	S (16)
ME-1 (E.C.1.1.1.40) Malic enzyme-1 (22)	S,C (23,24)
GLO (E.C.4.4.1.5) Glyoxalase I (25)	C (26)

C = Cellogel electrophoresis; S = Starch gel electrophoresis; M = Microabsorption test

RESULTS

Examples and ideograms of the t(1;6) translocation chromosomes and their normal homologs are shown in Fig. 1.

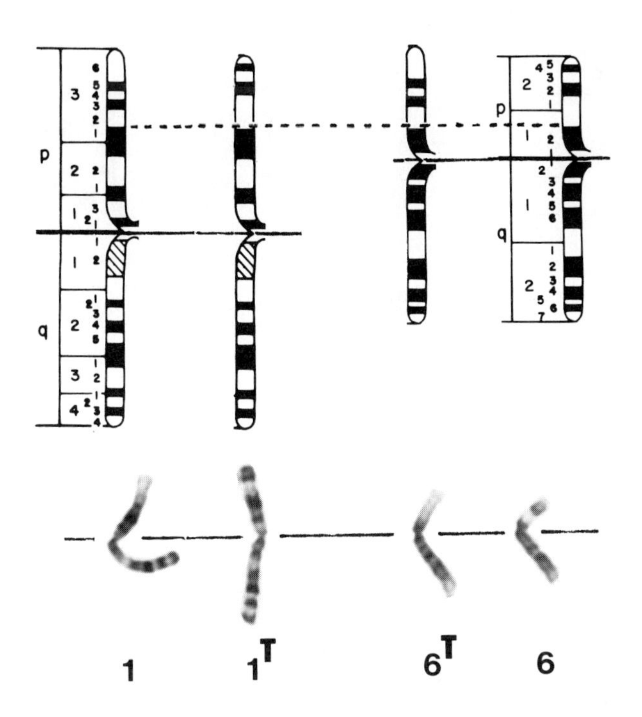

Fig. 1. The dotted line indicates the points of chromosome breakage and reunion that were interpreted to lie at the border of bands 1p31 and 1p32 in 1p3100, and at the border of bands 6p12 and 6p20 in 6p2100 (27,28). Solid lines mark position of the centromeres.

Chinese hamster/human hybrid clones retaining either 1^T (consisting of chromosome regions 6pter→6p2100::1p3200→1qter) or 6^T (consisting of 1pter →1p3200::6p2100→6qter) after having lost the normal chromosomes 1 and 6, should be useful for regional gene assignments on both chromosomes 1 and 6.

A representative hybrid cell karyotype is shown in Figure 2.

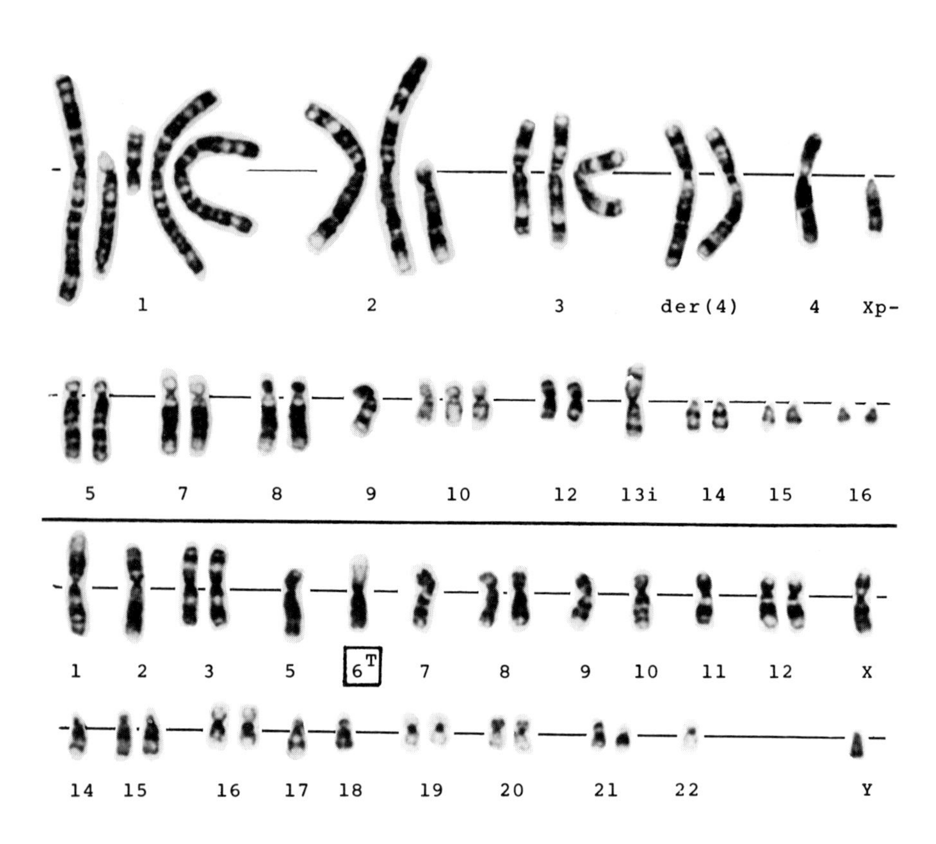

Fig. 2. The Chinese hamster (clone 380-6) chromosomes are arranged in the top 2 rows. The breakage involving CH chromosomes 1 and 2 is peculiar to this cell, otherwise, chromosomes were remarkably stable in these hybrid lines. The human chromosomes in the bottom 2 rows are structurally intact. Regarding the relevant chromosomes, 1 and 6^T have been retained and 1^T and 6 have been lost. Therefore, this hybrid clone is informative for regional mapping of chromosome 6, but not of chromosome 1.

The human parental fibroblast strain TH-5 containing the t(1;6) translocation possessed the HLA-A specificities A2 and A3, as determined by direct microcytotoxicity assay

(29) and confirmed by absorption (19). The TH-5/380-6 hybrids were tested for expression of HLA-A2 and -A3 antigens by absorption of the respective alloantisera, as well as for HLA-A1 not present on the TH-5 fibroblasts. Quantitative aspects of the results, indicating a possible gene dose effect for HLA antigen density on the cell surface have been described elsewhere (5).

Expression of isozymes in hybrids and controls. On starch gels stained for PGM activity, the slow-migrating PGM-1 bands developed first and were scored long before the PGM-3 bands started to appear (Fig. 3.).

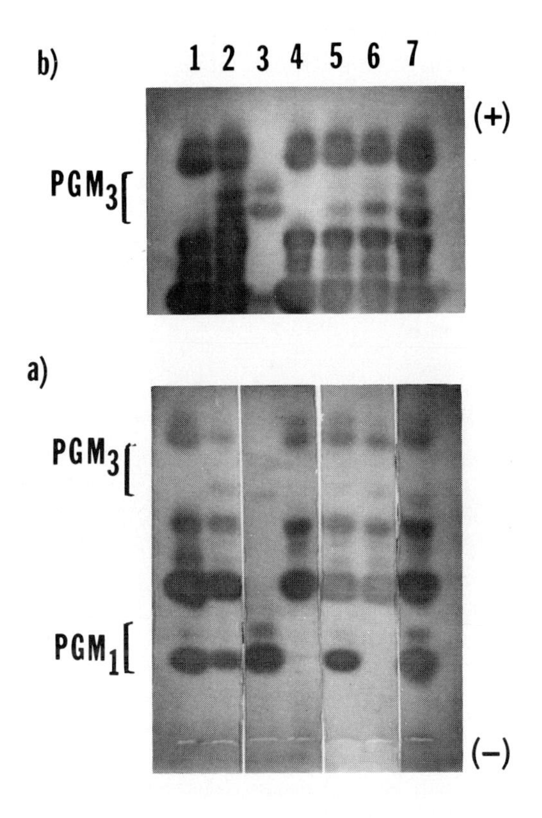

Fig. 3. Composite of photographs taken at 50 min (a) and at 135 min (b) after application of the PGM staining solution/agar mixture. Control lysates are in channels [3] (TH-5 human fibroblasts),[4](380-6 Chinese hamster cells)and [7](TH-5 and 380-6 mix). The hybrid cell line in

channel[1] is PGM-1 positive and PGM-3 negative. The hybrid in channel[6] is PGM-1 negative and PGM-3 positive. The hybrids in [2] and [5] are positive for PGM-1 and for PGM-3.

Hybrid clones expressing human PGD produced a gel pattern of 3 bands. There were a single heterodimeric band and a weak human band in addition to the single Chinese hamster band.

The pattern for ENO-1 was similar. "Positive" hybrid clones produced 3 bands of enzymatic activity. A single (fastest migrating) Chinese hamster band, one intermediate band and a single human band. "Negative" hybrid clones had only the Chinese hamster band.

The ME-1 activity in human TH-5 fibroblasts consisted of a single band migrating faster than the 380-6 single band. Hybrids containing the human gene for ME-1 produced a five-band pattern including 3 intermediate (heteropolymeric) bands.

In contrast, no intermediate bands were observed in hybrids positive for human UMPK which is consistent with a monomeric structure of this enzyme. The TH-5 and 380-6 single bands were separated by cellogel electrophoresis (Fig. 4).

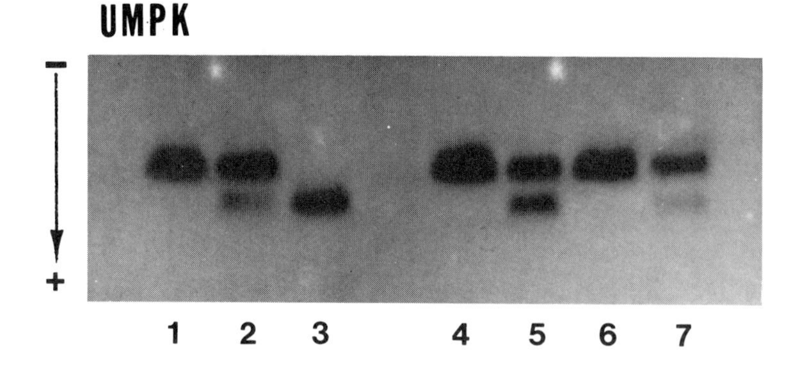

Fig. 4. Human and Chinese hamster UMPK activities separated by cellogel electrophoresis. Cell extracts applied: [1] Chinese hamster 380-6. [2] Mix of 380-6 and TH-5. [3] Human TH-5 fibroblasts. [4 - 7] TH-5/380-6 hybrid clones: [4] and [6] are negative and [5] and [7] are positive for human UMPK.

The glyoxalase studies were not informative with respect to mapping of the GLO gene to one of the translocation chromosomes. The TH-5 fibroblasts were heterozygous at the GLO locus. The slower migrating band segregated together with the normal chromosome 6, while the faster moving band could not be distinguished from the Chinese hamster enzyme in clones containing either the 1^T or the 6^T chromosome alone. The expression of both alleles, including an intermediate presumably heterodimeric band, in the t(1;6) human fibroblasts ruled out the possibility that the break in band 6p2100, leading to the translocation, has inactivated the GLO allele.

Correlations between the presence of the relevant chromosomes and expression of human HLA-A antigens and human isozymes in 15 hybrid clones, are summarized in Table 1.

DISCUSSION

The results as summarized in Table 1 indicate that both syntenic groups of genes - on the short arm of chromosome 1 and on chromosome 6 - have been disrupted by the breaks leading to the t(1;6)(p3200;p2100) translocation. On chromosome 1, PGM-1 maps proximal and UMPK, ENO-1 and PGD map distal to the break in 1p3200. On chromosome 6, PGM-3 and ME-1 map proximal and HLA-A maps distal to the break in 6p2100. When these results are combined with published linkage data,based on different approaches such as family and teratoma studies, more comprehensive and precise regional gene mapping can be accomplished.

In an attempt to correlate the genetic linkage map with the cytological map based on banded metaphase chromosomes, it is first of all necessary to construct more accurate ideograms of banded chromosomes than those provided by the Paris Conference (1971) on Standardization in Human Cytogenetics (Fig. 1). We have constructed ideograms of chromosomes 6 and 1 by measuring the relative widths of all major and minor bands that are distinguishable on GTG-banded early metaphase chromosomes and taking the average values of measurements on 10 chromosomes. The relative staining intensities of the bands were indicated not only by white and black, but also by 3 shades of grey (Figures 5 and 6).

Linkage map distances in centimorgans (cM) are based on cross-over frequencies between homologous chromatids paired in meiotic prophase. The apparent similarities between chromomere and G-banding patterns on pachytene chromosomes and G-banding patterns on mitotic prophase chromosomes (30) suggests that chromosome condensation follows a similar

TABLE 1

SEGREGATION OF t(1;6) TRANSLOCATION CHROMOSOMES AND HOMOLOGS, HLA-A ANTIGENS AND HUMAN ENZYME PHENOTYPES IN 15 TH-5/380-6 HYBRID CLONES

NO OF HYBRID CLONES[1]	HUMAN TRANSLOCATION CHROMOSOMES & HOMOLOGS[2]				HLA-A ALLELES[3]		HUMAN ENZYMES ON CHROMOSOME 1				HUMAN ENZYMES ON CHROM. 6	
	1	1^T	6	6^T	2	3	PGM_1	PGD	ENO-1[4]	UMPK[4]	PGM_3	ME-1
6	-	+	-	-	+	-	+	-	-	-	-	-
1	+	-	+	+	-	+	+	+	n.d.	n.d.	+	+
3	-	-	+	+	-	+	-	+	+	+	+	+
5	+	-	-	+	-	-	+	+	+	+	+	+

[1] Primary and secondary hybrid clones derived from at least 4 independent fusion events.

[2] "+" indicates the presence of respective human chromosome in at least 50% of metaphases.
"-" indicates the absence of chromosome from respective clone.

[3] "+" indicates Absorption Dosage 50 (AD_{50}) values between 2.0 and 14.3 x 10^3 (i.e. presence of antigen)
"-" indicates AD_{50} values greater than 10^5 (i.e. absence of antigen).

[4] Analyzed in only 2 clones from each class. "n.d." = not done.

course in meiosis and in mitosis. For the purpose of this comparison, we have made the simplifying assumption that cross-overs are distributed evenly among pachytene chromosomes. Thus, chromosome 6, which represents 6% of the autosomal haploid complement, should contain 6% of the autosomal genetic length, which amounts to 150 cM in the male (27). Chromosome 1, representing 9% of the autosomal haploid complement, should contain 225 cM in the male. Estimates based on the observed numbers of chiasmata on the respective bivalents identified in male diakinesis (31) are not significantly different: 133 cM for chromosome 6, and 100 cM for the short arm of chromosome 1. These chiasma counts by Hultén (31) involved a relatively small number of cells, and the standard deviations were large. Therefore, at this time, we prefer to use the earlier estimates that were based on 50 as the average number of chiasmata in human males (27).

Having decided on the number of centimorgans assigned to chromosomes 1 and 6, one is faced with the question of their distribution along the chromosome. In the absence of convincing evidence to the contrary, we have assumed, for the construction of the chromosomal maps, that cross-overs are distributed evenly along the individual chromosomes resulting in a linear relationship between cytological and genetic maps.

Map of chromosome 6. The genetic distance between HLA and PGM-3 is 15 cM in males as established by family studies (Report of the Committee on the genetic constitution of chromosome 6 (25)). In our TH-5/380-6 hybrid clones, HLA-A2 segregated with the 1^T chromosome and HLA-A3 with the normal chromosome 6, while clones with 6^T alone did not express an HLA-A antigen. In contrast, PGM-3 was expressed in hybrids containing 6^T but not in those containing 1^T. Thus, the break in 6p2100 has separated the gene for PGM-3 from the HLA region. In Figure 5, brackets spanning the cytological equivalent of 15 cM indicate the regional localizations of HLA and PGM-3. The HLA-A gene tested for in these hybrid clones is part of the major histocompatibility complex (MHC) in man which presently comprises 10 genes within about 2 cM (25). Although HLA-A is the most distal locus with respect to PGM-3 and the centromere (32), we think it unlikely that the break has occurred within the MHC, for the following reasons:

From the proportion of tumors heterozygous for PGM-3 among benign ovarian teratomas found in hosts that were heterozygous for PGM-3, Ott and colleagues have calculated a distance of 17 cM (95% confidence limits: 7 to 34 cM) bet-

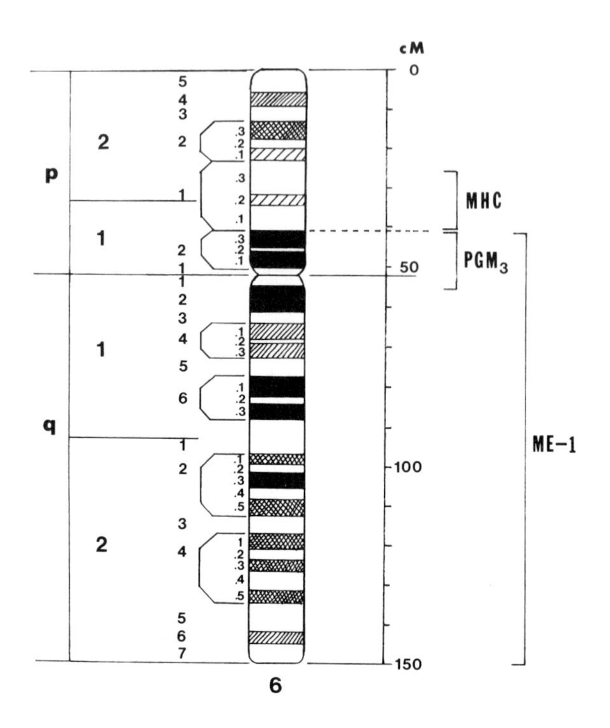

Fig. 5. Map of chromosome 6. p - short arm; q - long arm; cM - centimorgan; MHC - major histocompatibility complex; PGM_3 - phosphoglucomutase 3; ME-1 - malic enzyme-1.

ween the PGM-3 locus and the centromere (33,34). Since for all gene linkages in man the recombination frequencies have been found to be 1.55 to 1.74 times higher in females than in males (35), the centromere - PGM-3 distance in males should be less than 17 cM, probably around 10 to 11 cM. The equivalent of 10 cM from the centromere into the long arm of chromosome 6 is an unlikely location for PGM-3 as it would be more than 15 cM from the breakpoint in 6p2100. Therefore, it appears more likely that PGM-3 and MHC are both on the short arm. In this case the precise position for PGM-3 would be in band 6p12.3 near the breakpoint. The MHC should be 25 cM from the centromere, within the distal subband (p21.3) of the lightly staining region 6p21 (Fig. 5). The regional localization of the MHC in band 6p21 is consistent with its exclusion from region 6pter→6p22 by deletion (or translocation) mapping (25). The finding that the MHC maps in a region staining lightly by G-banding is consistent with the notion that structural genes are more often located in lightly-staining bands, while darkly-staining bands may contain more reiterated sequences (36). The MHC in man is homologous to the H-2 complex in the mouse which maps on chromosome 17 approximately 18 map units from the centromere (37) in a similar conspicuously light G-banded region (band 17B, Nesbitt and Francke (38)).

The gene for ME-1 has been assigned to chromosome 6 by somatic cell hybridization (22) and no family linkage data are available. We have established its regional localization in 6p2100→6pter.

Map of chromosome 1. Figure 6 depicts the early metaphase chromosome map of chromosome 1 (for males) - assuming a linear relationship between cytological and linkage map distances. The centromere position is at 108 cM from the end of the short arm (pter). The t(1;6) translocation breakpoint in 1p3200 should be 62 cM from the p terminal end and 46 cM from the centromere. We have found that PGM-1 maps proximal and that UMPK, ENO-1 and PGD map distal to the breakpoint. However, UMPK is linked to PGM-1 in males, with approximately 10 cM between them, according to the most recent combined linkage data from a number of laboratories (39). This allows for precise mapping of PGM-1 in band 1p31, and UMPK in band 1p32. The latter confirms previous data that UMPK is located in 1p32 distal to PGM-1 (14). However, the localization of PGM-1 proximal to the breakpoint in 1p3200, i.e. in band 1p31, disagrees with the most recent gene map (38) which places PGM-1 in region 1p32→1p33.

Douglas and colleagues (11) observed a spontaneous deletion of material from the short arm of a chromosome 1 in a hybrid clone after 50 cell passages. The deletion was interpreted as terminal with the break in band 1p33. Human PGD and PGM-1 were not expressed and the respective gene loci were therefore assigned to the missing segment 1p33→1pter.

Burgerhout and colleagues, after having induced chromosome breaks by X irradiation of human/Chinese hamster hybrid cells, found two clones with apparent deletions of the region distal to band 1p32. Both were negative for human PGD, but only one of them was negative for PGM-1 (10). However, more recent results from fusion of a translocation with break in band 1p32 indicated the location of PGM-1 (and UMPK) proximal to the breakpoint (14).

The same t(1;17)(p32;p13) translocation had previously been used by Grzeschik to produce human/mouse hybrids. He also found that PGM-1 segregated with the translocation

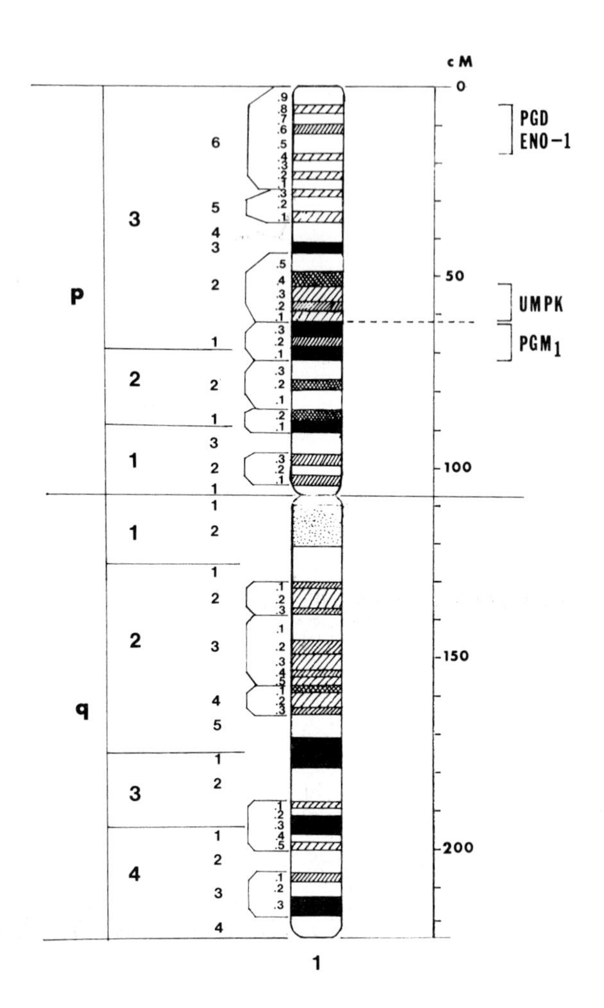

Fig. 6. Map of chromosome 1. p - short arm; q - long arm; cM - centimorgan; PGD - 6-Phosphogluconate dehydrogenase ENO-1 - Enolase-1; UMPK - Uridine monophosphate kinase; PGM_1 - Phosphoglucomutase-1.

product containing region 1p32→1qter and with gene markers on the long arm of chromosome 1 and had been separated from gene loci on the distal part of 1p (40).

In a patient with myelofibrosis, Marsh and colleagues (41) have found a clone of hematopoietic precursor cells which contained a complex rearrangement between chromosomes

1, 4 and 7 including a deletion of region 1p32→1pter. The Rhesus blood group locus had been lost, but the PGM-1 allele on the respective chromosome 1 had been retained.

It appears that the more recent evidence and that based on more systematic studies is consistent with our results indicating a more proximal location of PGM-1. This location is consistent with family studies indicating a map distance of 37 cM between PGM-1 and the heterochromatin region (1qh) on the long arm of chromosome 1 (band 1q21 on Fig. 6) (39). Although 1p31 appears as a dark band on the Paris Conference (1971) ideogram, on early metaphase chromosomes this band is found to resolve into dark and light staining subbands (Fig. 6). The exact location of PGM-1 in this region is presently under investigation using additional hybrid clones with breaks in this area.

The assignment of PGD and ENO-1 to region 1pter → 1p3200 is consistent with published reports (39). The genetic distance between PGM-1 and PGD is 55 cM in males (39). Since the region distal to the translocation breakpoint comprises 62 cM, the closely linked PGD and ENO-1 loci can be accommodated in band 1p36 without having to assume a relatively increased cross-over frequency in the distal part of 1p, as has been suggested earlier by Cook and colleagues (42).

We have demonstrated that regional gene localization data derived from somatic cell hybridization of human cells with a translocation can be combined with linkage data derived from family studies, or from benign ovarian teratomas, in the construction of a comprehensive human chromosome map. Once the feasability is established and the underlying assumptions are either verified or corrected, this approach should be particularly useful in mapping genes that are not expressed in cultured cells.

ACKNOWLEDGMENTS

The expert technical assistance of N. Busby, M. G. Brown and A. Pellegrino, the construction of the ideograms by N. Oliver and P. Olson-Wells, and the typing of the manuscript by S. Benson are gratefully acknowledged. This work was supported by U.S. PHS research grants GM 21110, CA 16071 and AI 10180, and by postdoctoral training grant GM 00054. This is publication number 1307 from the Department of Molecular Immunology, Scripps Clinic and Research Foundation, La Jolla.

REFERENCES

1. Ruddle, F.H. and Creagan, R.P. (1975) Annu. Rev. Genet. 9, 407.
2. Human gene mapping 3. Baltimore Conference 1975,(1976) Birth Defects: Original Article Series Vol. XII, No. 7, The National Foundation, New York.
3. Francke, U., Busby, N., Shaw, D., Hansen, S. and Brown, M.G. (1976) Somatic Cell Genet. 2, 27.
4. Croce, C.M., Koprowski, H. and Eagle, H. (1972) Proc. Nat. Acad. Sci. USA 69, 1953.
5. Francke, U. and Pellegrino, M.A. (1977) Proc. Nat. Acad. Sci. USA 74, 1147.
6. Giblett, E.R., Chen, S.H., Anderson, J.E. and Lewis, M. (1974), New Haven Conference, 1973, Birth Defects: Original Article Series X, No. 3, p. 91, The National Foundation, New York.
7. Meera Khan, P., Doppert, B.A., Hagemeijer, A. and Westerveld, A.(1974), New Haven Conference, 1973, Birth Defects: Original Article Series X, No. 3, p. 130, The National Foundation, New York.
8. Grzeschik, K.H. (1976) Baltimore Conference, 1975, Birth Defects: Original Article Series XII, No. 7, The National Foundation, New York, p. 142.
9. Ruddle, F., Ricciuti, F., McMorris, F.A., Tischfield, T., Creagan, R., Darlington G. and Chen, T. (1972) Science 176, 1429.
10. Burgerhout, W., v.Someren, H. and Bootsma, D. (1973) Humangenetik 20, 159.
11. Douglas, G.R., McAlpine, P.J. and Hamerton, J.L.
12. Meera Khan, P. and Rattazzi, M.c. (1968) Biochem. Genet. 2, 231.
13. Giblett, E.R., Anderson, J.E., Lewis, M. and Kaita, H. (1975) Rotterdam Conference, 1974, Birth Defects: Original Article Series Vol. XI, No. 3, p. 159, The National Foundation, New York.
14. Burgerhout, W.G. and Jongsma, A.P.M. (1976) Baltimore Conference, 1975, Birth Defects: Original Article Series Vol. XII, No. 7, p. 101, The National Foundation, New York.
15. Teng, Y.-S., Chen, S.-H. and Giblett, E.R. (1976) Am. J. Hum. Genet. 28, 138.
16. Spencer, N., Hopkinson, D.A. and Harris, H. (1964) Nature 204, 742.
17. Lamm, L.U., Friedrich, U., Peterson, G.B., Jorgensen, J., Nielsen, J., Therkelsen, A.J. and Kissmeyer-Nielsen, F. (1974) Hum. Hered. 24, 273.

18. v. Someren, H., Westerveld, A., Hagemeijer, A., Mees, J.R., Meera Khan, P. and Zaalberg, O.B. (1974) Proc. Nat. Acad. Sci. USA 71, 962.
19. Pellegrino, M.A., Ferrone, S. and Pellegrino, A. (1972) Proc. Soc. Exp. Biol. and Med. 139, 484.
20. Lamm, L.U., Svejgaard, A. and Kissmeyer-Nielsen, F. (1971) Nature new Biol. 231, 109.
21. Jongsma, A., v. Someren, H., Westerveld, A., Hagemeijer, A. and Pearson, P. (1973) Humangenetik 20, 195.
22. Chen, T.R., McMorris, F.A., Creagan, R., Ricciuti, F., Tischfield, J. and Ruddle, F.H. (1973) Am. J. Hum. Genet. 25, 200.
23. Cohen, P.T.W. and Omenn, G.S. (1972) Biochem. Genet. 7, 303.
24. v.Someren, H., v.Henegouwen, H.B., Los, W., Wurzer-Figurelli, E., Doppert, B.A,Vervloet, M. and Meera Khan, P. (1974) Humangenetik 25, 189.
25. Bodmer, W.F. (1976) Baltimore Conference, 1975, Birth Defects: Original Article Series Vol. XII, No. 7, The National Foundation, New York.
26. Meera Khan, P. and Doppert, B.A. (1976) Hum. Genet. 34, 53.
27. Paris Conference 1971 (1972),,Birth Defects: Original Article Series VIII, No. 7, The National Foundation, New York.
28. Paris Conference 1971 Supplement (1975), Birth Defects: Original Article Series XI, No. 9, The National Foundation, New York.
29. Ferrone, S., Pellegrino, M.A. and Reisfeld, R.A. (1971) J. Immunol. 107, 613.
30. Luciani, J.M., Morazzani, M.-R. and Stahl, A. (1975) Chromosoma 52, 275.
31. Hultén, M. (1974) Hereditas 76, 55.
32. Lamm, L.U., Kissmeyer-Nielsen, F., Svejgaard, A., Bruun-Petersen, G., Thorsby, E., Mayr, W. and Högman, C. (1972) Tissue Antigens 2, 205.
33. Ott, J., Hecht, F., Linder, D., Lovrien, E.W., Kaiser McCaw, B. (1976) Baltimore Conference, 1975, Birth Defects: Original Article Series Vol. XII, No. 7, The National Foundation, New York.
34. Ott, J., Linder, D., Kaiser McCaw, B., Lovrien, E.W., and Hecht, F. (1976) Ann. Hum. Genet. Lond. 40, 191.
35. Weitkamp, L.R., Guttormsen, S.A. and Greendyke, R.M. (1971) Am. J. Hum. Genet. 23, 462.
36. Hoehn, H. (1975) Am. J. Hum. Genet. 27, 676.

37. Womack, J.E. (1976) Mouse Newsletter 54, 6.
38. Nesbitt, M.N. and Francke, U. (1973) Chromosoma 41, 145.
39. Hamerton, J.L., Bodmer, W.F., Cook, P.J.L., McAlpine, P.J. and Rivas, M. (1976) Baltimore Conference, 1975 Birth Defects: Original Article Series Vol. XII, No. 7, The National Foundation, New York.
40. Grzeschik, K.-H. (1975) Cytogenet. Cell Genet. 14, 342.
41 Marsh, W.L., Chaganti, R.S.K., Gardner, F.H., Mayer, K., Nowell, P.C. and German, J. (1974) Science 183, 966.
42. Cook, P.J.L., Robson, E.B., Buckton, K.E., Jacobs, P.A. and Polani, P.E. (1974) Ann. Hum. Genet. 37, 261.

IN SITU HYBRIDIZATION FOR THE STUDY OF CHROMOSOME STRUCTURE AND FUNCTION

Mary Lou Pardue, J. Jose Bonner*, Judith Lengyel+, and Allan Spradling#

Department of Biology, Massachusetts Institute of Technology, Cambridge, Massachusetts 02139

INTRODUCTION

In situ hybridization using a known nucleic acid as the ^{3}H-labeled probe can be used to map sequences complementary to that nucleic acid on chromosomes in a cytological preparation. Conversely, *in situ* hybridization to known chromosomal regions can be used to help identify ^{3}H-labeled nucleic acids and to resolve components of mixed populations, such as the RNA transcripts of a particular cell type. For organisms with large, well-differentiated chromosomes such an analysis of RNA has a high level of resolution as well as the potential to relate directly to other genetic data.

When *in situ* hybridization is used to analyze RNA populations several factors determine whether any particular RNA species will be detected. Two of these factors, the number of molecules of that RNA present in the mixture and the amount of ^{3}H-uridine which is incorporated into that RNA during the labeling period, would also influence detection by other techniques, such as gel electrophoresis. In addition, detection of a particular RNA species by *in situ* hybridization is influenced by the number of complementary DNA sequences in the chromosomes. It

*Present address: Dept. of Biochemistry and Biophysics
University of California Medical School
San Francisco, CA 94143

+Present address: Dept. of Biology
University of California
Los Angeles, CA 95616

#Present address: Dept. of Zoology
Indiana University
Bloomington, IN 47401

is possible that the amount of hybridization is also influenced by the accessibility of the complementary DNA sequences and that such accessibility may change with the degree of condensation of the chromosomes. This possibility makes it important that comparisons between RNA populations be done by hybridization to identical chromosome preparations. For loci which are to be studied quantitatively it is necessary to avoid both saturation of hybridization at that locus and overexposure of the autoradiographic emulsion above it. If either of these situations occurs the number of autoradiographic silver grains above that locus will not reflect the relative levels of the complementary sequence in the RNA preparations being studied.

If the limitations of the technique are carefully taken into account, in situ hybridization is particularly useful for the analysis of mixed populations of RNA. The technique is most powerful when polytene chromosomes can be used as the assay system. These giant multistranded chromosomes appear to be banded, probably because the chromotids making up each chromosome are so precisely aligned that localized foldings in the individual nucleoprotein strands join to form bands. The pattern of this banding is the same in each polytene tissue (7, 9) and is probably the reflection of a fundamental structure of the chromatid which may exist even in non-polytene nuclei. The exact relation between this fundamental structure and the units of genetic function is still unclear (10); however the banding patterns produced on the polytene chromosomes allow the genome to be mapped cytologically with a resolution that has not been possible in higher organisms lacking such chromosomes. In situ hybridization to polytene chromosomes allows a ^{3}H-labeled nucleic acid to be mapped to a chromosomal region which can be defined with a high degree of precision. In addition polyteny produces high local concentrations of sequences. Thus even a gene sequence present only once per haploid complement should be relatively easy to detect by in situ hybridization. Although we have not measured the reiteration of the sequences detected in the work discussed here, we have done experiments with model systems (Fig. 1) from which we calculate that single copy sequences could be detected under the conditions used.

Polytene chromosomes are especially favorable for cytological studies not only because of their large size but also because they are essentially interphase nuclei. These chromosomes do not decondense during stages of DNA transcription and replication. Thus it is possible to

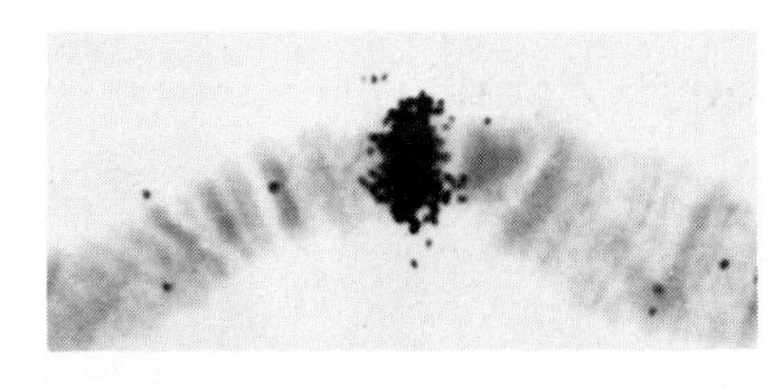

Fig. 1. Autoradiograph of a polytene chromosome showing in situ hybridization to DNA of low repetition frequency. The ^{3}H-cRNA used was transcribed in vitro from a segment of D. melanogaster DNA isolated and amplified by insertion into the Escherichia coli plasmid pSC101 (33). The hybrid plasmid so produced, pDm2, maps to polytene region 84D and reassociation kinetic studies show a repetition frequency of 3.5 for this segment of D. melanogaster DNA (33). The in situ hybridization was carried out in a buffer of 0.3 M NaCl, 0.01 M Tris; pH 6.8, at 65° C for 10 hrs. The same conditions have been used for the other in situ hybridization experiments described in this paper. Slide stained with Giemsa. Autoradiographic exposure 38 d. X 1400.

visualize genetic activity and to relate it to specific chromosomal sites.

One type of evidence for genetic activity which may be visualized on polytene chromosomes is puffing, a localized alteration in the structure of the polytene chromosome. Puffs behave in ways expected of specific gene transcription. Individual puffs appear and disappear at predictable times in the life cycle of the insect (3). Some puffs are tissue-specific: others are common to all polytene tissues (11). In several cases the correlation between the presence of a specific polytene puff and the occurrence of a particular cytoplasmic protein suggests a causal relationship (8, 5, 16, 17, 32).

Another type of evidence for genetic activity comes from autoradiography of polytene nuclei after ^{3}H-uridine incorporation. We refer to such preparations as transcription autoradiographs although it is possible that they may reflect storage as well as transcription. Transcription autoradiographs of polytene nuclei show that puffs are sites of active uridine incorporation; however many non-puffed sites are also labeled with uridine (25, 26). At least one of the regions which do not puff but do label with ^{3}H-uridine is known to produce cytoplasmic RNA in these cells; region 39DE contains the genes coding for histones (24).

Studies on patterns of puffing in the polytene chromosomes of Drosophila melanogaster have identified two sets of puffs whose appearance can be manipulated experimentally. One set of puffs, the "heat shock" puffs, are induced when larvae or larval salivary glands are moved from their normal growth temperature of 25° C to 37° C. The other set of puffs, the ecdysone puffs, are controlled by increases in the level of the steroid hormone ecdysone. A specific sequence of alterations in the puffing pattern is induced normally by the increase in ecdysone that preceeds molting (6, 1). The same alterations can be induced prematurely by injection of ecdysone into younger larvae (15, 11, 3).

Puffing is associated with local evidence of the incorporation of ^{3}H-uridine into RNA and appears therefore to be a morphological expression of gene transcription. Thus it seems likely that both heat shock and ecdysone are affecting transcription at specific chromosome loci. Although only a few *D. melanogaster* tissues have polytene chromosomes which are sufficiently large for puffing analyses, it is possible to study transcriptional changes in other polytene as well as diploid tissues by *in situ* hybridization of ^{3}H-labeled RNA from these tissues to cytological preparations of salivary gland chromosomes.

RESPONSE TO HEAT SHOCK: A SMALL SET OF COORDINATELY CONTROLLED GENES

Under the conditions of a heat shock (a shift from 25° C to 37° C) some nine new puffs are induced on the chromosomes within a few minutes (27, 28, 2). This is apparently a response to the metabolic disturbance due to the temperature shift. The same set of puffs can be induced by uncouplers of oxidative phosphorylation (27, 2, 19) or by recovery from anaerobiosis (28, 14). Consideration of the treatments which can cause induction of these heat shock puffs suggests that they might contain genes directing some aspect of energy metabolism but there is as yet no direct proof of this. Although we do not understand the biological significance of the heat shock response it does provide a small set of coordinately controlled genes which can be manipulated experimentally. Studies on such a system should help illuminate some general principles governing interactions between genes.

The response of the salivary gland cells to heat shock is apparently a general one shared by all of the *D. melanogaster* tissues which have been examined (Fig. 2). Imaginal discs, adult tissues, and cultured cell lines resemble salivary glands in the heat-induced changes in the synthesis of both RNA (30, 22, 12) and protein (32, 22, 21). RNAs which hybridize *in situ* with the six largest heat shock polytene puffs have been found associated with polysomes after heat shock, demonstrating directly the

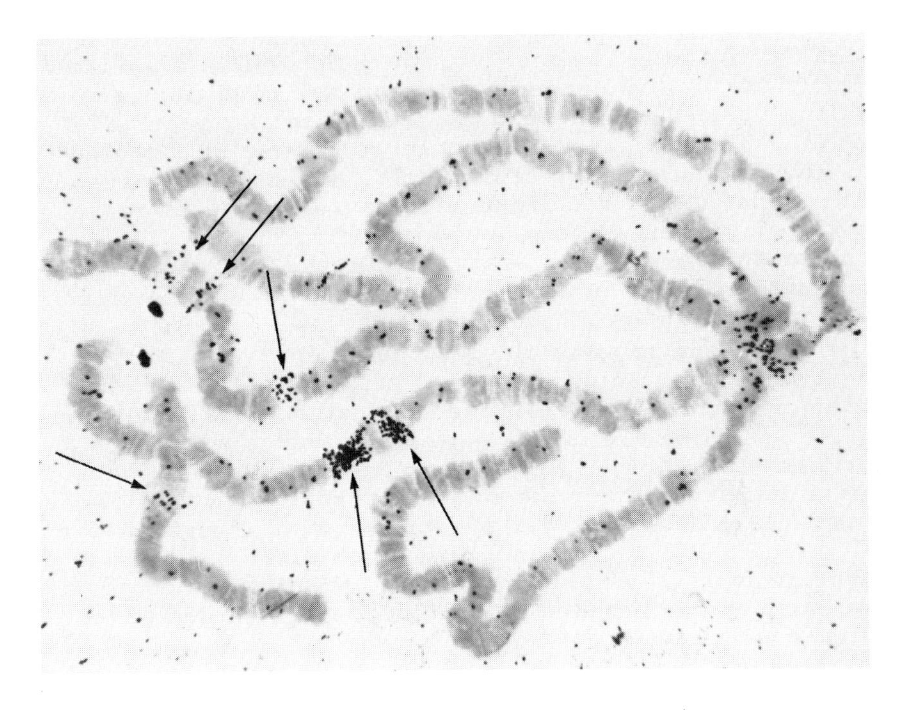

Fig. 2. In situ hybridization of poly $(A)^+$ cytoplasmic RNA from heat shocked D. melanogaster cultured cells. Cells were shifted from 25° C to 37° C. Five min. later ^{3}H-uridine was added and the incubation continued for an hour. The RNA hybridizes in situ to six of the nine puffs induced in salivary gland polytene chromosomes by the temperature shift. These puffs, 63BC, 64F, 67B, 87A, 87C and 95D, are indicated by arrows. Hybridization to the puff at 93D is seen only with poly $(A)^-$ cytoplasmic RNA. The two smallest puffs, 33B and 70A, bind only a small amount of ^{3}H-RNA in these experiments. 1.5×10^4 cpm/slide. Slide stained with Giemsa. Exposure 172 d. X 650.

linkage between chromosomal regions in which puffing can be induced and protein synthesis (22, 31). It is interesting that the level of hybridization of cytoplasmic RNA from diploid tissues to the various heat shock loci is approximately proportional to the size of the heat shock-induced puff at each of these loci on polytene chromosomes.

Despite the evidence that the nine heat shock loci are coordinately controlled, the expression of some members of the set can be differentially modulated. This is clearly illustrated by the locus in region 93D. When the heat shock is given under certain culture conditions region 93D is induced well beyond the level normally seen at 37° C (12). The increase in induction at 93D can be detected in RNA extracted from either diploid or polytene tissues. The same culture conditions also cause a large increase in the size of the puff produced in region 93D on polytene chromosomes, showing again the parallel between polytene puff size and amount of cytoplasmic RNA in both polytene and diploid tissues.

As in the polytene tissues, the primary effect of the heat shock in diploid tissues is likely to be at the level of transcription. When cells from an established _D. melanogaster_ cell line are incubated for 20 minutes with ^{3}H-uridine at 25° C the ^{3}H-nuclear RNA hybridizes to many sites distributed over the polytene chromosomes. However ^{3}H-nuclear RNA from cells incubated with ^{3}H-uridine for 20 minutes at 37° C gives a different pattern of _in situ_ hybridization (Fig. 3) even when the labeling period begins as early as two minutes after the cells are shifted to the higher temperature. The ^{3}H-nuclear RNA made at 37° C

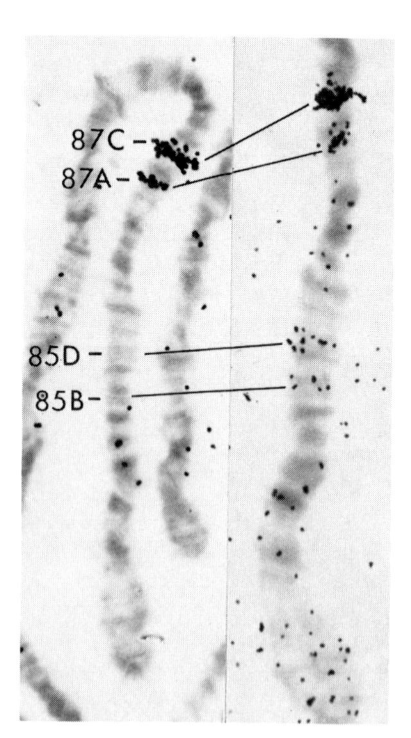

Fig. 3. _In situ_ hybridization of poly (A)$^+$ nuclear and cytoplasmic heat shock RNA. Cultured _D. melanogaster_ cells were shifted to 37° C and labeled with ^{3}H-uridine for 20 min. Slides stained with Giemsa. Exposure 46 d. X 1050. a) Hybridization by poly (A)$^+$ cytoplasmic RNA. Regions 87A and 87C are heavily labeled. b) Hybridization by poly (A)$^+$ nuclear RNA larger than 20S on a sucrose gradient. Regions 87A and 87C are the most heavily labeled. Significant hybridization is also seen over 85B and 85D. 85B and 85D are labeled in experiments with cytoplasmic poly (A)$^+$ RNA made during longer heat shocks.

hybridizes heavily to the same polytene bands which bind the cytoplasmic RNA produced under conditions of heat shock (20). In addition, poly $(A)^+$ ^{3}H-nuclear RNA prepared from heat shocked cells shows significant hybridization to a number of other bands which have not yet been seen to hybridize with poly $(A)^+$ cytoplasmic RNA preparations from the same cells (Lengyel and Pardue, in preparation). It is possible that some of the RNAs hybridizing to this second set of bands may never leave the nucleus. However competition experiments using unlabeled cytoplasmic RNA will be necessary to study this point since the apparent absence of these ^{3}H-RNAs in cytoplasmic fractions might also be due to dilution of the radioactive RNA with pre-existing unlabeled RNA.

The apparent change in RNA transcription in both diploid and polytene tissues, although dramatic, is not complete. Several types of RNA which were transcribed at 25° C continue to be produced at 37° C. These include the precursor to 19S and 26S ribosomal RNA (20), 5S RNA, and histone mRNAs (30). Although heat shock does not stop the transcription of the large ribosomal RNA precursor, it apparently does stop further processing and transport to the cytoplasm (20). The 5S RNA made at 37° is larger than normal _D. melanogaster_ 5S by some 15 nucleotides at the 3' end (29). It is possible that this, too, is the result of an effect on processing although 5S RNA is not known to have such a precursor.

With a coordinated set of genes such as the heat shock loci it is of particular interest to determine whether the chromosomal arrangement of the individual coding sequences gives insight into the integration of the response. The average polytene band in _D. melanogaster_ contains about 25, 000 nucleotide pairs per chromatid. Since this is much more DNA than would be necessary to encode the sequence of the average polypeptide it has often been suggested that a chromomere may be an operon coding for several related proteins. We have studied this possibility for the genes of the heat shock response by _in situ_ hybridization with individual RNA species (31).

The majority of the polyadenylated RNA produced at 37° C can be resolved into six bands (A1 through A6) by polyacrylamide gel electrophoresis. (See Table 1). These RNA sequences are almost entirely polysome-associated and are in the size range which might be expected of RNAs coding for the new proteins produced at 37° C.

The results of _in situ_ hybridization by individual RNA bands eluted from polyacrylamide gels are given in Table 1. The data have been arranged to show the order in which

	63C	67B	87A	87C	95D	56F	39E	Ch.
a)								
—	0	0	.76	4.8	0	0	0	0
A1	7.3	0	3.1	7.4	0	0	0	0.98
A2	3.3	0	31	57	2.5	0	0	15
A3	0.36	0	7.3	12	4.7	0	0	4.5
—	0	0	8.3	15	2.5	0	0	4.1
A5, A6	0	10	5.8	15	0.51	0	0	7.4
—	0	3.6	2.5	5.1	0	0	0	4.7
b)								
—	1.1	0	5.7	10	1.3	0	0	7.2
A4	2.2	3.3	4.4	19	1.8	0	0	18
A5, A6	0	7.0	3.1	7.7	0	0	0	2.6

Table 1. In situ hybridization of purified RNA fractions to polytene chromosomes. Each of the two blocks of entries in the table represents the results of in situ hybridization of the fractions, eluted from a single lane of a polyacrylamide gel, of RNA labeled in heat shocked cells. The gel fractions are listed in the order in which they ran on the gel (shown in the photograph on the left). Fractions which did not show the presence of any prominent bands of RNA are denoted by a line. The results are expressed as the mean number of grains (based on an analysis of at least 20 fields) at each site, normalized to a 60 day exposure period. Ch. =chromocenter. (From Spradling, et al., 1977. Reprinted with permission.) a) 20S and 13S RNA labeled at 37° C was purified by two cycles of sucrose gradient centrifugation followed by electrophoresis on 7M urea gels. The indicated fractions were eluted and hybridized. b) RNA labeled after a 30 min. heat shock at 37° C, conditions which result in a significant level of A4 labeling, was eluted after electrophoresis as above. Little A4 RNA was detected after the longer heat shock used in experiment a.

the RNA fractions ran on the polyacrylamide gel as well as the polytene bands to which each hybridized. Although each fraction hybridizes, to some extent, with more than one polytene band, much of the hybridization appears to represent the spread of RNA molecules in the polyacrylamide gel. For each polytene band hybridization is maximum with a single fraction of RNA and falls off rapidly with RNA fractions taken increasingly further away on the gel. (The one exception to this is region 87C which shows relative peaks of hybridization with two RNA fractions which are well separated on the gel. This intriguing result, obtained with RNA which appears during the first half hour at 37° C but is not detectable after longer heat shocks, is discussed below.)

If we consider only those gel fractions in Table 1 which show peaks in the level of hybridization with a polytene band, three of the heat shock loci show a one-to-one relationship with specific bands. Polytene region 63BC hybridizes with RNA A1, region 95D hybridizes with A3, and region 67B hybridizes with the combined A5-A6 fraction. In an experiment in which A5 and A6 RNAs were separated more completely only the A5 fraction hybridized with 67B.

Regions 87A and 87C appear to share a sequence. Both show maximum hybridization with RNA fraction A2. It is possible that gel fraction A2 is a mixture of two or more RNAs. However all fractions of the polyacrylamide gel show some hybridization to regions 87A and 87C. Although the level of the hybridization decreases as the RNA fraction is taken from parts of the gel more distant from A2, the ratio of the hybridization of 87A to that at 87C remains 1:2. This failure to obtain any indication that the RNAs which hybridize to 87A and 87C might be separable strengthens the possibility that regions 87A and 87C are binding the same RNA.

There is some evidence that region 87C may code for a second RNA in addition to the one apparently shared with 87A. One RNA band, A4, is detected on polyacrylamide gels only after heat shocks at 35° C or after short exposures to 37° C. When the A4 RNA fraction was used for *in situ* hybridization no new polytene bands were labeled. Instead the ratio of hybridization at region 87C to that at 87A increased from 2:1 to 4:1, suggesting that the A4 RNA fraction binds sequences at 87C which are not shared by 87A. Because of the unexpected spread of other sequences complementary to 87A and 87C throughout the gel our conclusions about both these regions remain tentative.

Chromosomal puffs originate from single or double bands although a number of surrounding bands become involved later as the puff enlarges (7). We have used in situ hybridization to confirm the early cytological observations on puff structure and to unambiguously identify the regions complementary to cytoplasmic RNA in the major heat shock puffs (Figs. 4 and 5).

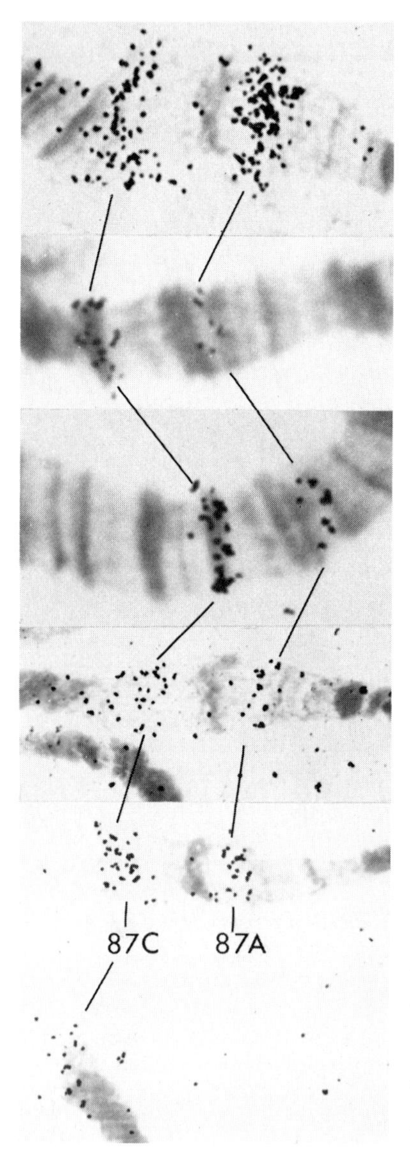

Fig. 4. Regions 87A and 87C. Analysis of ^{3}H-RNA, made during 37° heat shock, by transcription autoradiography (a) and by in situ hybridization (b-e). a) Salivary glands were labeled in vitro with 40 μCi/ml ^{3}H-uridine at 37° for 10 min. following a 10 min. preincubation at 37°. They were then fixed and prepared for autoradiography. In both regions 87A and 87C the ^{3}H-RNA is concentrated over one end of the puff. The region near the heavy bands of 87B between the two puffs shows no label. Giemsa stain. Exposure 5 d. X 1500. b) In situ hybridization of poly (A)$^+$ RNA prepared from heat shocked imaginal discs. Hybridization in region 87A is scattered over the region of fine bands between 87A4 and 87B1. Hybridization in region 87C is directly over the heavy band in 87C1. 6 x 10^3 cpm/slide. Exposure 14 d. X 1950. c) In situ hybridization with polyacrylamide gel band A4 prepared from heat shocked cultured cells. The A4 RNA hybridized to 87A and 87C in a ratio of 1:4 instead of the 1:2 ratio seen with other isolated RNA fractions in the same experiment. The distribution of the in situ hybrid is undetectably different from that seen with unfractioned poly (A)$^+$ RNA in (b). 4.2 x

10^3 cpm/slide. Exposure 52 d. X 1950. d, e) In situ hybridization of poly $(A)^+$ cytoplasmic RNA from heat shocked cultured cells to heat shock-induced puffs on polytene chromosome 3. The distribution of the hybrid closely parallels the distribution of label in transcription autoradiograms of heat shocked polytene nuclei (a). In (e) the chromosome has broken in the 87C region and shows hybridization over both ends of the broken chromosome. 1.5×10^4 cpm/slide. Exposure 37 d. X 850.

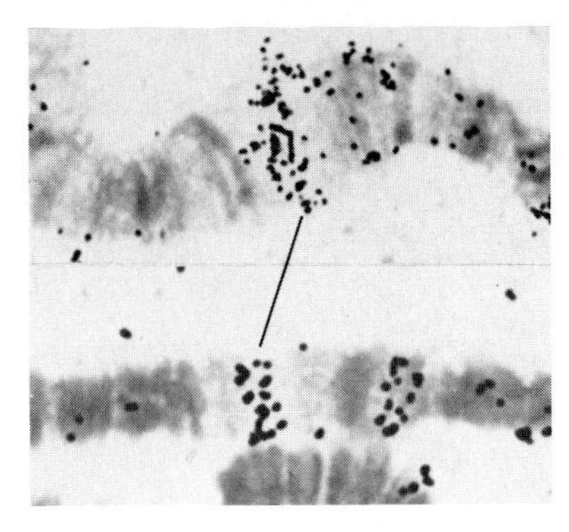

Fig. 5. Region 63BC. Analysis of ^{3}H-RNA, made during 37° heat shock, by transcription autoradiography (a) and by in situ hybridization (b). a) Salivary glands were incubated as described for Fig. 4a, fixed and prepared for autoradiography. There is a major site of labeling in the center of puff 63BC (indicated by the bar). Other label is detected over 63D and 63E. Giemsa stain. Exposure 5 d. X 2000. b) In situ hybridization of ^{3}H-RNA from heat shocked cultured cells. The principal region of hybridization in 63BC corresponds to the band of label on the puff in (a). Giemsa stain. 1.5×10^4 cpm/slide. Exposure 37 d. X 1750.

The sequences giving rise to these puffs were determined by hybridizing ^{3}H-labeled cytoplasmic RNA from heat shocked cultured cells to non-puffed chromosomes. One unexpected finding was that the relation between the polytene band which hybridized with cytoplasmic RNA and the bands which were involved in the puffing varied from puff to puff. In some cases, such as the puff at 63BC, the sequences which hybridized the ^{3}H-RNA were toward the center of the group of bands which eventually made up the puff. In other cases, such as the puff at 87C, the sequences which hybridized ^{3}H-RNA were very close to one end of the set of bands which eventually became involved in the puff.

When a region puffs the band of origin becomes very faint if it is visible at all. The area covered by sequences binding cytoplasmic RNA becomes much larger. However, when the chromosome preparation is well stretched it can be seen that the hybridization does not cover the entire area of the puff in any of the regions studied. The distribution of ^{3}H-RNA over the puff appears much the same whether the preparation was made by *in situ* hybridization with cytoplasmic RNA, by *in situ* hybridization with heterogeneous nuclear RNA, or by transcription autoradiography of ^{3}H-uridine incorporation directly on the chromosomes. Longer autoradiographic exposures of either *in situ* hybridization or transcription autoradiographs show broader distributions of ^{3}H-RNA, suggesting a gradient of compaction of the transcribed DNA sequences spreading from the band of origin into the puff.

Our observations on the major heat shock puffs are consistent with the accepted model of a puff arising by a localized uncoiling of sequences. However they give no explanation for the apparently non-transcribed sequences which are included in the puff. Nor do they explain why the puff may develop symmetrically around the transcribed region in some cases and quite asymmetrically in others.

RESPONSE TO ECDYSONE: HORMONALLY REGULATED GENES

In *D. melanogaster* many developmental processes are closely regulated by hormones. Increases in the level of the steroid hormone ecdysone are thought to trigger each molt. Ecdysone concentration affects most, if not all, of the larval tissues; the specificity of the response is dependent on the target tissue and not on the hormonal stimulus.

One of the earliest detectable effects of ecdysone is a change in the pattern of puffing of the polytene chromosomes in salivary glands (11, 3). Mid-third instar larvae show the beginning of the prepupal puffing sequence within 15 minutes after being injected with ecdysone. Ecdysone apparently acts as a trigger; it initiates a sequence of changes in the patterns of puffing of the polytene chromosomes but its presence is not required for the completion of the later parts of the sequence (4). Berendes (11) studied the early stages of the prepupal puffing sequence in several polytene tissues; salivary gland, malpighian tubules, and midintestine cells. He found the ecdysone-induced puffs were the same in all tissues, suggesting that these early puffs, at least, represent a generalized response to the hormone.

Ecdysone also initiates changes in diploid tissues. It induces metamorphosis in imaginal discs both *in vivo* and *in vitro* (23). Imaginal disc cells do not have polytene nuclei but it is possible to study ecdysome-induced changes in their RNA populations by *in situ* hybridization of ^{3}H-RNA from these diploid cells to polytene chromosomes from salivary glands.

Experiments using cytoplasmic RNA labeled in wing discs during the first few hours of ecdysone stimulation show the induction of a polyadenylated RNA species which hybridizes to polytene band 67B11 (13). Our early experiments also detected six other loci, 12E, 31A, 53C, 63B, 66B, and 92B, showing low levels of hybridization with cytoplasmic ^{3}H-RNA from ecdysone-stimulated wing discs. Later experiments showed that some RNA binding to those six regions is also produced in wing disc cells which have not been exposed to ecdysone. However ecdysone treatment specifically increases the level of ^{3}H-RNA binding to 12E, 63B, 66B, and 92B as compared to the levels of ^{3}H-RNA binding to other regions which are not affected by ecdysone.

The *in situ* experiments were done with RNA labeled during the first four hours of ecdysone induction of the wing discs. This is the same time period during which the early puffs are active in polytene tissues, the period for which Berendes (11) reported that the three polytene tissues he studied had identical puffing patterns. Despite the generality of the early response to ecdysone in different polytene tissues, the early response to ecdysone in wing discs results in a different set of changes in the cytoplasmic RNA of these diploid cells. Cytoplasmic ^{3}H-RNA from ecdysone-stimulated wing disc cells does not hybridize with polytene regions which puff during the early puff stages (e.g. 74EF, 75B). Nor has cytoplasmic ^{3}H-RNA from wing disc cells, synthesized either before or after ecdysone stimulation, been detected binding to the regions of the two major polytene puffs (25AC, 68C) which regress when ecdysone is added. The band which hybridizes ecdysone-stimulated RNA (but not control RNA) from the disc cells (67B11) does not puff during the early response to ecdysone stimulation, and also does not hybridize salivary gland RNA. Thus the initial response to ecdysone detected in the imaginal disc appears to be quite different from the response seen in the polytene tissues.

Due to the complex nature of RNA processing, it is quite probable that the genetic activity visualized directly on polytene chromosomes as either puffs or ^{3}H-uridine incorporation does not give an exact representation of the RNA population in the cytoplasm. For that reason we have examined the pattern of *in situ* hybridization of ^{3}H-RNA from the cytoplasm of salivary glands incubated with

^{3}H-uridine in the presence or in the absence of β-ecdysone (Bonner and Pardue, in preparation). Grain counts were made over region 68C, which regresses when ecdysone is added, and 74EF and 75B, which are induced as an early response to ecdysone, as well as over several control sites which show no effect of ecdysone. The results show that for these RNAs at least, the induction of a puff does result in the presence of ^{3}H-RNA in the cytoplasm which can be detected by *in situ* hybridization. ^{3}H-RNA hybridizing to 68C is found in the cytoplasm of glands labeled without ecdysone. After ecdysone stimulation the puff at 68C regresses and newly synthesized cytoplasmic RNA no longer hybridizes to this region. Cytoplasmic ^{3}H-RNA hybridizing to 74EF and 75B is seen only after the puffs have been induced by ecdysone.

All of the polytene puffs which have been studied thus far do produce RNAs which appear in the cytoplasm. The developmental puffs 68C, 74EF and 75B in this study, the major heat shock puffs discussed earlier, and Balbiani Ring 2 in *Chironomus tentans* (18), have all been shown to bind ^{3}H-labeled cytoplasmic RNA synthesized while the puffs were present. Thus *in situ* hybridization with cytoplasmic RNA reflects at least some of the changes in gene activity which can be detected by studies of puffing in polytene chromosomes. These studies on the cytoplasmic RNAs of salivary glands support the conclusion that wing discs respond to ecdysone in a different way from salivary glands, malpighian tubules, or midintestine cells.

ACKNOWLEDGEMENTS

This work has been supported by grants from the National Institutes of Health and the National Science Foundation. This paper will also be presented at the Joint Seminar and Workshop on the Aspects of Chromosome Organization and Function, Montevideo, Uraguay, February 1-5, 1977.

REFERENCES

1. Ashburner, M. (1967) Chromosoma (Berl.) 21, 398.
2. Ashburner, M. (1970) Chromosoma (Berl.) 31, 356.
3. Ashburner, M. (1972) In: Developmental studies on giant chromosomes. Results and Problems in Cellular Differentiation. Vol. 4. (W. Beermann, ed.) Springer-Verlag, Berlin, Heidelberg, New York, p. 101.
4. Ashburner, M., Chihara, C., Metzer, P., and Richards, G. (1974) Cold Spring Harbor Symp. Quant. Biol. 38, 655.
5. Baudisch, W., and Panitz, R. (1968) Exp. Cell Res. 49, 470.
6. Becker, H. J. (1962) Chromosoma (Berl.) 13, 341.
7. Beermann, W. (1952) Chromosoma (Berl.) 5, 139.
8. Beermann, W. (1961) Chromosoma (Berl.) 12, 1.
9. Beermann, W. (1962) Riesenchromosomen. Protoplasmologia, Handbuch der Protoplasmaforsch. Band IVd. Springer (Wein).
10. Beermann, W. (1972) In: Developmental studies on giant chromosomes. Results and Problems in Cellular Differentiation. Vol. 4. (W. Beermann, ed.) Springer-Verlag, Berlin, Heidelberg, New York, p. 1.
11. Berendes, H. D. (1967) Chromosoma (Berl.) 22, 274.
12. Bonner, J. J., and Pardue, M. L. (1976a) Cell 8, 43.
13. Bonner, J. J., and Pardue, M. L. (1976b) Chromosoma (Berl.) 58, 87.
14. Breugel, F. M. A. van (1966) Genetica 37, 17
15. Clever, U., and Karlson, P. (1960) Exp. Cell Res. 20, 623.
16. Grossbach, U. (1974) Cold Spring Harbor Symp. Quant. Biol. 38, 619.
17. Korge, G. (1975) Proc. Nat. Acad. Sci., U.S.A. 72, 4550.
18. Lambert, B. (1974) Cold Spring Harbor Symp. Quant. Biol. 38, 637.
19. Leenders, H. J., and Berendes, H. D. (1972) Chromosoma (Berl.) 37, 433.
20. Lengyel, J. A., and Pardue, M. L. (1975) J. Cell Biol. 67, 240a.
21. Lewis, M., Helmsing, P. J., and Ashburner, M. (1975) Proc. Nat. Acad. Sci., U.S.A. 72, 3604.
22. McKenzie, S. L., Henikoff, S., and Meselson, M, (1975) Proc. Nat. Acad. Sci., U.S.A. 72, 1117.

23. Oberlander, H. (1972) In: The biology of imaginal disks. Results and Problems in Cell Differentiation Vol. 5. (H. Ursprung and R. Nothiger, eds.) Springer-Verlag, Berlin, Heidelberg, New York, p. 155.
24. Pardue, M. L., Weinberg, E., Kedes, L. H. and Birnstiel, M. L. (1972) J. Cell Biol. 55, 199a.
25. Pelling, C. (1959) Nature 184, 655.
26. Pelling, C. (1964) Chromosoma (Berl.) 15, 122.
27. Ritossa, F. (1962) Experentia (Basel) 18, 571.
28. Ritossa, F. (1964) Exp. Cell Res. 35, 601.
29. Rubin, G. M., and Hogness, D. M. (1975) Cell 6, 207.
30. Spradling, A., Penman, S., and Pardue, M. L. (1975) Cell 4, 395.
31. Spradling, A., Pardue, M. L., and Penman, S. (1977) J. Mol. Biol. 109, 559.
32. Tissieres, A., Mitchell, H. K., and Tracy, U. M. (1974) J. Mol. Biol. 84, 389.
33. Wensink, P. C., Finnegan, D. J., Donelson, J. E., and Hogness, D. S. (1974) Cell 3, 315.

"MAPPING" WITH ANTIBODIES TO NONHISTONE CHROMOSOMAL PROTEINS: A BRIEF REVIEW

Sarah C.R. Elgin, Lee M. Silver, and Leslie Serunian

The Biological Laboratories, Harvard University
Cambridge, MA 02138

ABSTRACT. The *in situ* distribution of nonhistone chromosomal proteins (NHC proteins) in *Drosophila* polytene chromosomes is being studied using an indirect immunofluorescent "staining" technique (12). Formaldehyde fixation is used to prevent extraction or rearrangement of the proteins during preparation and staining of the chromosomes. Results obtained using antisera against total NHC proteins demonstrate that these proteins, purified from isolated chromatin, are indeed preferentially associated with the DNA in chromosomes *in vivo*. Recent studies using antisera against three molecular weight subfractions of the NHC proteins indicate that while some of the NHC proteins may be widely distributed in chromatin, others have a limited and specific distribution which may be indicative of their function. A highly selective staining pattern is obtained using antiserum against subfraction ρ; puffs (loci very active in RNA synthesis) and many nonpuffed chromomeres which are known to puff at other times during the third larval instar or prepupal stages are brightly fluorescent. In *Drosophila* new RNA synthesis can be induced at specific chromomeres, such as 87A and 87C1, by heat shock treatment; these loci, previously stained at low levels, are stained brightly using the ρ serum after heat shock (L. M. Silver and S.C.R. Elgin, manuscript in preparation). The sum of the available evidence suggests that a particular chromatin structure or configuration, indicated by staining using the ρ serum, is a necessary but not sufficient condition for gene activity as indicated by puffing. This correlation is being studied in greater detail.

The immunofluorescent "mapping" technique should be useful in studying the NHC proteins in two ways. First, the distribution patterns obtained using serum against individual NHC proteins isolated by two-dimensional gel electrophoresis may suggest the functions of these proteins, as above. Second, the distribution patterns of proteins isolated by use of other, more biological, properties (such as enzyme activity) may be obtained to test hypotheses concerning the role of such proteins in chromosome structure and function. Both approaches are being used in current work.

INTRODUCTION

During the last several years much has been learned about the structure of chromatin, both in terms of the DNA sequence organization and in terms of histone-DNA interactions. However, many mysteries remain. In particular, the question of the differences in chromatin structure between those regions which are accessible to transcription by RNA polymerase and those regions which are not has not been resolved. It is reasonable to hypothesize that this functional difference, and many others, is a consequence of differences in chromatin structure which reflect differences in the proteins associated with the DNA (see reference 4 for review). It now appears possible that such differences are superimposed upon the fundamental chromatin nucleosome structure which is the consequence of histone-DNA interactions (7,16, 6,10). Such differences could involve (among other possibilities) the association of different nonhistone chromosomal proteins (NHC proteins) with a given region of the chromatin fiber. In order to study this question, and others concerning the structural and functional roles of NHC proteins, we have developed a technique analogous to that of *in situ* hybridization to determine the distribution of NHC proteins in *Drosophila* polytene chromosomes using indirect immunofluorescence (12). By producing a cytological visualization of the distribution of specific macromolecular constituents, such mapping techniques have provided a considerable amount of new information on the organization of chromosomes and have aided in the correlation of biochemical, genetic, and cytological data. In this paper we will briefly review the work to date from our laboratory and others using indirect immunofluorescence to map the distribution of NHC proteins in polytene chromosomes.

IMMUNOFLUORESCENT STAINING OF POLYTENE CHROMOSOMES

Polytene chromosomes of *Drosophila* have been "stained" using antibodies prepared against chromosomal proteins by the following technique. The dissected salivary glands of third instar larvae are fixed in buffered 2% formaldehyde after brief incubation in buffered 0.5% Nonidet P40 (Shell Chemicals). The glands are then incubated in 45% acetic acid - 10 mM $MgCl_2$ and squashed. Following removal of the cover slip, the chromosome squash is washed and then treated first with the test sera prepared in rabbits and subsequently with a fluorescein-conjugated IgG faction of goat (anti-rabbit gamma globulin) serum (Miles Laboratories, Inc.). The preparation is then mounted for viewing with incident ultra-violet illumination (see references 12 and 13 for a detailed description of the procedure). For this study antibodies against Drosophila embryo chromatin, total Drosophila embryo NHC proteins prepared from isolated chromatin by two different techniques, and three molecular weight subfractions of the NHC proteins of Drosophila have been obtained from rabbits (5). In the latter case the antibodies were obtained using SDS slab gel fractions as described by Tjian et al. (15). Several control experiments have indicated that the presence of SDS does not interfere with the production of (at least some) antibodies which react specifically with the native protein (14,15).

Typical results obtained on "staining" Drosophila polytene chromosomes using antisera against Drosophila embryo total NHC proteins are given in Figure 1. It should be noted that the chromosomes are prominently stained, with little staining of the cytoplasmic material, confirming that the proteins so isolated are indeed preferentially associated with the DNA in chromosomes in vivo. It is also apparent that the distribution of the NHC proteins detected by the staining technique is nonuniform. The resolution possible approaches the detection of chromomeres, potential units of gene organization (see reference 8 for review). That the staining is the consequence of a specific antigen: antibody interaction involving the chromosomal proteins is confirmed by several control experiments. If antiserum from pre-immunization animals or phosphate buffered saline is used as the primary reagent, no staining is observed. If the anti-NHC protein sera are absorbed with chromatin before use, staining is abolished, while the presence of Drosophila DNA or yeast RNA does not alter the staining pattern (12). Some caution should be exercised in interpreting the staining pattern; since the chromosomes are fixed with formaldehyde prior to staining it is possible that some antigenic sites are altered or obscured from the antibody probe.

SPECIFICITY OF DISTRIBUTION PATTERNS

The results obtained above imply that individual NHC proteins will possess specific patterns of distribution; determination of these patterns could help both to suggest the functional role of various NHC proteins and to test hypotheses concerning individual NHC proteins. To explore this concept of a limited and specific distribution we have prepared antibodies against three molecular weight subfractions of the NHC proteins (13). In each case the fluorescence distribution pattern obtained is distinct and reproducible (L.M. Sil-

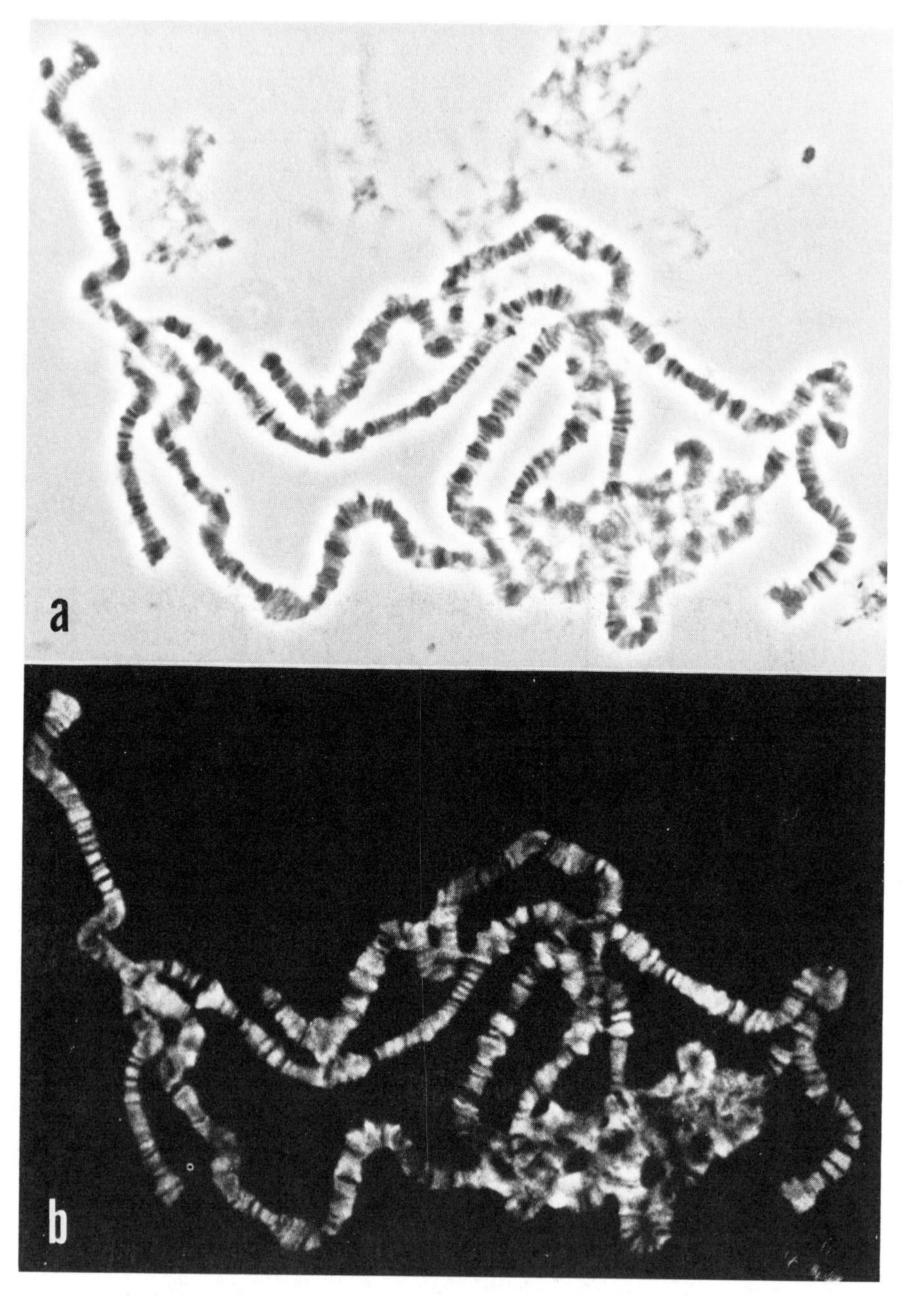

Fig. 1. Polytene chromosome spread prepared using formaldehyde fixation and stained using antiserum against total *Drosophila* embryo NHC proteins as described in reference 12 viewed by a) phase contrast and b) UV dark field optics.

ver and S.C.R. Elgin, manuscript in preparation). In particular, a highly limited pattern of prominent staining is obtained using a serum prepared against the subfraction ρ of NHC proteins (80,000 - 110,000 daltons molecular weight) (see Figure 2). In general it is observed that the puffs (loci highly active in RNA synthesis) and other chromomeres which are known to puff at some time during the third larval instar or prepupal stages are brightly fluorescent. In a detailed analysis of chromosome 3, an 88% correlation was observed; of 49 bands classified as intensely staining, 43 are reported to form major puffs (3). Conversely, all puffs are stained. Most of the staining pattern is stable; however, a few loci are brightly stained only when they are visibly puffed (L.M. Silver and S.C.R. Elgin, manuscript in preparation).

To confirm the relationship between positive ρ staining and gene activity as indicated by puffing we have examined the staining obtained using ρ serum at bands 87A and 87B-C1. These loci, not normally active during these developmental stages, are induced to puff by heat shock (elevation of the temperature from 25°C to 37°C). The staining of polytene chromosomes from larvae subjected to heat shock has been compared to that from normal larvae; a portion of the chromosome arm 3R is shown in Figure 3. In addition, 3H-uridine incorporation following heat shock has been monitored using autoradiography. The loci 87A and 87B-C1,stained only at very low levels in polytene chromosomes from normal larvae, are very brightly stained in polytene chromosomes from heat-shocked larvae. Other loci normally active at this stage of development show reduced levels of transcriptional activity following heat shock, but remain stained using the ρ serum. These results imply that the particular chromatin structure indicated by positive staining is a necessary but not sufficient characteristic of the active gene configuration (L.M. Silver and S.C.R. Elgin, manuscript in preparation).

A limited and specific staining pattern has also been observed by Alfagame et al (1,2) using antibodies prepared against D1, a small Drosophila nonhistone chromosomal protein. Using an independently developed similar staining procedure, they observe bright fluorescence at bands 81F-82A, 83C-E and two regions on chromosome 4. This staining pattern corre-

Fig. 2 (next page). Polytene chromosome spread stained using antiserum against fraction ρ of the NHC proteins, viewed by a) phase contrast and b) UV dark field optics.

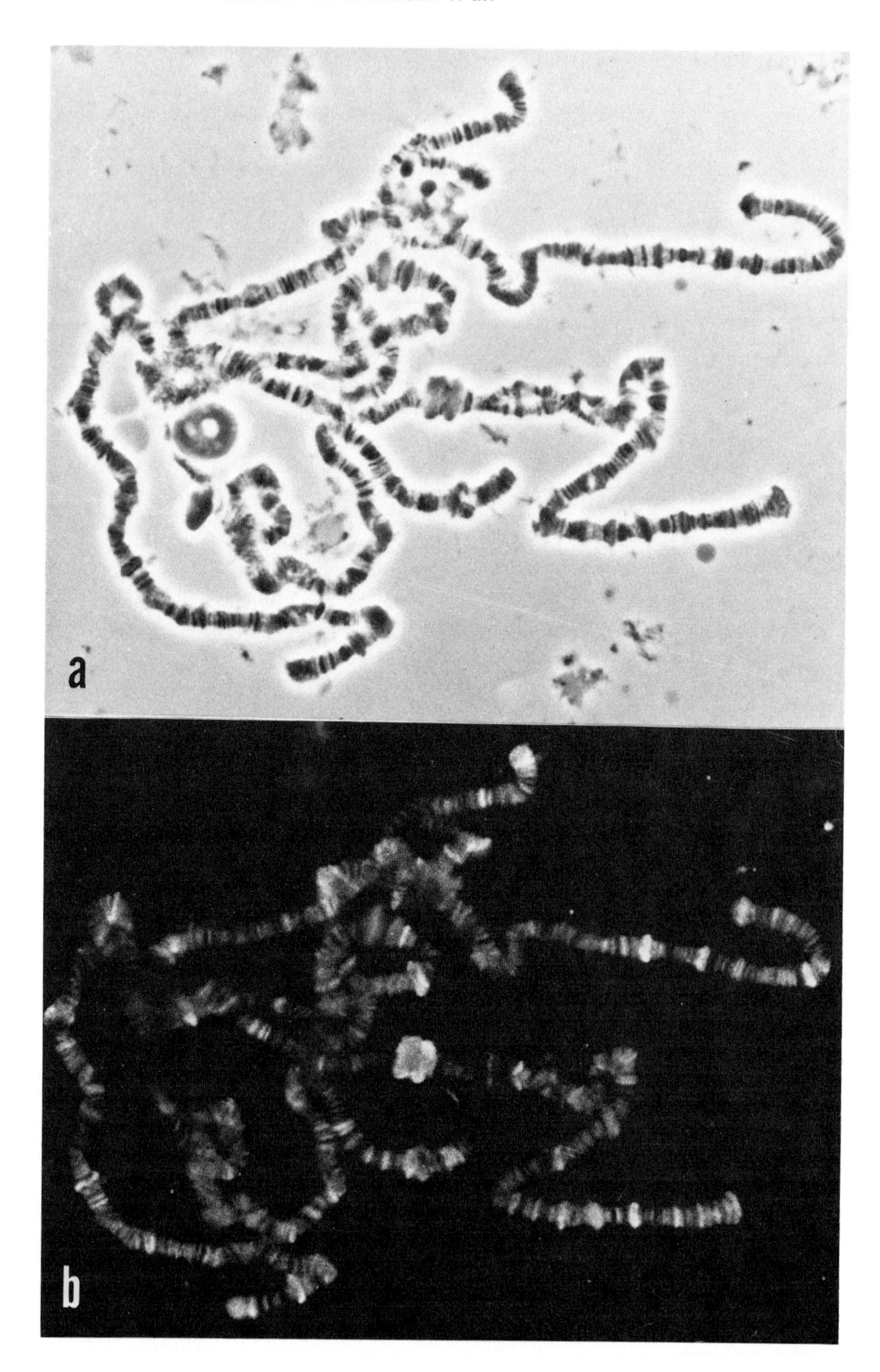
a
b

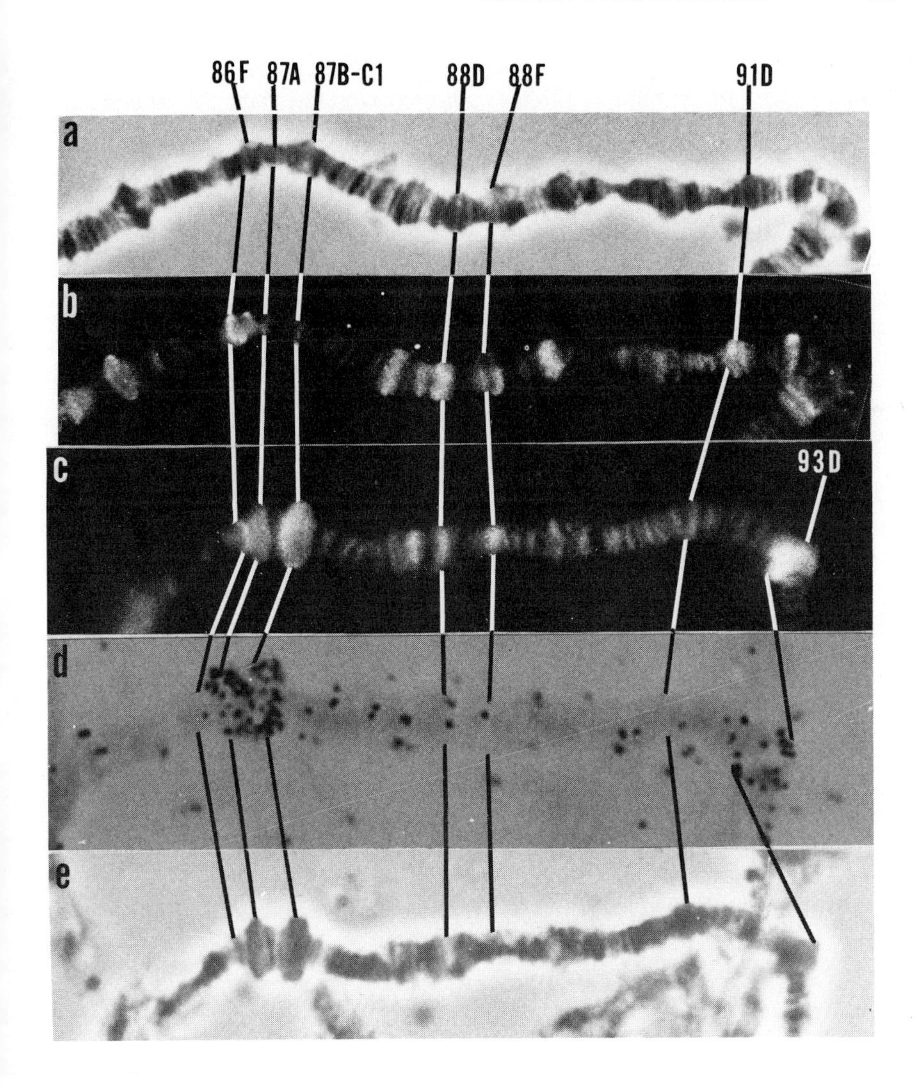

Fig. 3. Staining pattern of a central region of chromosome 3R using anti-ρ serum. 3a and 3b, phase contrast and fluorescent views of a chromosome from a larva at 25°; 3c, 3d, and 3e, fluorescent, autoradiographic, and phase contrast views of a chromosome from a larva at 37° (heat shocked).

lates well with the pattern of intense staining obtained with quinacrine or Hoechst 33258, suggesting that D1 is associated with certain DNA sequences having unique features of base composition or secondary structure. The staining pattern obtained using antibodies against D2, a second _Drosophila_ NHC protein, is distinctly different.

In a third study of this type Plagens, Greenleaf and Bautz (9) have examined the fluorescent staining pattern obtained on polytene chromosomes using antibodies prepared against purified _Drosophila_ RNA polymerase II. Both puffs normally present and those induced by heat shock are prominently stained using this sera. Many other loci are also stained; however, the pattern is distinct from that observed following staining using serum prepared against histone H1.

In these analyses the question may be raised as to whether or not the difference between a positive and a negative result reflects a difference in the distribution of an antigenic determinant (presence or absence of a particular protein) or a difference in the accessibility of the antigenic determinant to the antibody probe. In general, the former appears more likely since a) most NHC proteins are present in amounts too low to permit them to be associated with all chromomeres (although there certainly are exceptions to this) and b) many loci can be shown to be stained using anti-histone serum, indicating a general accessibility of the chromatin fiber to an antibody probe under these conditions. For example, the locus 87A which is intensely stained using the ρ serum only in chromosomes from larvae which have been heat-shocked, is stained using anti-H1 serum and anti-H3 serum in chromosomes from normal larvae (L.M. Silver and S.C.R. Elgin, manuscript in preparation).

FUTURE APPLICATIONS

The work to date, summarized above, suggests that this mapping technique will aid us in the study of NHC proteins in three ways. First, the distribution pattern observed by fluorescent antibody staining may correlate with patterns of functional significance, such as puffing, and hence suggest a role for these proteins. Second, the pattern may correlate with other structural features, such as DNA sequence, and thereby confirm the _in vivo_ structural association. Third, the technique can be used to explore a further parameter in the characterization of the mode of activity of known enzymes with known or suggested chromosomal roles. While the polytene chromosomes appear to be the most advantageous substrate for such work, it has been shown that lampbrush chromosomes can be stained by analogous immunofluorescence techniques(11); it is possible that such staining of metaphase chromosomes would yield interesting and useful results. The use of such cytological techniques in conjunction with genetic techniques should help to insure that results obtained by biochemical characterization of isolated chromatin in vitro can be correlated with the characteristics of chromosomes in vivo, and thus should help us work towards realistic models of the structure of chromatin in the active and inactive states.

ACKNOWLEDGEMENTS

Work in the authors' laboratory is supported by grants from the NIH and the American Cancer Society. L.M. Silver is supported by an NIH predoctoral training grant to the Committee on Higher Degrees in Biophysics; S.C.R. Elgin is supported by an NIH Research Career Development Award.

REFERENCES

1. Alfageme, C.R., Rudkin, G.T., and Cohen, L.H. (1976) Proc. Nat. Acad. Sci. USA 73, 2038-2042.
2. Alfageme, C.R., Rudkin, G.T., and Cohen, L.H. (1976) J. Cell Biol. 70, 298a.
3. Ashburner, M. (1972) In "Developmental Studies on Giant Chromosomes," W. Beerman, ed. (New York: Springer-Verlag), pp. 101-151.
4. Elgin, S.C.R., and Weintraub, H. (1975) Ann. Rev. Biochem. 44, 725-774.
5. Elgin, S.C.R., Silver, L.M., and Wu, C.E.C. (1977) In "International Symposium on Molecular Biology of the Mammalian Genetic Apparatus." P.O.P. T'so, ed. (Amsterdam: Asso. Scientific Publishers), in press.
6. Kuo, M.T., Sahasrabuddhe, C.G., and Saunders, G.F. (1976) Proc. Nat. Acad. Sci. USA 73, 1572-1575.
7. Lacy, E., and Axel, R. (1975) Proc. Nat. Acad. Sci. USA 72, 3978-3982.
8. Lefevre, G.,Jr. (1974) Ann. Rev. Genetics 8, 51-62.
9. Plagens, U., Greenleaf, A.L., and Bautz, E.K.F. (1976) Chromosoma 59, 157-65.
10. Reeves,R. (1976) Science 194, 529-532.
11. Scott,S.E.M., and Sommerville,J. (1974) Nature 250, 680-2.
12. Silver, L.M., and Elgin, S.C.R. (1976) Proc. Nat. Acad. Sci. USA 73, 423-427.
13. Silver, L.M., Wu, C.E.C., and Elgin,S.C.R. (1977) In "Methods in Chromosomal Protein Research," G. Stein,

J. Stein, and L. Kleinsmith, eds. (New York: Academic Press), in press.

14. Stumph, W.E., Elgin, S.C.R., and Hood, L. (1974) J. Immunology 113, 1752-1756.
15. Tjian, R., Stinchcomb, D., and Losick, R. (1974) J. Biol. Chem. 250, 8824-8828.
16. Weintraub, H., and Groudine, M. (1976) Science 193, 848-856.

FLOW-FLUOROMETRIC DNA CONTENT DIFFERENCES AS A FUNCTION OF CHROMOSOME CONSTITUTION IN RESTING HUMAN LYMPHOCYTES

Holger Hoehn and James Callis

Division of Genetic Pathology and Center for Inherited Diseases, University of Washington
Seattle, Washington 98195

ABSTRACT. Lymphocytes were isolated from heparinized venous blood and stained with the fluorescent intercalating dye propidium iodide in a hypotonic citrate solution. Their fluorescence was measured at rates of 2-3 x 10^4/min in a flow-microfluorometer of conventional design. Lymphocytes from patients with the 45,XO Turner syndrome were found to have between 2-3% less fluorescence than lymphocytes from 46,XX females. Conversely, the amount of fluorescence emitted by 47,XXY male lymphocytes was between 2-3% higher than that of 46,XY lymphocytes. When groups of euploid males and females were compared, a mean 1.6% difference in fluorescence intensity was observed. This difference is very close to the 1.7% expected on the basis of the size difference between the X and Y chromosomes, but the coefficient of variation of these measurements was high (± 0.6%). Further studies identified instrumental instability and non-stoichiometry of dye binding as major sources of variance in flow-fluorometric DNA content measurements. The problem of instrumental instability was controlled by employing chicken erythrocytes as internal standard for each lymphocyte sample. Likewise, problems with non-stoichiometry were minimized by allowing samples to equilibrate for 24 hours in the staining solution. While elimination of obvious instrumental and preparational noise considerably reduced the coefficients of variation of DNA content measurements among groups of males or females, a residual variance on the order of 0.3% was consistently found. Since samples with either low or high DNA content within their respective groups exhibited closely corresponding values at repeat venipuncture, we interpret this residual variance to reflect intrinsic, truly biological DNA content differences between euploid individuals of like sex. Preliminary chromosome banding studies suggest an association between polymorphism for constitutive heterochromatin and the extent of DNA content variation detected by flow-fluorometry. The extent of this variation is sufficiently small to permit the diagnosis of aneuploidy of larger chromosomes, but may prevent flow-fluorometric diagnosis of trisomy 21.

INTRODUCTION

Flow-microfluorometry (FMF) is an extremely convenient technique for ultra-rapid quantitation of a variety of cellular components (1-3). Its potential with respect to mammalian cytogenetics has recently been epitomized by reports which demonstrate the feasibility of flow-karyotyping (4,5). While this application of the FMF technique offers the exciting possibilities of recognizing numerical as well as structural chromosome changes on the basis of an "itemized" DNA profile of the entire complement, much work remains to achieve the resolution and reproducibility necessary to be competitive with conventional banded chromosome preparations. Like conventional cytogenetics, flow-karyotyping requires rather lengthy cell culture and chromosome preparation steps. In contrast, we have attempted to improve the resolution of the FMF technique with respect to the more traditional determination of total cellular DNA, which has the advantage of not requiring cell culture and chromosome preparation. This approach has the proven potential of identifying gross aneuploidies in malignant cell populations (6-8). While even with increased sensitivity measurements on interphase cells cannot possibly detect structural chromosome changes (unless extensive duplication-deficiency is involved), the rapid diagnosis of constitutional aneuploidy in itself should be useful, since nearly half of the growing requests for postnatal diagnostic cytogenetics concern the distinction between euploidy and aneuploidy. In prenatal cytogenetics the rapid diagnosis of aneuploidy is clearly attractive, since the exclusion of age-related aneuploidy is the target in over 80% of the cases, and speed matters most. In this article we will first review our initial experience with the FMF diagnosis of aneuploidy (9) and then discuss our attempts to define possible sources of DNA content variability between euploid individuals encountered during these studies.

FLOW-FLUOROMETRIC DIAGNOSIS OF ANEUPLOIDY

Human lymphocytes were purified from 5-10 cc of heparinized venous blood by a standard Ficoll-hypaque centrifugation technique (10). Aliquots of 0.5 x 10^6 lymphocytes in 100 λ of PBS were added to 2 ml of staining solution consisting of 0.5 μg/ml propidium iodide in 0.1% sodium citrate at pH 7.0 (11). Initially, measurements were made after 2 hours staining time, which was extended to a minimum of 24 hours in the more recent experiments.

Figure 1. Principle of the FMF system and data analysis. The symbols L and E in panels A-C illustrate data expected from cell populations with DNA contents on the order of, respectively, 6 and 2 pico gm/nucleus. See text for other details.

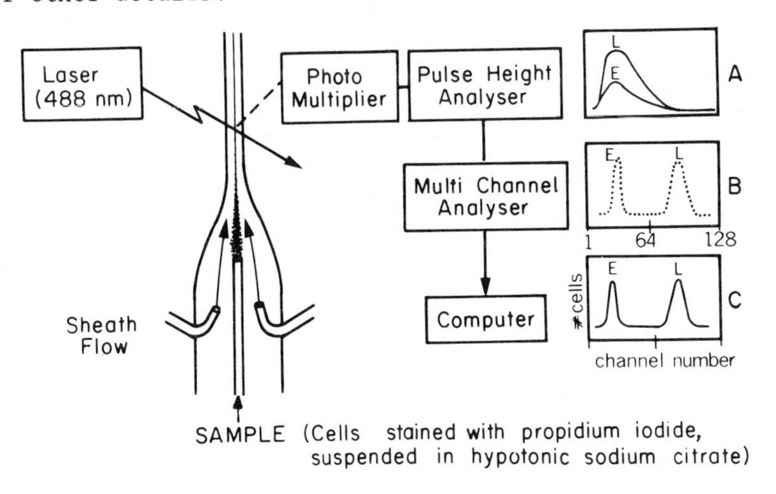

Figure 1 shows the basic design of our flow-fluorometer, including data collection and processing. The crucial parameter in FMF measurements is the amplitude of the fluorescent signal (A). Under stoichiometric conditions of dye binding to nuclear DNA, the signal height from a single cell as it crosses the laser beam is proportional to the total amount of DNA in that cell. The signals from an average of 2-3 x 10^4 lymphocytes are collectively displayed in the form of fluorescence pulse height histograms (B). The channel position of the mean of such a histogram therefore reflects the average DNA content of the entire lymphocyte population. Assuming that the DNA content measurements of resting human lymphocytes are normally distributed, the raw histogram data points can be fitted to Gaussian curves by standard fitting procedures (12,13). The histogram means obtained from these computations (C) serve as "DNA content" values throughout this report.

Figure 2A displays a representative histogram from an euploid 46,XY lymphocyte preparation. Typically, such histograms are highly symmetrical, with coefficients of variation between 3 and 5%. In Figure 2B, the horizontal display scale has been expanded in the area of interest to visually enhance the 1.5% difference observed between a 46,XY (left)

Figure 2. Examples of actual fluorescence pulse height histograms (alphabetical characters) and their respective best fits to Gaussian distributions (solid or broken lines). Ordinate: number of cells x 10^2 except B. Abscissa: channel number; proportional to fluorescence intensity and (under conditions of stoichiometric fluorochrome binding) DNA content.

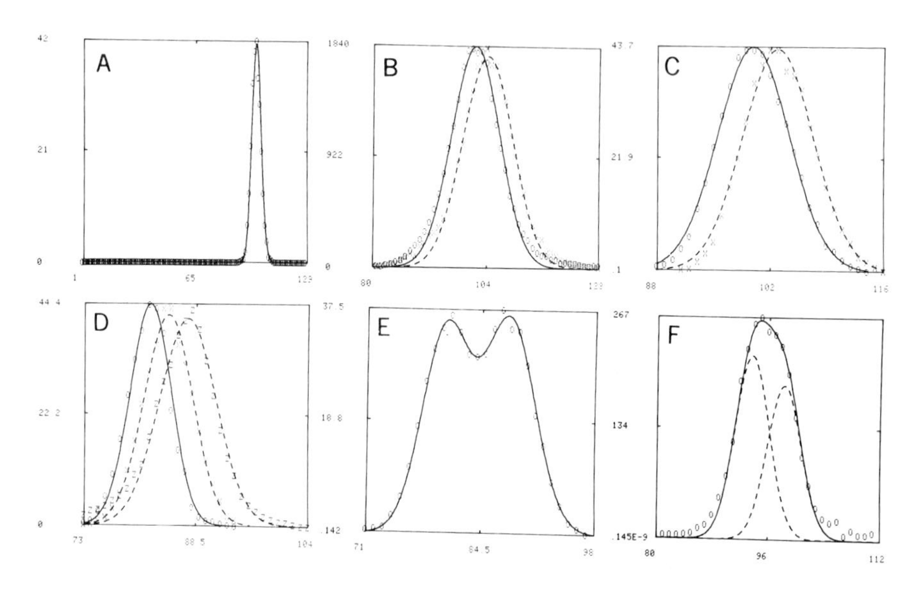

and a 46,XX (right) lymphocyte DNA content distribution. Such a difference is in fact expected because the Y chromosome is considerably smaller than the corresponding second X chromosome in female cells. When we compared 46,XY and 47,XXY lymphocytes (Figure 2C), the two distributions were still further apart (difference between computed means: 2.6%). Similarly, DNA histogram means between 46,XX, 47,XXX and 48,XXXX lymphocytes (Figure 2D) show the expected stepwise differences. Finally, a mixture between 45,XO and 49,XXXXY lymphocytes was found to exhibit bimodality (Figure 2E) with each peak reflecting the respective components (as can be demonstrated by curve-fitting on the basis of expected 8.7% difference between the two means). Such bimodality can only appear when the difference between the two components is greater than the sum of the coefficients of variation of each individual sample. Our lowest coefficients of variation in these experiments were on the order of 3%, which is not small enough to resolve the two populations in 45,XO/46,XX mosaicism. However, histograms from such individuals, even though they appear to be unimodal, may have slightly increased coefficients of variation and, under quantitative scrutiny, might reveal skewness with the possibility of finding best fits for two subpopulations differing by 2.5% in their respective means. Figure 2F depicts such a histogram from a patient with XO/XX mosaicism. Conventional cytogenetics revealed 55% XO metaphases, and curve fitting resolved two corresponding subpopulations (dashed lines) from the definitely skewed histogram.

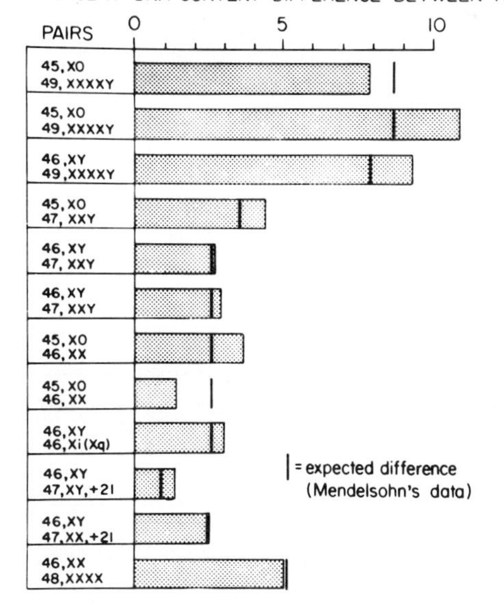

Figure 3. Examples of flow-fluorometric DNA content comparisons between pairs of cytogenetically distinct human lymphocytes. The horizontal bars illustrate the percentage of DNA-content difference observed by the FMF measurements. The solid line indicates the percentage of difference expected from cytophotometric DNA content determinations of individual human metaphase chromosomes (14).

Figure 3 summarizes an entire series of comparisons between cytogenetically distinct lymphocyte samples (9). In most instances, the differences between pairs were in close agreement with those predicted from their respective chromosome constitutions (14). Some of these pairs, however, differed either by considerably less or by more than expected, indicating that FMF measurements in these instances failed to accurately reflect DNA content. The possible reasons for such inaccuracy were explored in the following experiments.

DNA-CONTENT VARIATION AMONG EUPLOID INDIVIDUALS

In two series of 6 pairs each, we compared isolated lymphocytes from 12 cytogenetically confirmed 46,XY males and as many 46,XX females. The average DNA content difference between the two groups was 1.6 ± 0.5%, which agrees well with the expected value of 1.7%. However, inspection of the individual data (Figure 4) shows that while mean FMF distributions differed by more than 2% between pair 5/4, others (e.g. 15/16; 9/10 and 2/1) showed less than 1% difference between their respective histogram means. What is the reason for this variability? Could this variation reflect variant amounts of highly repetitive, simple sequence DNA between certain individuals (15)? C-banding of conventionally prepared metaphases from the male/female pairs with either large or small DNA content differences in Figure 4 failed to reveal striking associations between the amount of heterochromatin and the degree of DNA content difference. We then selected males with known large amounts of C-banding positive material on chromosomes 1, 9, 16 and Y and contrasted these with lymphocyte donors whose metaphases had revealed only "minus" C-banding variants and small Y chromosomes. In the course of these experiments, instrumental instability surfaced as a major obstacle in probing

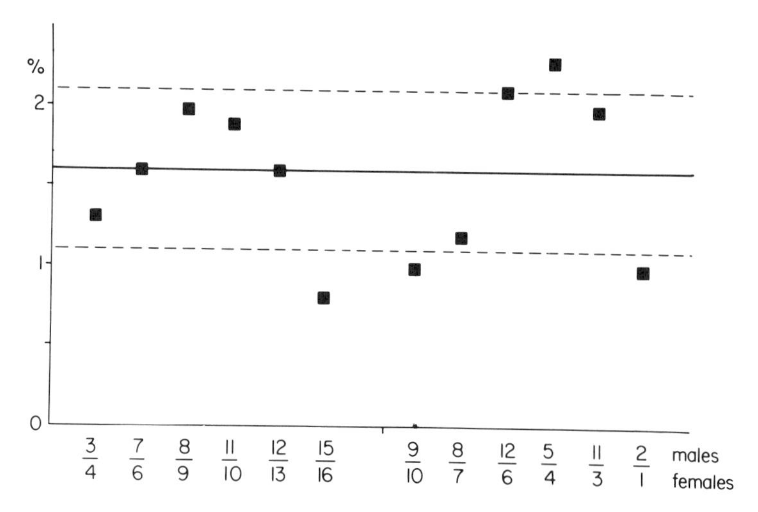

Figure 4. Percent DNA content differences between pairs of euploid male and female lymphocytes (arabic numerals). Mean difference & standard deviation shown in solid and broken lines.

such small DNA content differences between individuals. It became increasingly clear that instrumental noise had to be eliminated before attempts at definition of true (biologic) variability could be made. Our approach to the elimination of extrinsic noise focused on the search for particles suitable for inclusion with lymphocytes in every measurement. Of all possibilities (e.g., polystyrene spheres, calf thymocytes) chicken erythrocytes were found to hold greatest promise as "ideal" internal standard.

USE OF CHICKEN ERYTHROCYTES AS INTERNAL STANDARD

Chicken erythrocytes are readily available, nucleated cells with approximately one third of the mammalian DNA content (16). Accordingly, after staining with the hypotonic solution of propidium iodide, their fluorescence signals are much smaller than those of human lymphocytes. Figure 5 illustrates the performance of chicken erythrocytes as internal reference for lymphocyte DNA content determinations. In Figure 5A, L-histograms show different means, even though both are derived from euploid 46,XY lymphocytes. Noting that the respective E histograms were neither congruent, we computed the ratio between L and E means for each sample and found this parameter to be virtually unchanged (2.829 as opposed to 2.825), and the visually less extensive shift of the E as opposed to the L histogram is explained by scale factors. We can therefore conclude that the difference between the L histograms in this example reflects instrumental variation rather than variation in DNA content. We encounter such shifts more frequently during

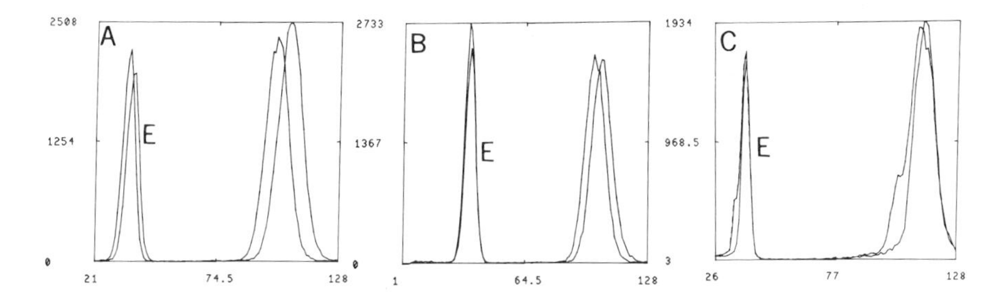

Figure 5. Detection of instrumental noise in FMF measurements by internal standardization with chicken erythrocytes (E). Ordinate and abscissa as in Figure 2.

the initial warm-up phase of the instrument, when laser power fluctuations, beam wander and instability in the electronics of the detection system seem to occur. In contrast, Figure 5B illustrates the comparison of two L-samples with a genuine DNA content difference (left L-histogram; 45,XO; right L-histogram; 46,XX). While the L-histogram means are apart by 2.4% (corresponding closely to the 2.5% value expected for a single X chromosome), the E-histograms overlap each other very closely, indicating a stable performance of the instrument during this comparison. Finally, Figure 5C illustrates malfunction of the instrument due to obstruction of the flow capillary by cell clumps. A "shoulder" appears at the left base of both the E- and L-histograms, presumably caused by change of the laminar flow profile. After cleaning of the flow channel and readjustment, both histograms regain their smooth and symmetrical appearance.

Prior to the introduction of the E-standard the only means for checking the stability of the instrument consisted of multiple measurements of a given sample over a prolonged period of time. This was done with most samples, and the results described in the previous sections were obtained under conditions of such apparently stable performance. However, the examples shown in Figure 5 illustrate that the integrity of lymphocyte DNA content distributions, as well as their reproducibility, can be much more reliably monitored by inclusion of chicken erythrocytes in all FMF measurements. In addition, such internal standardization considerably improves our basis for investigation of truly biological DNA content variations between euploid individuals (17).

SOURCES OF INTERINDIVIDUAL DNA CONTENT VARIATION

As discussed in the previous paragraph, <u>instrumental instability</u> has to be considered as one of the major sources of non-biological DNA content variation observed among euploid individuals in our initial FMF measurements. Clearly, without internal standardization, this instability (residing mainly in the rather complex nature of the instrumentation) severely affects the reproducibility of DNA content measurements. The introduction of the chicken erythrocyte standard and the use of the <u>L/E ratio</u> as a relative measure of DNA content proved to be highly effective for the control of instrumental noise in our measurements. However, even with internal standardization, consecutive measurements of independently stained L-samples continued to reveal a con-

Figure 6. Male/female DNA content differences represented by L/E ratios. Open bars: male samples; solid bars: female samples. Aliquots of lymphocytes from 8 individuals were stained during 6 successive days. 1R and 2R refer to repeat measurements, after 6 hours, of the day 1 and day 2 samples. See text for details.

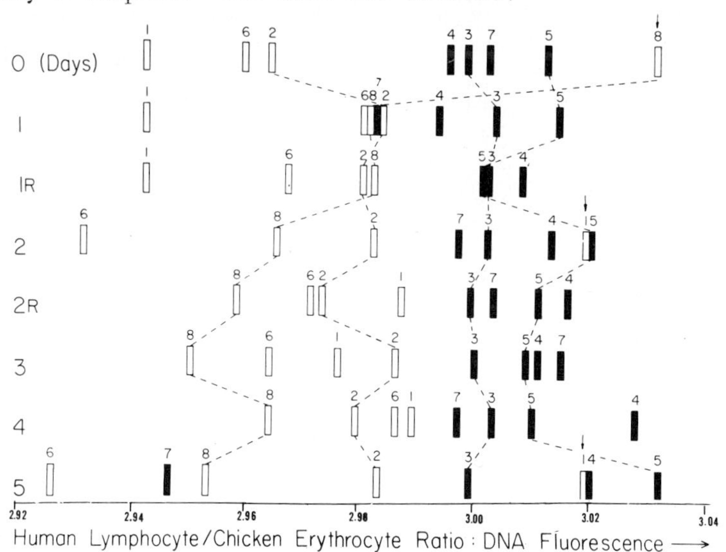

siderable degree of variability. In the example shown in Figure 6, aliquots from 4 male and 4 female lymphocyte preparations were assayed over a period of five days. The distinction between males and females is generally excellent and consistent until day 5, when the purified lymphocytes showed signs of clumping and decay and correspondingly less clustering of L/E ratios.

Certain samples, however, exhibited a rather irregular behavior of the L/E ratio at much earlier stages. Sample #8, for instance, had an extremely high L/E ratio before gradually assuming a lower position within the male range. Conversely, Sample #1 showed low L/E ratios during the first three measurements but fluctuated thereafter within the higher range of the male samples, to be found even among female L/E ratios in two instances (days 2 and 5). Other samples (e.g., #3 and, to a lesser extent, #2 and 5) showed relatively stable L/E ratios throughout the entire series of experiments. These assays were done within 2-8 hours after staining of the L-sample with propidium iodide. We have since observed that much of the extreme fluctuations (e.g., samples #1 and 8 in Figure 6) can be avoided by allowing at least 24 hours (at 4°C) for equilibration in the hypotonic staining solution.

Another source of non-stoichiometry of dye binding could arise from variations in the binding constants or dye-nucleotide ratio. Fortunately, the high equilibrium constant and the large excess of dye assures that greater than 99.9% of the binding sites will be occupied. Non-stoichiometry due to selective availability of DNA for dye binding (18) is unlikely to be of importance in resting lymphocyte preparations with presumably quite uniform states of chromatin. However, the unfixed lymphocytes, over a period of days, experience a slight continual increase in dye binding, which is probably associated with a gradual unfolding of chromatin and/or loss of histones or non-histone proteins in the hypotonic milieu. Fixation (19) renders dye binding more stable, but enhances clumping and cell loss. Until a satisfactory fixation protocol has been developed, we therefore restrict our FMF analysis to comparisons between samples which have been stained and assayed at closely corresponding times.

Given the hazards of instrumental instability and non-stoichiometry, does there remain any evidence for truly intrinsic DNA content variation between euploid individuals of a given sex? The experiments displayed in Figure 7 were designed to answer this question. In a series of 17 male and 15 female individuals we encountered substantial overlap between male and female L/E ratios. Since the original set (A) had been assayed within only 2-8 hours after staining, non-stoichiometry probably accounts for most of these inconsistencies. This interpretation is strengthened by the asymmetric appearance of the individual histograms of samples #12, 9, 40 and 39 in Panel A (17). Individuals with perfectly symmetrical and narrow histograms from both extremes of the male and female set (#34 and 10, as well as 37 and 18, respectively) were asked to donate a repeat blood specimen. As shown in B, the original male samples #34 (low L/E ratio) and 10 (high L/E ratio), when assayed together with 4 other male and 6 other female specimens, tended to reoccupy their distant positions. The B-series measurements were performed after proper equilibration (24 hours at 4°C) in the staining solution. The repeat analysis of blood samples from females #37 and 18 of the original series in Panel A gave a similar result: as shown in Panel C, samples P (34; low L/E ratio) and O (18; high L/E ratio) assumed almost identical extreme positions upon reanalysis.

Figure 7. Male (open bars) and female (solid bars) lymphocyte L/E ratios. A comparison of 32 samples assayed within 2-8 hours after staining. Arrows identify individuals (arabic numerals) from whom repeat blood specimens were obtained and analyzed in Series B and C, respectively. The original and repeat L/E ratios were connected by dashed lines to illustrate relative preservation of extreme positions.

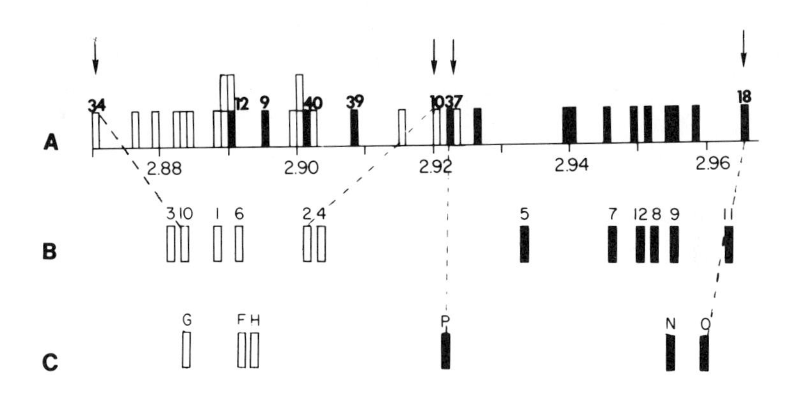

The remaining coefficient of variation among both males and females in Panel B is on the order of 0.3%. Since this is the lowest possible value obtained in many experiments carried out under careful exclusion of instrumental noise and maximal reduction of non-stoichiometry of dye binding, we believe that at least some of this residual variation reflects intrinsic DNA content differences between individuals.

COMPARISON BETWEEN CONVENTIONAL CYTOGENETICS AND FLOW CYTOGENETIC DATA

If we assume that the residual coefficient of variation of 0.3% between euploid individuals of like sex reflects intrinsic DNA content variation, two hypotheses can be tested. (1) The residual variation is due to mosaicism and (2) it is due to variant amounts of highly repetitive, simple sequence DNA as visualized by the C-banding procedure. For example, the consistently low L/E ratio of female #37 of Figure 7A might be explained by the discovery of XO/XX mosaicism. Conversely, the relatively high L/E ratios of individual 18 (Figure 7A) could be due to a mixture of 46,XX and 47,XXX cells. Over 200 metaphase counts in each instance ruled out that substantial mosaicism of T-lymphocytes accounts for the observed extreme DNA content values. However, when metaphases from males #34 and #10 (Figure 7A) were compared with respect to their C-banding patterns, a substantial difference in the amount of C-banding material was found (Figure 8). While male #34 has a Y chromosome of small to average size, the Y chromosome of individual #10 is comparable in size to human chromosome 18 and exhibits more than double the amount of constitutive heterochromatin found in the Y chromosome of individual 34. While the Y chromosome difference is most conspicuous, the comparison of C-banding patterns between females 37 and 18 with respect to the heterochromatin region of the chromosomes 1, 9 and 16 (Figure 8C and D) likewise suggests the presence of somewhat larger amounts of C-type heterochromatin in individual 18.

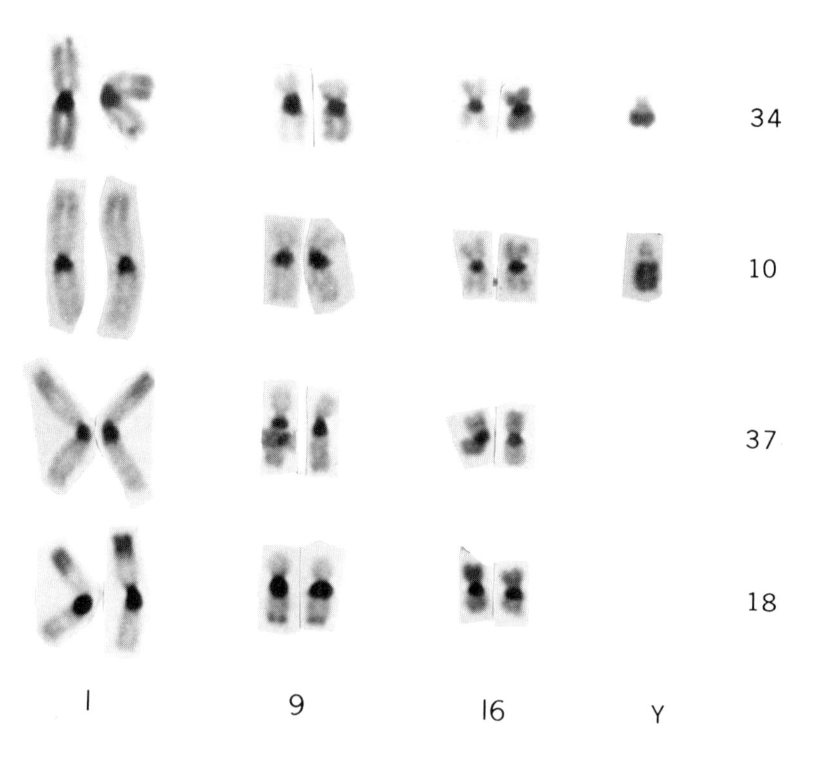

Figure 8. Partial C-banded metaphases (20) from 4 individuals (Figure 7) with either low or high L/E ratios.

These preliminary findings are consistent with the hypothesis of association between residual DNA content variation in the FMF assay and C-band polymorphism as revealed by conventional cytogenetics. If confirmed, such an association would limit the utility of the FMF technique for the diagnosis of aneuploidies involving small chromosomes, such as trisomy 21, because of the relatively high background of intrinsic variation. On the other hand, it would provide a convenient method for the quantitation of C-type heterochromatin polymorphism among individuals which cytogeneticists would certainly welcome.

The studies of Mayall et al. (21) suggest yet another type of DNA variation for which a correlation with C-banding polymorphism cannot be demonstrated. Possibly, this type of variation resides in regions of intercalary heterochromatin (22). It would therefore be of utmost importance to find means of differentiating, with specific fluorochromes, between the various classes of genetically relevant and less relevant chromatin. Adaptation to the FMF system of the fluorescent antibody techniques developed by Miller's group (23), for example, might be a promising avenue of research. In exploring such possibilities, however, one should not lose sight of the fact that what we and others using the FMF technology operationally define as "DNA-content" fluorescence is, in essence, merely a determination of available binding sites for a given fluorochrome which is itself highly base pair specific. Even though we have shown that with a uniform chromatin system such as the resting human lymphocyte the availability of binding sites closely corresponds to chromosome constitution, we predict formidable difficulties with selective quantitation of different types of chromatin. This is so because all of these techniques involve the selective denaturation of chromatin, and attendant risk of non-stoichiometry.

ACKNOWLEDGEMENTS

These studies were supported by grants from the National Institutes of Health and by a grant from the Human Growth Foundation to Judith G. Hall, to whom we are indebted for blood specimens.

1. VanDilla, M.A., Trujillo, T.T., Mullaney, P.F. and Coulter, J.R. (1969) *Science* 163, 1213.
2. Crissman, H.A. and Steinkamp, J.A. (1973) *J. Cell Biol.* 59, 766.
3. Melamed, M.R. and Kementsky, L.A. (1975) *Int. Rev. Exp. Path.* 14, 206.
4. Gray, J.W., Carrano, A.V., Steinmetz, L.L., VanDilla, M.A., Moore, D.H., II., Mayall, B.H. and Mendelsohn, M.L. (1975) *Proc. Nat. Acad. Sci. (USA)* 72, 1231.
5. Stubblefield, E., Cram, S. and Deaven, L. (1975) *Exp. Cell Res.* 94, 464.
6. Stöhr, M. (1975) *Acta Cytol.* 19, 299.
7. Barlogie, B., Büchner, T., Hart, J.S., Ahearn, M.J. and Freireich, E.J. (1975) "Pulse-Cytophotometry" (C.A.M. Haanen, H.F.P. Hillen and J.M.C. Wessels, eds.) European Press Medikon, Ghent, p. 299.
8. Swartzendruber, D.E., Cram, L.S. and Lehman, J.M. (1976) *Cancer Res.* 36, 1894.
9. Callis, J. and Hoehn, H. (1976) *Am. J. Hum. Genet.* 28, 577.
10. Bøyum, A. (1968) *Scand. J. Clin. Lab. Invest.* 21, Suppl. 97.
11. Krishan, A. (1975) *J. Cell Biol.* 66, 188.
12. Smith, L.B. (1970) *Commun. Assoc. Comp. Machin.* 13, 625.
13. Knott, G.D. and Reece, D.K. (1972) *Proc. Online Int. Conf.* 1, 497.
14. Mendelsohn, M.L., Mayall, B.H., Bogart, E., Moore, D.H., II. and Perry P.H. (1973) *Science* 179, 1126.
15. Müller, H.J., Klinger, H.P. and Glasser, M. (1975) *Cytogenet. Cell Genet.* 15, 239.
16. Rees, H. and Jones, R.N. (1971) *Internat. Rev. Cytol.* 50, 53.
17. Hoehn, H., Johnston, P. and Callis, J. (1977) in preparation.
18. Latt, S.A. and Wohlleb, J.C. (1975) *Chromosoma* 52, 297.
19. Büchner, Th., Hiddeman, W., Schneider, R. and Kamanabroo, D. (1974) *Blut* 28, 191.
20. Sehested, J. (1974) *Humangenetik* 21, 55.
21. Mayall, B.H., Carrano, A.V. and Moore, D.H., II. (1977) this volume.
22. Comings, D.E. (1972) *Adv. Hum. Genet.* 3, 237.
23. Miller, O.J. (1977) this volume.

NATURALLY OCCURRING DNA/RNA HYBRIDS. I. NORMAL PATTERNS IN POLYTENE CHROMOSOMES

G. T. Rudkin* and B. D. Stollar†

*The Institute for Cancer Research, Fox Chase Cancer Center, Fox Chase, Philadelphia, Pennsylvania 19111;
†Department of Biochemistry, Tufts School of Medicine, Boston, Massachusetts 02111

ABSTRACT. An antiserum specific for RNA-DNA hybrids was used to probe for such hybrid molecules in acetic acid fixed polytene chromosomes by means of an indirect immunofluorescence technique. Hybrids were detected in chromosomes of salivary glands of the _giant_ phenotype _Drosophila melanogaster_ during the last 4 days of larval development and the first 8 hr of prepupal development (25°C). The most conspicuous positive reactions were in puffs, such as at three prominent sites on the right arm of chromosome 3, although other sites on the third and on other chromosomes also gave a positive reaction at some time (or times) in development. Many puffed regions displayed no hybrids while some regions in which no puff was visible gave positive reactions, in some places apparently in bands, in others seemingly in interband regions. A unique distribution of hybrid sites appears to be characteristic of a developmental stage.

INTRODUCTION

The development of an antiserum directed against DNA-RNA hybrids with a high degree of specificity (1) and its application to the polytene chromosomes of _Drosophila melanogaster_ utilizing an indirect immunofluorescence technique (2) has enabled us to reveal not only that hybrids do exist on chromosomes but that the distribution of their locations undergoes developmental changes. Since the most conspicuous sites we have seen so far are ones at which puffs (sites of transcription) occur, it is likely that they represent transcripts annealed to templates. However, since not all puffs give a positive reaction for hybrids, and some sites with no morphological puff do give a positive reaction and, finally, the sites at which the hybrids are found vary with stage of development, it has not yet been possible to assign a function at any particular occurrence of indigenous hybrid.

MATERIALS AND METHODS

The anti-hybrid antibodies, raised in rabbits by the injection of (poly rA)•(poly dT) complexed to methylated serum albumin have been described previously (1, 3). The indirect immunofluorescence method flagged the hybrid sites with rhodamine-tagged IgG from serum of a goat immunized against rabbit IgG, as described in earlier publications (2, 4).

The chromosomes were prepared from salivary glands of *Drosophila melanogaster* larvae dissected in a balanced salt solution (gland medium of Cohen *et al.*, 5), fixed in 45% aqueous acetic acid, squashed on an acid-cleaned glass slide under a siliconed cover slip and frozen on dry ice. The cover slip was snapped off and the slide was fixed briefly in 3 parts ethanol: 1 part acetic acid (v/v), then stored in 95% absolute ethanol at 4°C until used. The ethanol was replaced by phosphate buffered saline (PBS = 0.14 M NaCl, .01 M phosphate, pH 7.1 ± 0.1) before the immunological procedures were carried out.

The indirect immunofluorescence reaction was carried out as described earlier (2, 4), first with anti-hybrid rabbit serum, then, after washing with PBS, with the rhodamine labeled goat anti-rabbit IgG. The slides were mounted in PBS-glycerine and examined with a Zeiss epiilluminated (Ploem system) fluorescence microscope. Photomicrographs were taken on 35 mm Tri-X-Pan film at ASA 1600 and developed in Diafine developer.

RESULTS

The specificity of the antiserum for DNA/RNA hybrids has been documented in earlier publications (1, 3, 2) with respect to standard immunological procedures. It was further tested in the cytological context by the controls listed in Table 1. The first three items show that the anti-hybrid serum is required in order to evoke fluorescence in the chromosomes and the last three items show that treatment designed to remove naturally occurring hybrids from the chromosomes (as for hybridization *in situ*) prevents the development of chromosomal fluorescence. The demonstration that the hybridization of a purified rRNA (5S) to the chromosomes results in fluorescence only at the site of the 5S gene cluster has been published (2). It should be noted that in our hands there is usually some non-specific fluorescence of extra nuclear cell debris that is not affected by the control treatments listed in Table 1.

TABLE 1

DEMONSTRATION OF THE CYTOLOGICAL SPECIFICITY OF THE ANTI-HYBRID RABBIT SERUM. (See ref. 2 for the conditions for the last three items).

Control Treatment	Chromosome Fluorescence
Normal rabbit serum	none
Omit anti-hybrid serum	none
Absorb anti-hybrid serum with (poly rA)•(poly dT)	none
RNase pretreatment	reduced
Denature (Formamide, low salt)	reduced
RNase, Denature, RNase	none

Developmental patterns of fluorescent sites were observed in larvae expressing the sex-linked *giant* (*gt*) phenotype. They were grown from timed eggs collected from a mass cross of *gt* w^a (w^a = apricot eye color) virgin females by males carrying a deficiency for the *gt* locus [$Df(1)62g^{18}$]. Since the penetrance of the *gt* allele is extremely low (less than 5%), the sons from such a cross are phenotypically *non-giant* while the daughters, heterozygous for the mutant and for the deficiency, are phenotypically *giant*. The larval period of *giant* larvae is twice that of their *non-giant* siblings (or of wild type larvae), namely 8 days as compared to 4 days at 25°C. Slides were prepared daily through the extended 4 days of the larval period and hourly through the first 8 hr of the prepupal period. We will restrict our attention to the right arm of chromosome 3 (chromosome 3R).

The morphology of the chromosomes changed little during the extension of the larval period except for an additional replication cycle that increased chromosome size. Fig. 1 is from a larva on the second day after the *non-giants* had pupated, illustrating the morphology of the chromosomes and showing the maximum number of fluorescent sites observed during the extended larval period. There is a fluorescent puff in section 93D and a less conspicuous one in section 88. There are fluorescent bands at places where no puff is obvious in sections 91 and 83. It should be noted that several prominent puffs on other chromosomes do not fluoresce at this stage. It is also clear that the fluorescence in the puff at 93D and also in the one at 88 is localized to a narrow region within the puffed segment.

At the time of puparium formation, we observed very pale to negative reactions in the chromosomes. That nearly

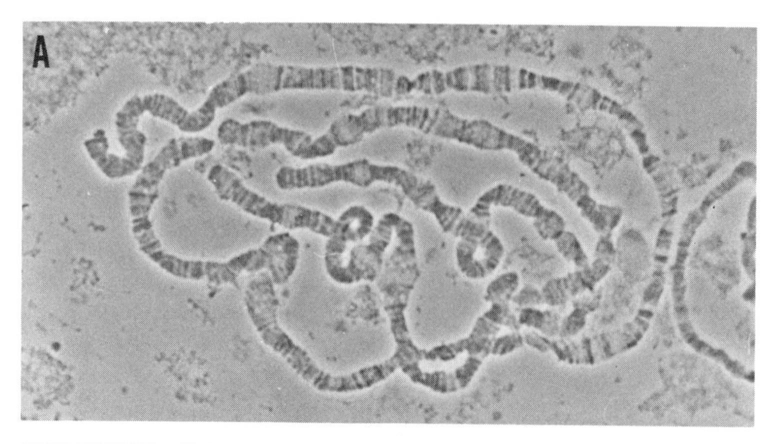

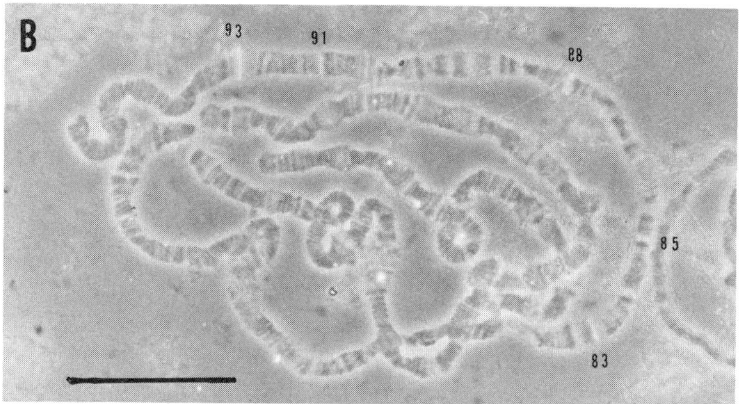

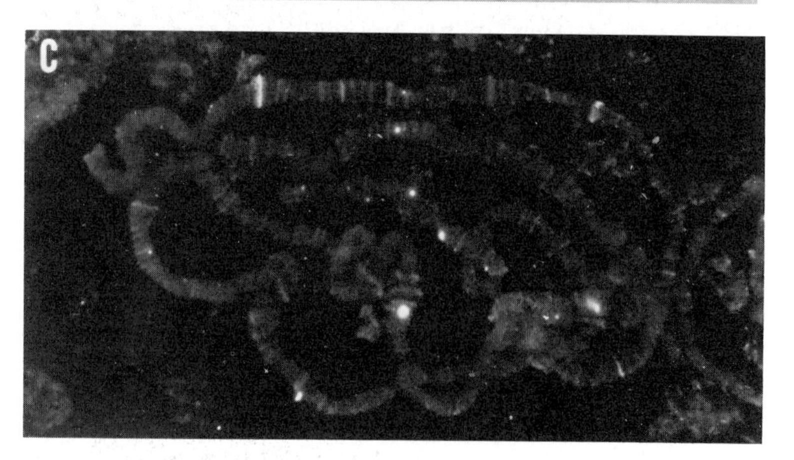

Fig. 1. Two days extension of larval period.

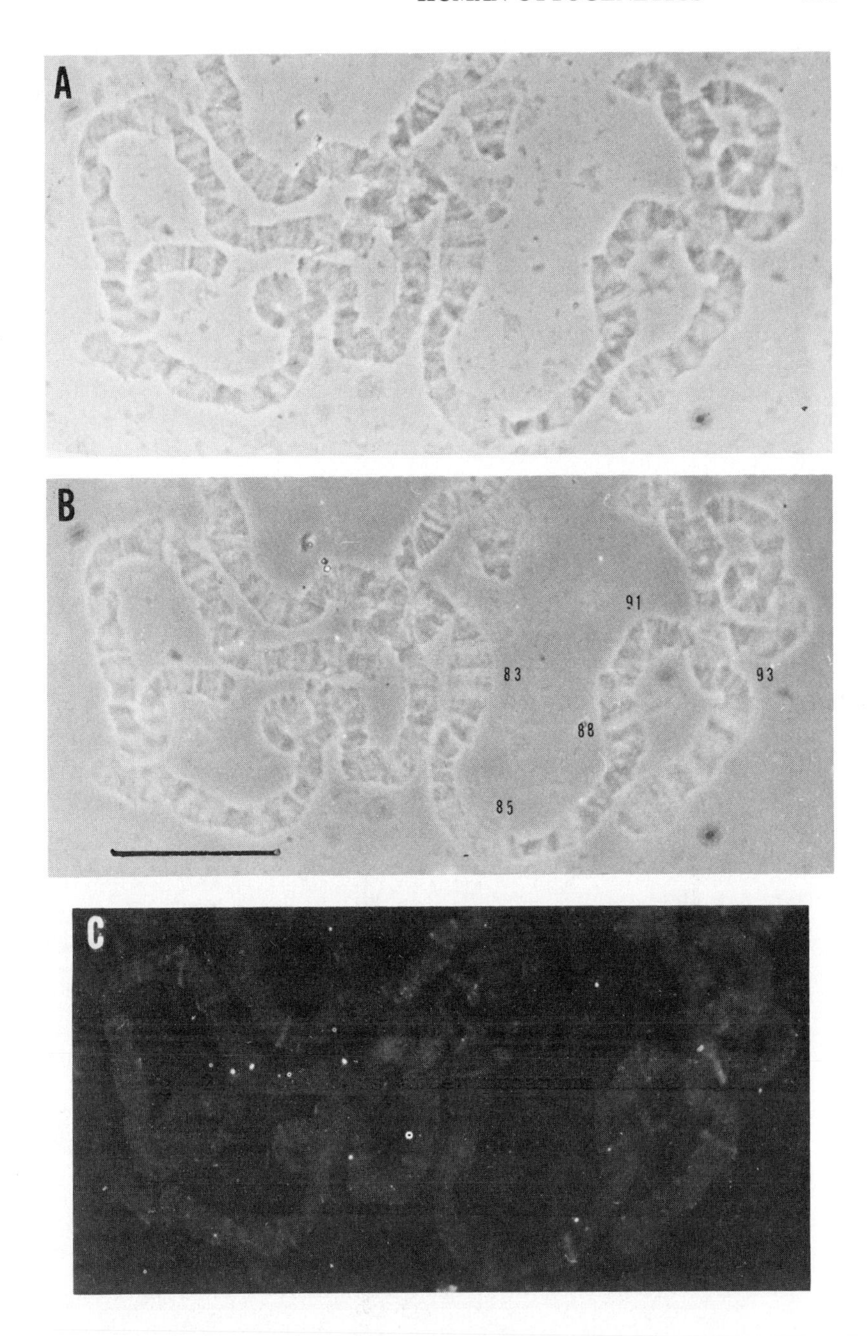

Fig. 2. Two hours after puparium formation.

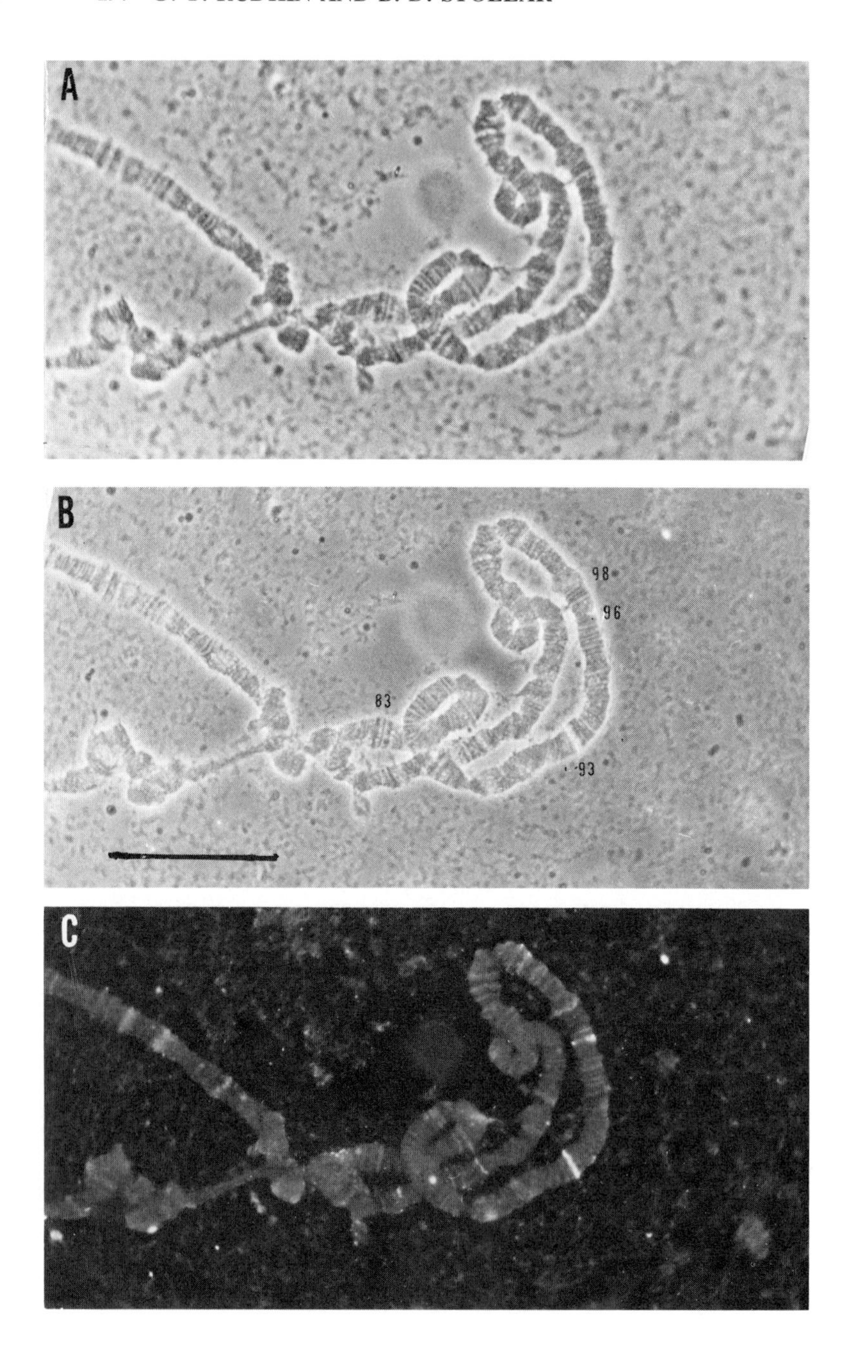

Fig. 3. Six hours after puparium formation.

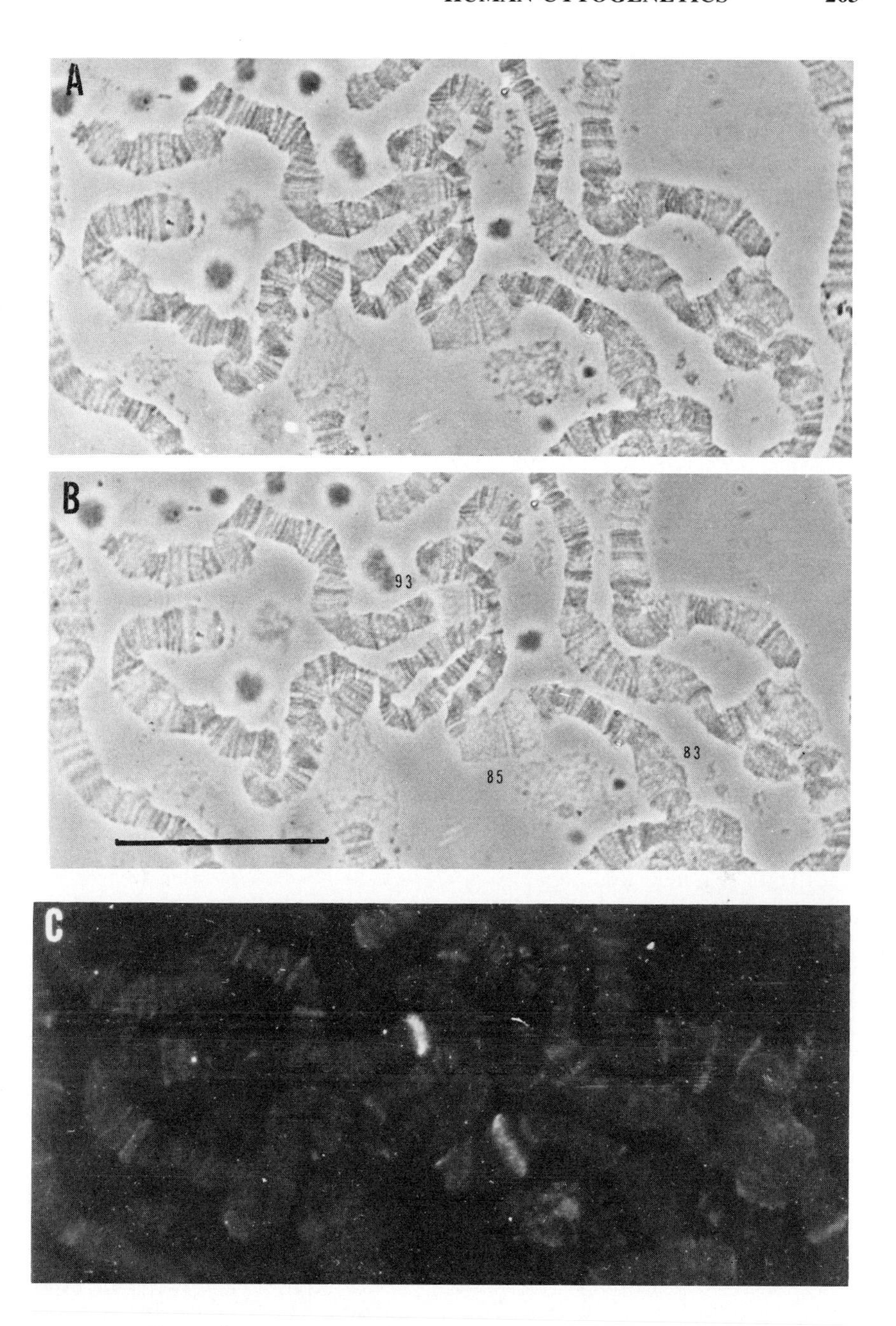

Fig. 4. Seven hours after puparium formation.

FIGURE LEGENDS

Figs. 1-4. Localization of rabbit anti-DNA/RNA-hybrid antibodies on chromosome 3R of polytene nuclei from Drosophila melanogaster salivary glands by detection with goat anti-rabbit-IgG fluorescent antibody. In each figure, the uppermost panel (A) is a photomicrograph of chromosomes taken with phase contrast optical conditions, the lowest panel (C) is a photomicrograph of the same field under fluorescence conditions, and the center panel (B) is taken with both phase contrast and fluorescence conditions on simultaneously. Zeiss 40X phase contrast objective, scale lines = 50 μ.

Fig. 1. Two days after non-giant larvae formed puparia. The tip of chromosome 3R is at the upper left, the chromocentric end at the lower right. The locations of sections 93, 91, 88, 85 and 83 are indicated. Each has a conspicuous fluorescent band which lies in a clearly puffed site (93, 88) or at a site of low density in the phase contrast image. Small round brightly fluorescent spots are artifactual. Puffed sites visible on other chromosomes are either non-fluorescent or very weakly so.

Fig. 2. Two hours after puparium formation. Chromosome 3R lies to the right in the figure and is crossed by 3L just proximal to the fluorescent band in section 93. The fluorescence is paler than in Fig. 1 but is clearly visible in sections 93 and 86.

Fig. 3. Six hours after puparium formation. Chromosome 3R lies to the right and is crossed by the X chromosome approximately in section 88. Even though the puff has nearly regressed in 93, there is a very bright band at the site. There are many more clearly fluorescent sites in this nucleus than in nuclei from younger prepupae. A case of "ectopic pairing" between chromosome 3R and the X chromosome is clearly fluorescent on the 3R side but not on the X.

Fig. 4. Seven hours after puparium formation. The tip of chromosome 3R is at the upper left, its chromocentral end at the lower right and a segment near the middle from about 94 to 86 is unpaired. Bright fluorescence remains in 93 and is also prominent in a puff in section 85. Note, however, that the proximal region including 82 and 83 is quite dark.

negative reaction persisted into the second hour after puparium formation, as shown in Fig. 2. By the fourth hour, strong fluorescence again appeared in the puff at 93D and at other sites along chromosome 3R. Fig. 3 shows a typical reaction 6 hr after puparium formation in which it is clear that 93D shows a very bright reaction in spite of the fact that its puff has regressed considerably in size: the fluorescent band appears to be again narrower than the ones in the earlier, more fully developed puff states. Sections 96 and 98 have conspicuous positive reactions and others may be seen along the chromosome. However, bands that fluoresced brightly near the base of the chromosome arm (in sections 81-83) in earlier larvae are not bright at this stage.

Finally, Fig. 4 displays a nucleus from a prepupa 1 hr older (7 hr post puparium formation). 93D and 85D are conspicuously bright, sections 81-83 continue to be negative and a very large number of other sites are clearly fluorescent.

A summary of preliminary results of the series is given in the matrix of Table 2. A dash indicates no significant fluorescence; "p" indicates that fluorescence was present in a puff in the appropriate chromosome section; "b" signifies band, "i", interband, in each case fluorescent; "bi" indicates ambiguity in the nature of the location of the fluorescence; + and ± indicate weak and marginal fluorescence, respectively, and parentheses (p) indicate slight enlargement that suggests puffing at a fluorescent place.

DISCUSSION

The controls leave little doubt concerning the specificity of the immunofluorescence reagents employed in these experiments. We are convinced that chromosomal fluorescence indicates the presence of DNA/RNA hybrid molecules. There is, however, room for a wide-ranging discussion concerning the significance of the hybrids we detect. But before addressing these interesting questions which concern function, it will be necessary to consider some "mechanical" issues concerning the origin of the hybrids.

The use of fixed material raises the question of fixation artifact, and it is necessary to ask whether the hybrids could have been annealed as a consequence of fixation. That seems unlikely since the fixation conditions used do not favor denaturation of chromosomal DNA and since fluorescent localizations are observed after prefixation with formaldehyde (as in ref. 2) which should preserve close molecular associations essentially in their *in vivo* state. It

TABLE 2

RNA-DNA HYBRIDS IN CHROMOSOME 3R *giant* *D. melanogaster*

Larval Extension (Days)	Chromosome Section										
	82	83	85D	85EF	86	90	93	94	95	96	97 98
0	–	–	–	p	–	–	p	p	–	–	–
1	–	–	–	p	–	–	–	–	–	–	–
2	p	bi	–	p	–	p	p	p	–	–	–
3	–	–	–	–	–	–	±	–	–	–	–
4	–	–	–	–	–	–	p	–	–	+	+
White Prepupa											
+ 0 hrs	–	–	–	–	–	–	(p)	–	–	–	–
+ 2 hrs	–	–	–	p	–	–	+	–	±	–	–
+ 3 hrs	–	–	p	p	–	p	bi	(p)	–	–	±
+ 4 hrs	–	–	–	–	–	–	+	–	±	±	–
+ 5 hrs	b	–	p	p	–	–	bi	–	–	–	–
+ 6 hrs	i	–	–	–	–	–	p	–	–	p	±p
+ 7 hrs	–	–	p	p	–	–	b/p	–	–	–	–
+ 8 hrs	–	–	–	–	+	–	–	±p	–	–	–

Preliminary localization of prominent sites of hybrid RNA/DNA molecules in acetic acid fixed chromosomes 3R of *Drosophila melanogaster*. Although specified only to section of chromosome, the hybrids were localized in the same chromosome bands within that section at the different stages. See text for explanation of the time dimension in the leftmost column.

could also be asked whether those hybrids detected are simply the least easily extracted by the fixative, whether we are really detecting places where blocking molecules (*e.g.*, protein) are removed or moved by the fixative, thus rendering the antigen accessible to the antibodies, or, finally, whether we are, for whatever reason, seeing all of the naturally occurring hybrids detectable by our probe. For the most part we have as yet little or no direct evidence bearing on these questions. The existence of the chromosomal hybrids has been established and we have yet to define precisely their distribution and functions.

The hybrids at puffed sites and at chromosome sites that appear to be interbands could be interpreted in two ways. On the one hand, they could consist of transcripts annealed to their templates, for which there is some precedent in studies of in vitro transcription (e.g., ref. 6). On the other hand, they could be "control RNA" base paired to the non-transcribed strands of active genes in the manner of Paigen (7) or Frenster (8) or Britten and Davidson (9) or Davidson et al. (10). The possibility that both phenomena exist cannot be excluded. It is clear that most places where very bright fluorescence is observed are regions where transcription has occurred in salivary gland nuclei, was occurring at the time of fixation, or would occur at a later time of development. Bright fluorescence implies the binding of much labeled antibody (and the absence of quenching) which suggests long stretches of exposed hybrid nucleic acid which, in turn, is most easily conceived in terms of transcription: control molecules would not have to be very long, whence their hybrids would bind a correspondingly small amount of antibody.

An alternative explanation for the occurrence of bright sites is the absence of a blocking agent such as a protein that otherwise would prevent antibody binding. Such agents could be thought of as proteins which are either released from sites where transcription is occurring or are so altered in their affinity for chromatin that they are easily removed by the fixative. Whatever mechanism is postulated, the patterns of hybrid formation at different genetic loci must reflect changes in the state, and presumably in the activity, of the genetic material.

The correlation between the occurrence of hybrids and transcription is not a firm one. When ^{3}H uridine is offered to salivary glands in vivo or in vitro, every puff site becomes radioactive, indicating that transcription continues whenever a morphological puff exists. However, only a few specific puffs of those in the nuclei of any one salivary gland have been observed to exhibit the bright fluorescence that indicates the presence of abundant hybrid molecules. It is premature to indulge in speculations about how that could come about until more information is available from experiments in which all of the pertinent factors are controlled at the same time. An obvious possibility is that transcripts are removed from non-fluorescent puff sites with high efficiency either in vivo or in the course of tissue preparation.

A regulatory role for RNA has also been proposed with respect to DNA replication (see review in ref. 11): short stretches of RNA hybridized to initiation sites on DNA act as primer. In view of the multitude of replicating units in polytene chromosomes (see 12 for review), there should be a very wide distribution of such sites. It is possible that some of the "background" chromosomal fluorescence arose from that source.

Attention has been given to RNA that appears tightly associated with chromatin, often called cRNA (for chromatin- or chromosomal-RNA), in several laboratories with a variety of results forthcoming. Some recent references are 13, from Bonner's group, 14, from Tanaka, and 15, from Infante's group. The relationship between their cRNA's and the hybrids reported here can only be guessed at the present time.

The thesis that RNA is involved in the control of chromosomal functions has received a boost from studies on eukaryote enzymes of the type called ribonuclease H which break down the RNA moiety of hybrid molecules (16, 17, 18, 19). Their chromosomal distribution should be detectable by immunofluorescence methods of the type used here and used by others to follow the distribution of RNA polymerases (20, 21). It is probably not coincidental that we have observed hybrid molecules at or near many of the sites that are activated by heat shock (see 22) and at which RNA polymerase II is said to be mobilized after a heat shock (21). The extensive distribution of mixed polymerases A and B throughout the genome (20) and polymerase B in interbands and puffs (21) in non-shocked larvae appears not to be directly related to the patterns observed here. A continuation of the studies described here should lead to an understanding of some of the functions of chromosomal RNA.

ACKNOWLEDGMENTS

We wish to thank Dr. C. R. Alfageme for the fluorescent antibody, Dr. L. Cohen for the use of a Zeiss epi-illuminator fluorescence attachment, and both of them for profitable discussions; and Ms. D. J. Hazler for technical assistance. This work was supported in part by grants from the National Science Foundation, the National Institutes of Health, and by an appropriation from the Commonwealth of Pennsylvania.

REFERENCES

1. Stollar, B.D. (1970) Science 169, 609.
2. Rudkin, G.T., and Stollar, B.D. (1977) Nature 265, 472.
3. Colby, C., Stollar, B.D., and Simon, M.I. (1971) Nature New Biol. 229, 172.
4. Alfageme, C.R., Rudkin, G.T., and Cohen, L.H. (1976) Proc. Nat. Acad. Sci. USA 73, 2038.

5. Cohen, L.H., and Gotchel, B.V. (1971) J. Biol. Chem. 246, 1841.
6. Howk, R.S., Williams, D.R., Haberman, A.B., Parks, W.P., and Scolnick, E.M. (1974) Cell 3, 15.
7. Paigen, K. (1962) J. Theoret. Biol. 3, 268.
8. Frenster, J.H. (1965) Nature 206, 4990.
9. Britten, R.J., and Davidson, E.H. (1969) Science 165, 349.
10. Davidson, E.H., Klein, W.H., and Britten, R.J. (1977) Devel. Biol. 55, 69.
11. Chargaff, E. (1976) in Prog. Nucl. Acid Res. Mol. Biol., Cohn, W.E., ed., Academic Press, New York, Vol. 16, 1.
12. Rudkin, G.T. (1972) in Results and Problems in Cell Differentiation, Beermann, W., ed., Springer, Berlin, Vol. 4, 58.
13. Sivolap, Y.M., and Bonner, J. (1971) Proc. Nat. Acad. Sci. USA 68, 387.
14. Tanaka, T., and Kanehisa, T. (1972) J. Biochem. 72, 1273.
15. Alfageme, C.R., and Infante, A.A. (1975) Exptl. Cell Res. 96, 263.
16. Stein, H., and Hausen, P. (1969) Science 166, 393.
17. Wyers, F., Huet, J., Sentenac, A., and Fromageot, P. (1976) Eur. J. Biochem. 69, 385.
18. Berkower, I., Leis, J., and Hurwitz, J. (1973) J. Biol. Chem. 248, 5914.
19. Büsen, W., Peters, J.H., and Hausen, P. (1977) Eur. J. Biochem. 74, 203.
20. Greenleaf, A.L., Plagens, U., and Bautz, E.K.F. (1976) in Molecular Mechanisms in the Control of Gene Expression, Nierlich, D.P., Rutter, W.J., and Fox, C.F., eds., Academic Press, New York, 249.
21. Greenleaf, A.L., Jamrich, M., and Bautz, E.K.F. (1977) J. Supramol. Struct., Suppl. 1, 277 (Abst.) [and these symposia].
22. Bonner, J.J., and Pardue, M.L. (1976) Chromosoma (Berl.) 58, 87.

HYBRIDIZATION IN SITU OF ^{125}I-cRNA TRANSCRIBED FROM SORTED METAPHASE CHROMOSOMES

Virginia L. Sawin,* Neal Scherberg,* and Anthony Carrano†

ABSTRACT. In this report, preliminary data is presented on the use of sorted chromosomes in the analysis of sequence homologies among the chromosomes of a single species. The fourteen morphologically distinct chromosomes of the Chinese hamster cell line M3-1 can be resolved by a flow microfluorimeter into nine groups (peaks) on the basis of DNA content. The chromosomes of a single peak can be separated from the remainder by an electronic cell sorter. For the present experiments, the DNA of 10^6 Peak A chromosomes (95% pure Chinese hamster chromosome #1) was denatured and purified by gradient sedimentation. ^{125}I-cRNA was transcribed from Peak A chromosome DNA or from whole Chinese hamster or *E. coli* DNA in an RNA polymerase reaction mixture containing carrier-free ^{125}I-CTP. Each cRNA was hybridized in situ to Chinese hamster cells, and its chromosomal location was determined by autoradiography.

The transcript from *E. coli* DNA showed no hybridization to Chinese hamster chromosomes, whereas that from Chinese hamster DNA hybridized extensively over the entire genome. When the DNA from Chinese hamster Peak A chromosomes was used as template, the ^{125}I-cRNA again hybridized to numerous sites throughout the Chinese hamster genome. Values of C_rt for the hybridization reactions were between 10^{-2} and 4×10^{-2} mole sec l^{-1} suggesting that the sites of hybridization represent highly reiterated sequences common to all the chromosomes.

INTRODUCTION

Various approaches have been taken in the search for information concerning the organization of eukaryotic chromosomes. One of these approaches at a molecular level involves the application of radiolabeled DNA or RNA to hybridization

*Department of Medicine and The Franklin McLean Memorial Research Institute, The University of Chicago, Chicago, Illinois 60637.
†Biomedical Research, Lawrence Livermore Laboratory, Livermore, California 94550.

studies in cytological preparations. Hybridization of specific DNA or RNA sequences to metaphase chromosomes immobilized on a glass slide, followed by autoradiography, has proved to be a useful technique for mapping the location of particular regions in the genome. The method has been used for the demonstration of fine structure, that is, specific gene localization. Gall and Pardue (5) and Jones (8) identified the location of the satellite DNA in the laboratory mouse. This highly repetitive fraction was found to be distributed in the heterochromatic regions near all the centromeres except in the Y chromosome. More recently, the technique has been used in mapping of the genes for various RNA fractions, i.e., 5S and ribosomal RNAs, in species ranging from Drosophila to man (4, 13, 15, 17).

Recent technological advances have allowed the separation of metaphase chromosomes from certain species on the basis of DNA content (3, 7). In this report we present preliminary data on the use of sorted chromosomes in the analysis of sequence homologies among the chromosomes of a single species. Procedures are described for the purification of the DNA from sorted chromosomes, and for the subsequent use of the DNA as template to produce a radiolabeled cRNA of high specific activity. Such cRNA was hybridized in situ to fixed metaphase chromosomes, and the sites of hybridization were identified by autoradiography. The procedures discussed here will be applied to mapping of the sites of hybridization of highly repetitive and moderately repetitive sequences transcribed from the #1 chromosome of Chinese hamster cells. Further development of the techniques may allow the identification and mapping of chromosomal rearrangements, such as translocations.

METHODS

Enzymes and nucleic acids. *E. coli* RNA polymerase (588 units/mg) was obtained from Sigma Chemical Company. *E. coli* DNA obtained from Sigma was further purified by banding in cesium sulfate, followed by dialysis. The DNA was heated at 100° C for 4 min prior to its use as template. Chinese hamster DNA was prepared from nuclei by standard procedures (10). Yeast RNA (Koch-Light Laboratories, Ltd.) was extracted with phenol twice before being used as carrier nucleic acid.

Isotopes and nucleotides. Iodine-125 (IMS-30, 80% free of stable iodine) was purchased from Amersham Corp. Nucleoside triphosphates were obtained from Pabst Laboratory. Trace contamination of ATP with CTP was checked by radioiodination of an amount of nucleotide 10 times the usual level in the reaction for the preparation of ^{125}I-CTP. The products of radioiodination were partially repurified and tested in reactions with *E. coli* RNA polymerase, DNA, and various nucleotide mixtures. The isotope was incorporated into RNA at 3% of the rate of an equal amount of ^{125}I-CTP. The polymerization was competitively inhibited by addition of ATP, but not CTP. The net upper limit of CTP contamination of the ATP was less than one part in 10^4.

Sorting of chromosomes. The procedures for preparation and sorting of Chinese hamster metaphase chromosomes have been reported previously (7). The chromosomes were prepared from the cell line M3-1 and sorted at the Lawrence Livermore Laboratory. The chromosomes of each peak were inspected by light microscopy after sorting. Peak A was found to consist of Chinese hamster chromosome #1 at 95% purity. The contaminant (5%) was chromosome #2. Quantities of 10^6 chromosomes were stored and shipped from Livermore frozen in PIPES buffer.

Purification of DNA. Approximately 10^6 sorted chromosomes were obtained and collected by centrifugation. They were lyophylized and redissolved in 0.4 ml of a solution containing 8 M guanidinium chloride, 1.0% sodium lauryl sarcosine and 0.02 M EDTA, pH 7.5. This mixture was heated at 80° for 3 min and layered directly on a cesium sulfate step gradient (1 ml 96% Cs_2SO_4; 3.3 ml 40% Cs_2SO_4). Centrifugation was at 35,000 rpm for 20 hours at 25° C in a Spinco SW 50. Fractions containing DNA were determined by the location of calf thymus marker DNA run in a parallel gradient. The DNA containing fractions were combined and dialyzed first against molar sodium chloride, then against water, and finally lyophylized to dryness. The DNA was dissolved in 100 λ of water prior to its use as a template in the transcription reaction.

Transcription of ^{125}I-cRNA. The CTP was iodinated in reaction mixtures containing 2.5 mCi ^{125}I and 23 nmol CTP by the procedure of Scherberg and Refetoff (14). The ^{125}I-CTP was stored at -20° in 50% ethanol. Prepared and stored in this manner, ^{125}I-CTP has been found to remain better than 95% undegraded after two months. Its specific activity, based on that of the ^{125}I, was estimated as 1760 Ci/mM.

The RNA polymerase reaction mixture contained 10 mM Tris-HCl, pH 8.0; 100 mM KCl; 5 mM $MgCl_2$; 1 mM $MnCl_2$; 0.5 mM each of ATP, GTP, and UTP; 1 mM dithiothreitol; 100 μCi ^{125}I-CTP; template DNA; and 1.7 μg *E. coli* RNA polymerase (1 unit) in a final volume of 0.05 ml. The reaction mixture was incubated 2 hrs at 30° and the reaction was terminated by the addition of 1 ml water and heating for 2 min at 100°.

The ^{125}I-cRNA was separated by adsorption to an 0.5 ml column of DEAE cellulose, followed by sequential elution with 0.5 ml portions of water, buffer A (0.125 M NaCl, 0.025 M sodium acetate, pH 5.0), and 4X buffer A. The ^{125}I-cRNA eluted by the 4X buffer A was mixed with 100 μg of carrier yeast RNA and precipitated by the addition of 2 volumes of ethanol. It was redissolved in 50% formamide/50% 4X SSC (SSC = 0.15 M NaCl, 0.015 M sodium citrate, pH 6.8) for use in hybridization reactions. The specific activity of a ^{125}I-cRNA containing 25% cytidine transcribed under these conditions would be 3 x 10^9 dpm/μg RNA (1.35 mCi/μg RNA).

Hybridization in situ. Metaphase spreads were prepared from log phase cultures of the Chinese hamster cell line V79, clone AL 162 (courtesy of Dr. A. Han, Argonne National Laboratory). The slides were processed by the procedure described by Pardue and Gall (12) for hybridization in situ and the chromosomes were denatured by treatment in 70% formamide/2X SSC at 70° for 10 min (Dr. Dale Steffensen, personal communication). Seven λ of ^{125}I-cRNA were added to each slide, spread with a coverslip and sealed with rubber cement. The slides were incubated at 40° C for 18 hours. Non-hybridized RNA was removed by treatment with RNAase and extensive washes in 2X SSC.

Slides were dipped in Ilford L4 emulsion diluted 1:1 with 2% glycerol and stored at 4° for 2 weeks (2). They were developed in Kodak D-19 developer and stained with Leishman-Giemsa for cytogenetic analysis and photography.

RESULTS

After purification, aliquots of the DNA from the sorted chromosomes were tested in a polymerase reaction mixture containing 1 μCi of ^{125}I-CTP. The reaction mixture was incubated for 30 min, and the amount of transcript was determined as the percent of acid precipitable counts (Table 1). In the reaction mixture containing no template DNA, 0.42% of the label was acid precipitable. When 0.0042 μg of Chinese hamster DNA was present, 20.7% of the label was incorporated into acid precipitable polymers. The purified DNA from 10^4 Peak A chromosomes incorporated 9.4% of the label and that from 2 x 10^4 Peak B chromosomes (chromosome #2, 90% purity) incorporated 7.1% into an acid precipitable product.

Reactions were then carried out in which ^{125}I-cRNA was transcribed for hybridization in situ (Table 2). Ten micrograms of *E. coli* DNA resulted in the incorporation of 5.1 x 10^7 cpm of label into cRNA (0.173 μg); 1.7 μg whole Chinese hamster DNA incorporated 3.2 x 10^7 cpm (0.0106 μg

TABLE 1

TEMPLATE FUNCTION OF DNA FROM SORTED METAPHASE CHROMOSOMES*

Template Data	^{125}I-cRNA (% cpm Acid Precipitated)
---	1.3
4.2×10^{-3} μg Chinese hamster	20.7
10^{4} Peak A chromosome	9.4
2×10^{4} Peak B chromosome	7.1

*Each mixture contained 1 μCi of ^{125}I-CTP. Peak A contained Chinese hamster chromosome #1, 95% pure, and Peak B contained chromosome #2, 90% pure.

TABLE 2

TRANSCRIPTION OF ^{125}I-cRNA FROM VARIOUS TEMPLATE DNAs*

Template DNA		^{125}I-cRNA		C_rt
Species	μg	cpm	μg	mole sec l^{-1}
E. coli	10	5.1×10^{7}	0.0173	4×10^{-2}
Chinese hamster	1.7	3.2×10^{7}	0.0106	2.4×10^{-2}
10^{5} Peak A chromosomes	0.053	1.3×10^{7}	0.0041	0.95×10^{-2}

*Each reaction mixture contained 0.1 mCi of ^{125}I-CTP. The mixtures were incubated for 2 hrs at 30° C, and the transcribed RNA was purified. The amount of RNA transcribed in each reaction was calculated from the specific activity of 3×10^{9} dpm/ug. This was used in the calculation of the C_rt values for the hybridization in situ of each cRNA.

[125]I-cRNA); and the DNA of 10^5 Peak A chromosomes (0.053 μg) resulted in the transcription of 1.3×10^7 cpm of ^{125}I-cRNA, which was approximately equal to 0.0041 μg of RNA. The micrograms of RNA produced were calculated from the counts incorporated when the specific activity of the cRNA was assumed to be 3×10^9 dpm/μg. These cRNAs were dissolved in 70 λ of formamide/SSC and hybridized to metaphase cells for 18 hours. The C_rt values for each reaction were calculated on the basis of specific activity of 3×10^9 dpm/μg for the transcript RNA. In mole sec l^{-1}, these values were 4×10^{-2} for the *E. coli* cRNA, 2.45×10^{-2} for the Chinese hamster cRNA, and 0.95×10^{-2} for the Peak A chromosome DNA.

After exposure of the photoemulsion for 2 weeks, the *E. coli* ^{125}I-cRNA did not produce any grains over the Chinese hamster chromosomes (Fig. 1, upper portion), whereas the Chinese hamster ^{125}I-cRNA hybridized extensively over the entire genome (Fig. 1, lower portion). Peak A ^{125}I-cRNA hybridized to the #1 chromosome as well as to numerous sites on all the other chromosomes (Fig. 2). Under the conditions employed, hybridization was nonselective for specific chromosomes.

DISCUSSION

Use of the flow sorter is potentially a more effective method for purification of specific chromosomes than previous procedures which depend upon differential sedimentation. Metaphase chromosomes stained and sorted in a flow microfluorimeter have been reported to be morphologically normal (7). It is evident from the results of this study that the sorting process does not alter the DNA in such a way as to inactivate its capacity to function as a template in the transcription of RNA in vitro. It was found that after separation of the DNA by density gradient sedimentation, significant amounts of RNA could be transcribed from the DNA recovered from the equivalent of as few as 10^4 sorted chromosomes.

When ^{125}I-cRNA was transcribed from sorted Peak A chromosomes and hybridized to Chinese hamster metaphase cells, it was found to hybridize to sites distributed throughout the entire genome. Transcripts from *E. coli* DNA did not bind to Chinese hamster chromosomes, although the *E. coli* cRNA was present in the hybridization mixture at approximately 1.5 times the concentration of the Chinese hamster transcript and at 4 times that of Peak A chromosome DNA.

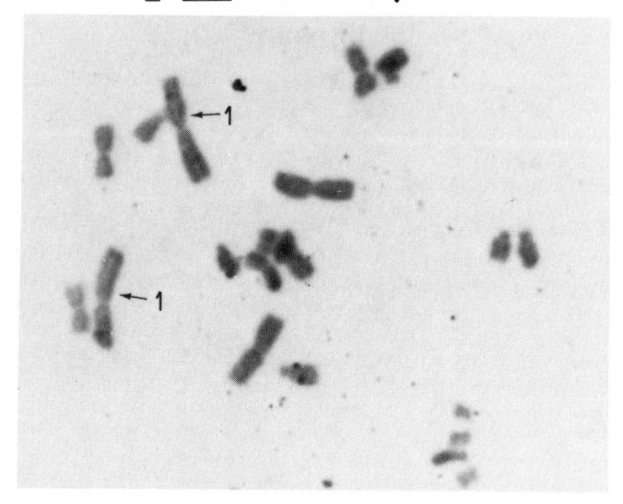

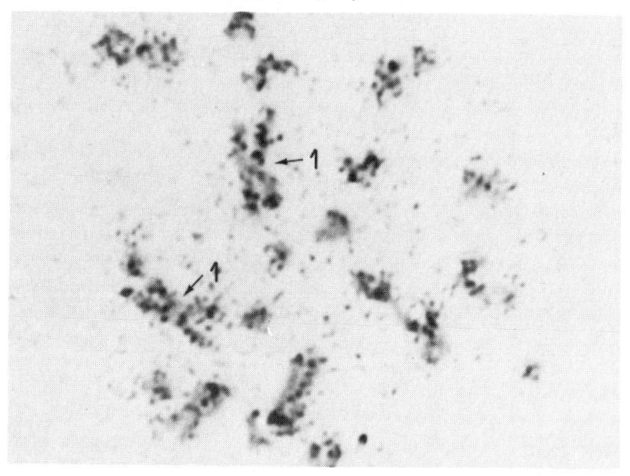

Fig. 1. Hybridization of ^{125}I-cRNA to metaphase cells of the Chinese hamster line V79. Arrows mark the #1 chromosomes in each metaphase. Exposure time was 2 weeks.

Upper: The ^{125}I-cRNA was transcribed from *E. coli* DNA.

Lower: The ^{125}I-cRNA was transcribed from whole Chinese hamster DNA.

Peak A chromosome DNA template

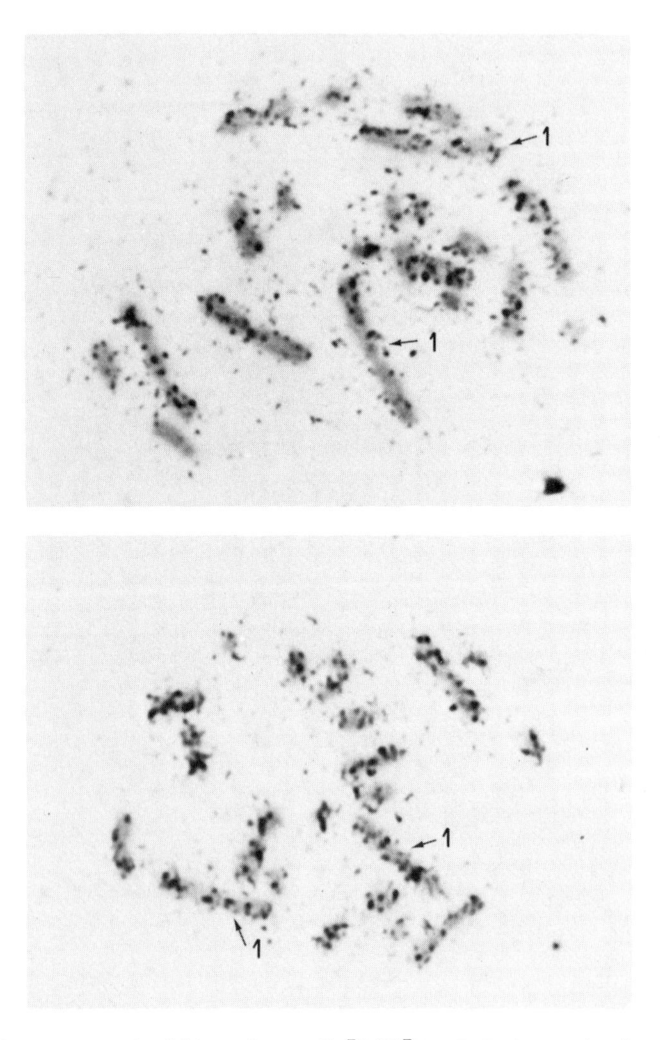

Fig. 2. Hybridization of ^{125}I-cRNA to metaphase cells of the Chinese hamster line V79. Arrows mark the #1 chromosomes of each metaphase. Exposure time was 2 weeks.

The ^{125}I-cRNA was transcribed from the DNA of sorted Peak A chromosomes.

This widespread annealing of Peak A ^{125}I-cRNA to sites throughout the genome might be an artifact of the techniques employed. Such artifacts could result from (1) the purity of the sorted chromosomes, (2) the chromosomal homology of the cell lines, or (3) the nature of the transcript RNA.

First, the sorted Peak A chromosomes consist of 95% pure chromosome #1. If the 5% impurity represents pieces of all the remaining chromosomes, hybridization to sites on other chromosomes might be expected. However, the 5% contaminant consisted of chromosome #2, and one would expect the sites of hybridization to be located on only these two chromosomes if the hybridization is chromosome specific.

Second, two different cell lines were used in this study; M3-1 was the source of the sorted chromosomes, and V79 was used as the source of metaphase cells for in situ hybridization. Both had been cultured for several years and chromosomal rearrangements had occurred (18). It is possible that the #1 chromosome from M3-1 contains translocations from many other chromosomes, or that the #1 of V79 had become translocated to numerous sites. Either of these series of events would account for the widespread hybridization of Peak A cRNA obtained in this study. Some minor rearrangement of chromosome #1 of one or both cell lines may have occurred, but it is unlikely that extensive rearrangement and exchanges had taken place. Two morphologically normal #1 chromosomes were present in each cell line, and the extensive rearrangements needed to explain the results would have been evident.

Third, the transcript RNA might not hybridize specifically. A factor which can affect specific pairing is the size of the polynucleotides. Sequences of the order of 6 to 10 nucleotides in length might hybridize randomly on a statistical basis. The low concentration of CTP (10^{-7} M) which we used was the rate-limiting factor and may have caused the polymerase to transcribe only very short lengths of cRNA. The sizes of the RNA transcripts were determined by polyacrylamide gel electrophoresis of the cRNA with *E. coli* f-met tRNA as a marker. It was found that 60% of the cRNAs recovered after preparation of the product for hybridization had lengths equal to or greater than that of the marker.

Although the chains described were of a satisfactory length, they may have been degraded rapidly, since the disintegration of ^{125}I is reported to cause multihit breakage of nucleic acid polymers (9, 11). In a test of this possibility, RNA was transcribed and stored at 4° C for 11 days. The chain length was again determined by electrophoresis; only a slight decrease was noted which did not exceed the extent expected from single-hit kinetics. It was also determined that RNA, transcribed under the experimental conditions, hybridized in liquid conditions specifically with its template species DNA (Sawin and Scherberg, unpublished results). These experiments demonstrated that the extensive hybridization of Peak A was not an artifact due to the size of the transcript RNA.

The most likely explanation is that the sites of hybridization represent highly reiterated sequences which are common to all of the chromosomes. The C_rt values for the hybridizations in situ in this study were between 10^{-2} and 4×10^{-2} mole sec l^{-1}. A study by Szabo et al. (16) on the kinetics of hybridizations in situ demonstrated that the reaction was consistent with the expected first order rate kinetics. The values of the rate constants were very similar to those obtained from analogous filter disc DNA/RNA hybridization experiments. At the concentrations used here, only the highly repetitive and moderately repetitive sequences would be expected to hybridize. The numerous sites of hybridization of the cRNA from a single chromosome may represent the loci of repetitive sequences common to all the chromosomes.

These results are comparable to those formerly reported by Arrighi (1) on the hybridization of total human repetitive DNA to numerous sites throughout the human genome. Hybridization appeared to be localized in the centromeric and telomeric regions, but did not appear to be restricted to specific chromosomes. Gosden et al. (6), however, separated human repetitive DNA into four satellite DNA fractions; each satellite could be shown to hybridize to specific sites on the chromosomes.

There are two possible approaches to the problem of obtaining a fraction that will be specific for its chromosomal template. First, the highly repetitive sequences can be removed by annealing of the Peak A cRNA to whole cell DNA, or to the DNA of all the sorted chromosomes except that of Peak A. Those sequences remaining should be chromosome specific.

A second approach is to subfractionate the Peak A chromosomal DNA (or its cRNA), on the basis of hybridization kinetics, into several highly repetitive fractions. By hybridizing each of these fractions separately to metaphase chromosomes, one may obtain a preparation that is specific for its template chromosome, or for a few sites only. An alternate method of isolating very highly repetitive sequences would be to prepare transcripts from each sorted group and to hybridize each separately to cells at extremely low C_rt values. In this procedure only the extremely repetitive sequences would be permitted to anneal, and it is possible that only a few sites of hybridization would be found for each chromosome. When a fraction specific for a single chromosome has been obtained, it will be possible to identify rearrangements of that chromosome in other cell lines. This would provide a molecular basis for the identification of translocations.

ACKNOWLEDGMENTS

We want to thank Ms. K. Burkhardt-Shultz and Mr. T. Merril for excellent technical assistance. We are particularly grateful to Dr. Janet Rowley for her continued support and advice throughout this investigation.

This work was supported by grants from the Leukemia Research Foundation, U.S.P.H. Training grant 05250, and A.C.S. Institutional grant 3016. A portion of the work was performed under the auspices of the U.S. Energy Research and Development Administration, Contract No. W 7405 ENG 48. The Franklin McLean Memorial Research Institute is operated by the University of Chicago for the U.S. Energy Research and Development Administration under Contract No. EY-76-C-02-0069.

REFERENCES

1. Arrighi, F.E., Saunders, P.P., Saunders, G.F. and Hsu, T.C. (1971) *Experientia* 27, 964.
2. Bogorich, R. (1972) *Autoradiography for Biologists*, Academic Press, N.Y., p. 63.
3. Carrano, A.V., Gray, J.W., Moore II, D.H., Minkler, J.L., Mayall, B.H., VanDilla, M.A. and Mendelsohn, M.L. (1976) *J. Histochem. Cytochem.* 24, 348.
4. Evans, H.J., Buckland, R.A. and Pardue, M.L. (1974) *Chromosoma* 48, 405.
5. Gall, J.G. and Pardue, M.L. (1975) *Methods in Enzymology* 21, 470.
6. Gosden, J.R., Buckland, R.A., Clayton, R.P. and Evans, H.J. (1975) *Exp. Cell Res.* 92, 148.
7. Gray, J.W., Carrano, A.V., Steinmetz, L.L., VanDilla, M.A., Moore II, D.H., Mayall, B.H. and Mendelsohn, M.L. (1975) *Proc. Nat. Acad. Sci.* 72, 1231.
8. Jones, K.W. (1970) *Nature (Lond.)* 225, 912.
9. Krisch, R. and Ley, R. (1974) *Intern. J. Radiation Biol.* 25, 21.
10. Miura, K. (1967) *Methods in Enzymology* 12, 543.
11. Painter, R.B., Young, B.R. and Burki, H.J. (1974) *Proc. Nat. Acad. Sci.* 71, 4836.

12. Pardue, M.L. and Gall, J.G. (1970) *Science* 168, 1356.
13. Prensky, W., Steffensen, D.M. and Hughes, W.L. (1973) *Proc. Nat. Acad. Sci.* 70, 1860.
14. Scherberg, N.H. and Refetoff, S. (1974) *Biochem. Biophys. Acta* 340, 446.
15. Steffensen, D.M., Duffey, P. and Prensky, W. (1974) *Nature* 252, 741.
16. Szabo, P., Elder, R. and Uhlenbeck, O. (1975) *Nucleic Acids Res.* 2, 647.
17. Wimber, D.E. and Steffensen, D.M. (1970) *Science* 170, 639.
18. Yu, C.K. (1963) *Can. J. Genet. Cytol.* 5, 307.

KINETIC ASPECTS OF *IN SITU* HYBRIDIZATION IN RELATION TO THE PROBLEM OF GENE LOCALIZATION

Paul Szabo, Loh-chung Yu and Wolf Prensky

Sloan-Kettering Institute for Cancer Research
New York, New York 10021

ABSTRACT. Different components of an RNA sample hybridize to chromosomes at rates consistent with their individual complexities and concentrations, *i.e.* the predominant species in a preparation anneals more rapidly than any other RNA present as a low concentration contaminant. The predominant species anneals to loci containing homologous DNA with predictable kinetics. Therefore identification of these sites is independent of the absolute number of grains associated with them. This permits the design of experiments incorporating analyses which can verify the validity of gene localization based on results from *in situ* hybridization.

INTRODUCTION

In situ molecular hybridization is a technique for finding the chromosomal location of DNA which is homologous to a given nucleic acid sequence. The basic procedures used in cytological annealing studies were developed by Gall and Pardue who successfully applied the technology of nitrocellulose filter hybridization to chromosome preparations (1). The techniques are analogous because in both cases hybridization is carried out between a labeled RNA probe in solution, and a heterogeneous mass of immobilized DNA. Szabo *et al.* (2,3) have found that the kinetics of RNA:DNA annealing on slides are similar to the kinetics of RNA:DNA annealing on filters as described by Birnstiel *et al.* (4), except that the absolute annealing rate is slower on slides (3). Birnstiel *et al.* have shown that the observed rate of RNA:DNA hybridization on filters is a function of the complexity and concentration of the RNA in the hybridization solution and is independent of the amount of DNA bound to the nitrocellulose except at extremely high DNA inputs (4). The similarities between the two experimental systems suggest that the quantitative concepts derived from studies of the kinetics of filter hybridization can be applied both to the design as well as to the interpretation of *in situ* localization studies. The application of these concepts will be especially useful in difficult experimental situations, where the correct interpretation of the results may otherwise be in considerable doubt.

In this presentation, we will demonstrate that human and Drosophila 5S RNAs hybridize to their respective chromosomes with essentially the same rate. This is consistent with the similarities in the size of the two molecules. The mass of homologous DNA in the two preparations has no effect on annealing rate.

We will show that the iodinated probe used in localizing the genes coding for human 5S RNA is primarily composed of 5S RNA. Fingerprinting studies demonstrated trace amounts of a low complexity contaminant and *in situ* hybridization detected the presence of small amounts of 18+28S RNA in the 5S RNA probe. The contaminants hybridize to a number of sites in the human genome, but at rates slower than the rate observed for the annealing of 5S RNA to the long arm of chromosome 1.

The data prove that the assignment of the 5S RNA genes to chromosome arm 1q by Steffensen *et al.* (5) was correct. First, the kinetic behaviour of human 5S RNA over chromosome 1q is similar to the kinetic behaviour of our standard, Drosophila 5S RNA hybridized to region 56F in salivary gland chromosomes of *D. melanogaster*. Second, the complementary sites of chromosome 1 have a larger proportion of total silver grains under conditions of low C_rt hybridization, *i.e.* conditions under which only the predominant species of RNA anneals to any appreciable extent. The criteria used here for confirming the location of the human 5S genes are independent of the absolute amount of label over the chromosome, and permit the segregation of label due to contaminating RNAs and other factors. The implications of these observations to other gene localization and to our ongoing histone experiments will be briefly discussed.

MATERIALS AND METHODS

Preparation of 5S RNA. *Drosophila melanogaster* 5S RNA was prepared from wild type (Canton S) larvae as described by Szabo *et al.* (3). The HeLa 5S RNA was a gift of Dr. T. Borun. The sample was prepared by preparative electrophoresis on 6% polyacrylamide gels and tested for homogeneity by disc gel electrophoresis on 15% polyacrylamide gels both before and after labeling with ^{125}I.

RNA was labeled with ^{125}I using Commerford's reaction as described by Prensky (6). The iodinated HeLa 5S RNA was purified by two passages through CF11 and one through DEAE cellulose. The Drosophila 5S RNA was purified by one passage through CF11 and chromatography on RPC-5 (3). The samples were precipitated with 2.5 volumes of ethanol after the addition of carrier RNA, either *E. coli* tRNA or polyA. The precipitate was collected by centrifugation, dried under vacuum and dissolved in the hybridization solution, 50% formamide: 2xSSC (pH 7.0) containing 10^{-4}M KI. The specific activity of the HeLa 5S RNA was 1.1 x 108 dpm/μg.

Fingerprint Analysis. T_1 fingerprints of ^{125}I 5S RNA were obtained according to the procedure of Barrell (7). Procedures used here were the same as those described by Robertson *et al*. (8).

Preparation of Slides. Dissected salivary glands from late third instar larvae of *Drosophila melanogaster* were swelled in 45% acetic acid for several minutes on a siliconized cover slip. The gland was squashed on a microscope slide with steady, even pressure. After freezing the slide on dry ice, the cover slip was removed and the slide was fixed in ethanol:acetic acid, 3:1, v/v for 10 minutes. The slides were dehydrated with two washes in 70% ethanol and one in 100% ethanol and then air dried.

Human metaphases were prepared from 3 day PHA stimulated whole blood cultures. 5 x 10^{-7} M colcemide was added to the cultures 90 minutes prior to harvesting, the cells were collected by centrifugation and resuspended in hypotonic solution (0.075M KCl). After 8 minutes incubation at 37°C, the cells were collected by centrifugation. The cells were fixed with methanol:acetic acid, 3:1, v/v for at least 15 minutes. They were then collected by centrifugation and resuspended in fresh fixative. One or two drops of suspended cells were pipetted onto glass slides and air dried. Slides were stored at -20°C.

In situ Hybridization. *In situ* RNA:DNA hybridization was done using the procedure described by Szabo *et al*. (3). Slides were treated with 0.1 mg/ml RNAase A for 1 hour at room temperature, washed with two changes of 2xSSC (pH 7.0), and denatured in 0.2N HCl for 20 min at room temperature. The slides were again washed in two changes of 2xSSC, dehydrated with two changes of 70%, and one change of 100% ethanol, and then air dried.

Solutions containing various concentrations of 5S RNA were added to the slides and these were then incubated at 40°C in petri dishes containing 10 ml of hybridization buffer. The HeLa 5S RNA was hybridized for 2 hours; the Drosophila 5S RNA, for 10 minutes to 3 hours. The final C_rt values (concentration of RNA in moles of nucleotides per liter x incubation time in seconds) for these reactions ranged from 2.4 x 10^{-4} to 1.2 x 10^{-2} moles.sec/l for the human chromosomes and 10^{-4} to 8×10^{-2} moles·sec/l for the Drosophila chromosomes.

At the end of the reaction excess RNA was washed off the slides with 5-6 changes of 2xSSC (pH 7.0). The slides were treated with RNAase A, as before, and washed with 6-7 additional changes of 2xSSC. The slides were autoradiographed and stained as described before (3). The human slides were exposed for 45 days. The Drosophila slides were exposed from 1 to 8 days, depending on experimental conditions.

Autoradiographic Data Recording and Analysis: An average of 50 metaphase spreads were scored on each human slide. We scored for the number of silver grains over chromosome arm 1q, containing the 5S RNA gene site (5) and over the D and G group chromosomes, containing the 18+28S RNA gene sites (9). The number of grains over 100 interphase cells per slide were also determined. For the Drosophila 5S RNA hybridizations, an average of 50 salivary gland chromosomes were examined per slide and the number of grains over the 5S RNA gene site, region 56F (10) was recorded.

The grain count from each slide was averaged; two slides were analyzed for each C_rt value. The grain counts were analyzed by a computer program described by Szabo *et al.* (2,3) which determined rate constants and saturation values of the reactions and generated the curves shown in Fig. 1.

Mayall *et al.* (11) have measured the amount of DNA in each human chromosome. Using their data chromosome arm 1q contains about 4.3%, the NOR chromosomes 13.5%, and the remainder of the genome 82.2% of the total metaphase chromosomal DNA. We assumed that random experimental noise is distributed as a function of DNA content over the chromosomes. We corrected the observed grains over chromsome arm 1q by the formula

$$X_{cr} = X_o - \left(\frac{4.3}{82.2}\right) Y$$

where X_o and X_{cr} are the observed and corrected grain number, and Y is the observed number of grains over the rest of the genome. Appropriate values were also used to correct the data for the NOR chromosomes.

RESULTS

Figure 1 compares the annealing rates of Drosophila and human 5S RNA to their respective chromosomes. Curve B shows the hybridization of highly purified *D. melanogaster* 5S RNA to region 56F of Drosophila salivary gland chromosomes. This region contains all the 5S RNA genes of the species

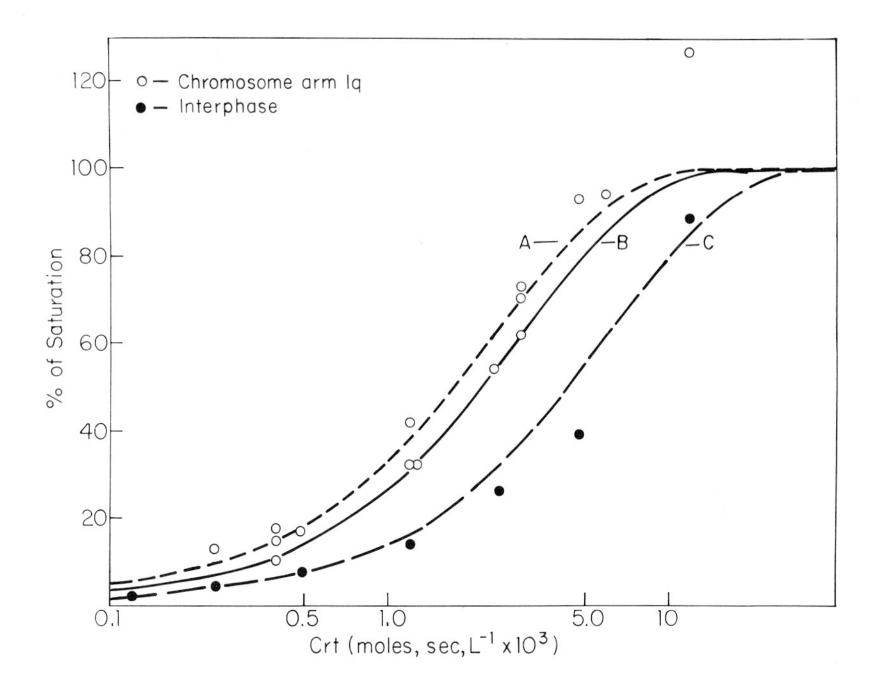

Figure 1. Kinetics of 5S RNA hybridization to chromsomes. ^{125}I human 5S RNA (1.1 x 10^8 dpm/μg) was hybridized to human metaphase preparations for a range of C_rt values. After autoradiography, the slides were analyzed. The data was fit to a first order reaction rate equation. Curve A represents the hybridization of human 5S RNA to chromosome arm 1q (-o); curve C, to human interphase cells (-o-). Curve B represents hybridization of Drosophila 5S as described in the text. The curve was derived from 32 individual data points which are not shown.

(10). The rate constant (k) is 310 l/mole·sec corresponding to a $C_rt_{1/2}$ of 2.2 x 10^{-3} mole·sec/l (3). Curve A shows the hybridization of human 5S RNA to human chromosome arm 1q. Both species of RNA exhibit similar cytological annealing rates to their respective homologous DNAs, an observation which is in accord with their expected behaviour in filter hybridization tests (4). Region 56F in Drosophila salivary gland polytene chromosomes contains about 500-1,000x more 5S DNA than is present in the analogous site on a human metaphase chromosome. The hybridization rate to the polytene chromosomes could be depressed by diffusion limitation. This has been observed by Birnstiel et al. (4) for the rate of hybridization of rRNA to nitrocellulose filters bearing increasing amounts of DNA. The slight displacement of curves A and B in Figure 1 could also be due to imprecision in the determination of the specific activities of the iodinated RNAs, and no significance can be attached to the observed differences.

Curve C in Figure 1 shows that hybridization of human 5S RNA to total human DNA of interphase cells is about 3x slower than hybridization to chromosome arm 1q. The suspicion arises that the 5S RNA probe consists not only of 5S RNA but also includes contaminating RNAs at lower concentrations. Since ^{125}I labeled RNA can be examined by "fingerprinting" (8), we investigated the oligonucleotide composition of our probe by this technique. Figures 2A and 2B present autoradiographs obtained following two-dimensional separation of ribonuclease T_1 products of iodinated human 5S RNA. Figure 2A is an autoradiograph exposed for "normal" periods of time. The pattern seen is that of HeLa cell 5S RNA (8, 12). Figure 2B is an overexposure of the same plate and reveals a number of oligonucleotides not previously detectable. These new oligonucleotides most likely represent a single low complexity contaminant present at a concentration much lower than 5S RNA itself. The presence of small amounts of completely random or of 18+28S ribosomal RNAs would mainly add haze to the area of the plate containing large oligonucleotides and their presence cannot therefore be detected. Thus, the fingerprint identifies the nature of the major components in the sample. It does not provide information to account for the non-5S DNA specific hybridization seen in our studies with this probe. Therefore, the shape and position of curve C in Figure 1 cannot be predicted by the fingerprint data.

However, the molecule detected in the fingerprint as the major constituent of the probe must anneal more rapidly than any other component. Thus, 5S DNA would be preferentially detected on slides hybridized to low C_rt values. Table 1 shows the change in grain distribution over chromosome arm 1q as the extent of hybridization is changed. At low C_rt values ($2.4 - 4.8 \times 10^{-4}$ moles·sec/l), about 20% of all silver grains are associated with the long arm of chromosome 1. Only 4.3 % of the grains are expected if distribution over chromsome arm 1q were random (based on its DNA content as measured by Mayall et al., 11). As the extent of hybridization is increased, the fraction of the total grains associated with the long arm of chromosome 1 decreases, i.e. a larger proportion of the resulting silver grains is over other sites. Some of these silver grains are due to non-

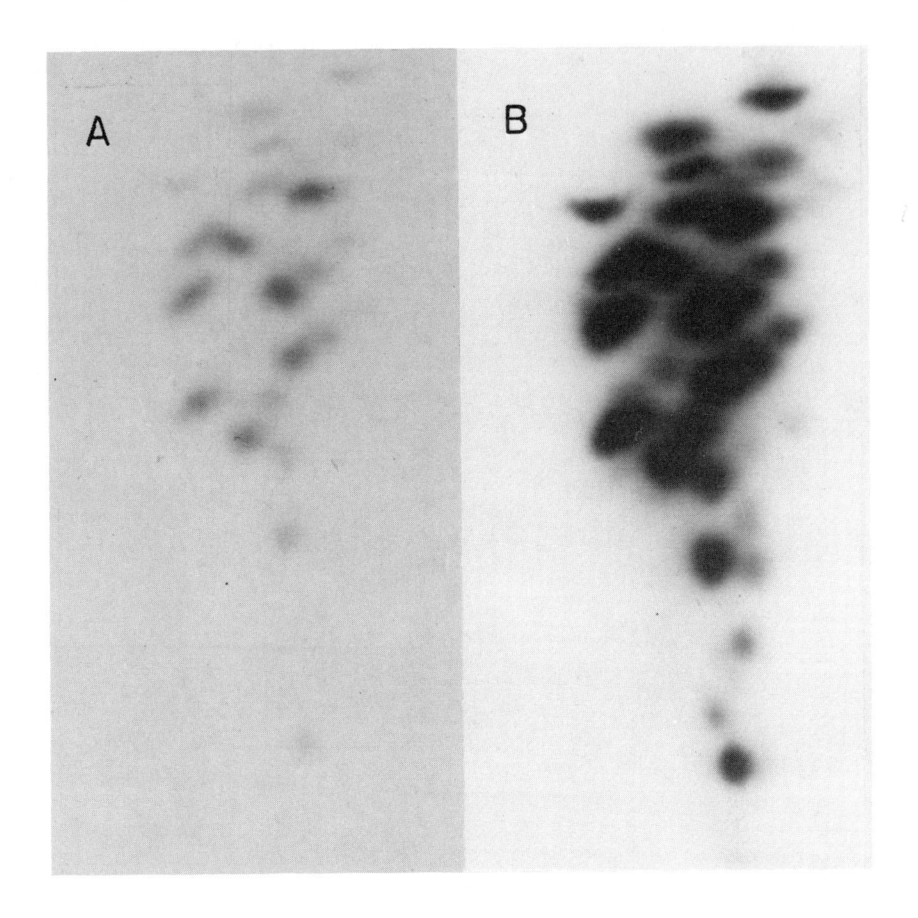

Figure 2. Fingerprint analysis of the human 5S RNA sample. ^{125}I 5S RNA was treated with T_1 ribonuclease. The resultant oligonucleotides were subjected to two-dimensional separation as described in Methods. From bottom left to bottom right, electrophoresis at pH 3.5 in pyridine acetate; homochromatography from bottom towards top of plate. (A) An autoradiogram exposed for a short period of time, showing the ^{125}I containing oligonucleotides of 5S RNA. (B) An overexposure of the same preparation showing the presence of new spots corresponding to a low complexity contaminant in the RNA sample.

TABLE 1

THE DISTRIBUTION OF SILVER GRAINS OVER HUMAN CHROMOSOMES AFTER HYBRIDIZATION WITH ^{125}I-5S RNA

Crt x 10^3 moles.sec/l	Average grains over chromosome arm 1q (corrected)	% of grains over chromosome arm 1q (observed)	% of grains over all nucleolar organizer regions (NOR) containing chromosomes (corrected)	% of grains over all chromosomes due to noise
0.24	.22	19.5	11.2	69.3
0.48	.47	25.7*	-	-
1.2	.7	17.8	14.6	67.6
2.4	.92	14.5*	-	-
4.8	1.58	14.5	34.4	51.1
12.0	2.16	6.1	37.7	56.2

* These values have not been corrected for background over 1q because NOR chromosomes were not scored for these points. Correction downward would have been by 2 to 3%.

specific background and increase as the number of input counts are increased. Others are caused by the hybridization of contaminant RNAs. Table 1 (fourth column) shows that label over the chromosomes containing the human rDNA sequences increased with C_rt. The hybridization rate of pure 18+28S rRNA to human cells *in situ* has been measured ($C_rt_{1/2}$ = 2 x 10^{-2}moles·sec/l, Szabo and Cotte, unpublished observations). Since the specific activity of 18+28S RNA must be the same as that of the 5S RNA, we can estimate that at the highest C_rt used here (1.2 x 10^{-2} mole·sec/l) we observed only 3% of the grains which would have resulted if 18+28S RNA had been hybridized to saturation. At this level of hybridization we observed 4.9 grains per haploid set of NOR chromosome and only 2.16 grains per chromosome arm 1q. On the basis of total grain number alone, one could assign the 5S RNA genes to the NOR containing chromsomes but the kinetic analysis precludes this possibility.

DISCUSSION

The most abundant RNA species in a sample is prominently visualized by fingerprint analysis and hybridizes at a rate commensurate with the complexity of the RNA. Such an RNA species can be assigned to those genetic loci which hybridize at the expected C_rt values. Therefore, as the C_rt is increased beyond experimental needs, a progressively smaller proportion of the silver grains will be due to specific hybridization by the most abundant species of RNA in the probe.

In situ hybridization studies are conducted in vast RNA excess, thus all eukaryotic RNA samples must be considered suspect with respect to homogeneity. This is especially troublesome when the major RNA is homologous to a low or unknown number of nucleotides in the genome. A number of steps can be taken to insure correct interpretation of autoradiographic results.

1. The probe should have a specific activity sufficient for detecting the expected number of nucleotides in the complementary DNA.
2. The RNA preparation should yield a non-random family of oligonucleotides following T_1 RNAse digestion and two-dimensional separation (fingerprinting).
3. The probe should show the proper hybridization kinetics in RNA excess annealing tests to DNA on nitrocellulose filters or in solution, *i.e.* the rate constant should be commensurate with the complexity of the RNA probe.

4. The specificity of the probe should be tested for its hybridization properties with DNA and chromosomes prepared from a number of different species.
5. In situ experiments should be designed on the basis of the data from step 3, taking into account that in situ annealing will be slower than DNA filter annealing rates (3). The identical probe can be used sequentially for both tests.
6. Grain distribution over individual chromosomes should be evaluated as a function of C_rt. Homologous sites should exhibit the expected hybridization kinetics.

All the above steps need not always be done in all studies. Some combination of these steps will be necessary, in cases in which precise localization will depend on a statistical analysis of the data rather than on a qualitative visualization of the hybrid region. We consider the last step to be an absolutely essential one when a new preparation of RNA is used the first time.

In the case of 5S RNA, we have carried out all of the above steps in different contexts. The molecule is highly conserved and therefore anneals to DNA as well as chromosomes from a number of different eukaryotic species. The kinetics of hybridization of the human 5S RNA to DNA and chromosomes obtained from a number of different mammalian species were indistinguishable from the kinetics observed with human material (our unpublished observations). This RNA also gave exceptionally clean autoradiographs when hybridized to chromosomes of the frog Rana pipiens. Our preliminary observations suggest that the greater the evolutionary distance between the human and the species whose chromosomes are challenged with the human 5S RNA probe, the smaller is the background noise over the hybridized metaphase cells. Since we cannot account for all the factors causing the noise seen over human chromosomes, we cannot explain the reasons for the reduced amount of noise seen over the chromosomes of other species. What is beyond doubt, however, is that the 5S RNA itself exhibits the specificities expected of such a molecule under a variety of conditions.

We have used the above guidelines for mapping the genes coding for H4 histone protein in man. The HeLa cell mRNA used in our studies translated exclusively into H4 protein (13). T_1 and RNAse A fingerprints of the two H4 mRNA's used for in situ studies were consistent with a molecule rich in guanines and about 400 nucleotides long. As with 5S RNA, less background was observed when the mRNA was hybridized to heterologous chromosomes. Because of its conserved sequence the HeLa mRNA could be used to localize H4 genes in a number of species, including those of the frog. The probe behaved as expected on filters. On the basis of its high specificity for chromosome 7 in the human genome at low $C_r$ts, we are provisionally assigning the H4 gene to that chromosome (Yu, Szabo, Borun and Prensky, in preparation).

ACKNOWLEDGEMENTS

The studies described herein were supported by N.C.I. Program Project grants: CA16599 and CA17085 and N.I.H. postdoctoral fellowhip GM05645 to L.-C.Y. We wish to thank Dr. T. Borun for 5S RNA and Robert Elder for aid in the computer analyses of the grain counts.

REFERENCES

1. Gall, J.G. and Pardue, M.L. (1969) *Proc. Nat. Acad. Sci. 63,* 378.
2. Szabo, P., Elder, R. and Uhlenbeck, O. (1975) *Nucleic Acids Res. 2*, 647.
3. Szabo, P., Elder, R., Steffensen, D.M. and Uhlenbeck, O.C. (1977) *J. Mol. Biol.* (in press).
4. Birnstiel, M.L., Sells, B.H. and Purdom, I.F. (1972) *J. Mol. Biol. 63,* 231.
5. Steffensen, D.M., Duffey, P. and Prensky, W. (1975) *Nature 252,*741.
6. Prensky, W. (1976) In *Methods in Cell Biology,* D.M. Prescott (ed.), vol. 13. Academic Press, New York, p. 51.
7. Barrell, B.G. (1971) In *Procedures in Nucleic Acid Research,* G.L. Cantoni and D.R. Davies (eds.), vol. 2. Harper and Row, New York, p. 751.
8. Robertson, H.D., Dickson, E., Model, P. and Prensky, W. (1973) *Proc. Nat. Acad. Sci. 70,* 3260.
9. Henderson, A., Warburton, D. and Atwood, K. (1972) *Proc. Nat. Acad. Sci. 69,* 3398.
10. Wimber, D.E. and Steffensen, D.M. (1970) *Science 170,* 639.
11. Mayall, B.H., Carrano, A.V., Moore, II, D.H., Ashworth, L.K., Bennett, D.E., Bogart, E., Littlepage, J.L., Minkler, J.L., Piluso, D.L. and Mendelsohn, M.L. (1975) In *Automation of cytogenetics,* M.L. Mendelsohn (ed.), Conf-751158. U.S. Energy Research and Development Administration, Washington, D.C., p. 133.

12. Dickson, E. (1976) Ph.D. Thesis, Rockefeller University New York.
13. Borun, T.W., Ajiro, K., Zweidler, A., Dolby, T.W. and Stephens, R.E. (1977) *J. Biol. Chem. 252*,173.

THE X AND Y CHROMOSOMES: MECHANISM OF SEX DETERMINATION

Susumu Ohno, Yukifumi Nagai and Salvatrice Ciccarese

Department of Biology, City of Hope National Medical Center, Duarte, California 91010

ABSTRACT: The embryonic plan of mammals is inherently feminine. While the Y chromosome initiates male development, its role is strictly limited to the determination of gonadal sex. The Y-linked gene, probably in multiple copies, apparently specifies a plasma membrane protein, or its sugar residues, that are serologically detectable as H-Y antigen. The plasma membrane presence of H-Y antigen diverts the embryonic indifferent gonad's inherent tendency to develop into an ovary and organizes a testis instead. Thus, in exceptional mammalian species and individuals, the expression of H-Y antigen strictly coincided with the possession of testicular tissue and neither with the presence of the Y chromosome nor with the masculine phenotype expression. Furthermore, in the Moscona-type rotation culture, free XY gonadal cells, lysostripped of their plasma membrane H-Y antigen sites, organized ovarian follicle-like aggregates. The expression of the H-Y gene appears to be under the control of an X-linked gene, and aside from testicular organization, neither the Y chromsome as a whole, nor the H-Y gene make any further contribution to male development.

All the extragonadal masculine developments are induced by testosterone normally synthesized by the H-Y antigen organized testis, and the responsiveness of all the target cell types to testosterone is mediated by a single species of nuclear-cytosol androgen-receptor protein controlled by the X-linked *testicular feminization* locus.

INTRODUCTION

In most mammalian species, sexual dimorphism is very pronounced; i.e., antlers of the stag, mane of the lion, crested neck of the stallion and the bull, beard of man. Since males, more often than females, are adorned with these obvious sex specific characters for display, males of the species tend to entertain an illusion of gradeur of being genetically better endowed. Nothing can be further from the truth. Artificial insemination by, which a bull of the exemplary phenotype and genetic background is yearly able to fertilize literally hundreds of cows, has contributed markedly to the rapid improvement of dairy breeds of cattle. Indeed, *Bovidae* in nature practice polygamy; only victors of intense combats, among males, gaining harems of females. No doubt, massive neck and shoulders that taper to narrow hind quarters are desirable combative traits to a bull, yet if these traits are genetically determined *sensus stricto*, the practice of polygamy or polygynous mating brings no benefit to the species, for the simple reason that these same traits, if manifested by his daughters, bring about disastrous consequences. Females of the species should be endowed with large enough hindquarters to house and permit the passage of a fetus as well as ample mammary glands to nurish their young.

The above then is reason enough for the strictly hormone-dependent development of all the extragonadal masculine traits in mammals. The dictate of natural selection has always been to keep a genetic difference between sexes at a necessary minimum. There obviously are two alternatives to achieve this end. 1) A majority of the fish and amphibians have opted to maintian the X and the Y or the Z and the W in a largely homologous state with each other. Thus, XY males and XX females differ genetically only with regard to alleles of a few sex determining loci. 2) Mammals, and possibly also birds, have opted to permit extensive genetic degeneration to affect the heterogametic sex specific chromosome; the Y or the W. Thus, the mammalian Y or the avian W came to retain only a few, very possibly one, key genes. In mammals, a sexual difference, with regard to the number of X chromosomes, has further been nullified, or minimized, by the X-inactivation mechanism which leaves only one active X in somatic cells of both sexes.

THE HORMONE INDEPENDENCE OF EARLY FEMININE DEVELOPMENT AND THE ANDROGEN-DEPENDENCE OF ALL THE EXTRAGONADAL MASCULINE DEVELOPMENT:

Mammalian fetuses, housed in the mother's womb, are in constant danger of being feminized by maternal steroid hormones. This apparently is the evolutionary reason that initial feminine differentiation of mammals became an autonomous process that depends neither on estradiol nor on progesterone. These female steroid hormones play really significant roles only after the female reaches puberty. A male hormone, testosterone, and its effective intracellular metabolite, on the other hand, are solely responsible for the induction of all extragonadal masculine development. The Y chromosome plays no role in the androgen responsiveness of mammalian fetuses. Thus, XX fetuses are as responsive to androgen as XY fetuses. Accordingly, if exposed to an exogenous source of androgen, XX fetuses readily aquire all the extragonadal masculine characteristics. Conversely, if deprived of an endogenous source of androgen by castration, XY fetuses automatically aquire all the extragonadal feminine characteristics (7).

The absence of sexual difference in the androgen responsiveness is attributable to the ubiquitous presence of the nuclear-cytosol androgen-receptor proteins in corresponding cell types of both sexes. The fact that an apparent point mutation of the X-linked *Tfm* locus of mice affects the nuclear-cytosol androgen-receptor protein of divergent target cell types in the identical way, and renders all the cell types totally nonresponsive to androgen, suggests that a single species of androgen-receptor protein mediates the androgen responsiveness of all the target cell types in the mammalian body (1, 10). How the same species of androgen-receptor protein manages to induce different sets of structural gene products in divergent target cell types has been discussed in some detail (10). In view of the above, the X-linked *Tfm* locus should be regarded as the primary regulatory locus of mammalian extragonadal sexual differentiation.

THE Y CHROMOSOME AND H-Y ANTIGEN

Although the Y chromosome of mammals, as a rule, is a miniscule element compared to other chromosomes, the amount of DNA it contains is still considerable; roughly 10,000,000 base pairs for the human or mouse Y. Now that the Y chromosome has revealed itself to be no omnipotent male determiner but a mere testis organizer, its DNA content is still disproportionately greater than its assigned task calls for. Indeed, a minute Y, representing no more than a tiny pericentric region of the ordinary human Y, has been found in three generations of perfectly fertile men (Harold Klinger, personal communication). It would appear that a bulk of the human Y, or the Y or any other mammalian species for that matter, is readily dispensable.

The importance of plasma membrane components in organogenesis has been well established since the classical experiment of Moscona (8). By definition, all the histocompatibility antigens, recognized as such by immunologists, are the plasma membrane components. Indeed the apparently Y-linked H-Y antigen was first recognized as a weak transplantation antigen that was the cause of consistent male skin graft rejections by females of the same inbred mouse strain (3). Subsequently, humoral H-Y antibody was raised by weekly injections of male lymphocytes to females of the same inbred strain, and the cytotoxicity test was developed to test its specificity as well as potency. The three types of cytotoxic targets are available in mice; 1) spermatozoa (5), 2) male epidermal cells (12), and 3) 8-cell XY embryos (6). Our favorite targets are male and female epideraml cells obtained from tails of BALB/c mice. In the presence of complements, a good H-Y antibody kills up to 70% of the male epidermal cells even at 1/48th dilution, while it does not touch female epidermal cells.

Once the cytotoxicity test for H-Y antigen was developed,

it became possible, by absorption, to test the presence or absence of H-Y antigen on cells of other mammalian species, and such a study revealed an extreme evolutionary conservation of this plasma membrane component (14). The plasma membrane component, so conserved, must have been performing an invariant function throughout adaptive radiations of mammals. That invariant function almost has to be the determination of primary or gonadal sex. Accordingly, the testis-organizing function of H-Y antigen was proposed (15).

INVARIANT ASSOCIATION BETWEEN THE POSSESSION OF TESTES AND THE H-Y ANTIGEN EXPRESSION IN EXCEPTIONAL MAMMALIAN SPECIES AND INDIVIDUALS:

If H-Y antigen is responsible for the testicular organization, in exceptional mammalian individuals whose gonadal sex does not coincide with the chromosomal sex, the expression of H-Y antigen should strictly coincide with the possession of testicular tissue and not with the presence of the Y chromosome. Indeed, XX males of man, the mouse and the dog expressed H-Y antigen and so did a human XX true hermaphrodite. Even XY women, with pure gonadal dysgenesis, expressed H-Y antigen if their streak gonads either contained residual seminiferous tubules or developed gonadal tumors of the testicular type. Since the above findings have been summarized by S.S. Wachtel (17), I shall only mention our findings on two exceptional rodent species belonging to the family *Microtinae*.

Both sexes of the mole-vole *(Ellobius lutescens)* are characterized by the XO sex chromosome constitution (2n=17) and claims to have identified a male specific segment on one or the other autosome have not been substantiated. This species has been enduring a 50% loss of zygotes since its inception represented by lethal XX- and OO-zygotes. Yet these XO males expressed a level of H-Y antigen quite comparable to that of ordinary XY males of any other mammalian species, while XO females of the same sepcies did not express this antigen (9). Thus, the minimal requirement for mammalian testicular organization is *sensus stricto*, the possession and expression of the H-Y gene and nothing more.

The wood lemming *(Myopus schisticolor)* is one of a few mammalian species whose sex ratio greatly deviates from 1 : 1 in females' favor. An X-linked mutant gene has recently been identified as the cause. This mutant gene, by suppressing the testis-organizing function of the paternally derived Y, produces fertile XY females. The progeny of such XY females include no males, but two types of females; XX and XY (4). H-Y antigen was expressed by XY males but not by XY females of this species (16). Thus, without the H-Y antigen expression, the Y chromosome is useless, and XY gonadal cells readily organize functional ovaries instead of testes.

The above indeed constitute an impressive set of circumstantial evicences that favor the proposed testis-organizing function of mammalian H-Y antigen. Yet the above also cast some doubt on the purported Y-linkage of the H-Y structural gene. Neither its X-linkage nor its autosomal inheritance can be vigorously excluded; the expression of H-Y gene normally requiring activation by a Y-linked gene. Nevertheless, S.S. Wachtel and myself currently favor the existence in multiple copies, of the order of 10 rather than 100, of the H-Y structural gene on the pericentric region of mammalian Y chromosomes, the X-chromosome carrying a modulator of these multiple H-Y gene copies. The above view is schematically illustrated in Figure 1.

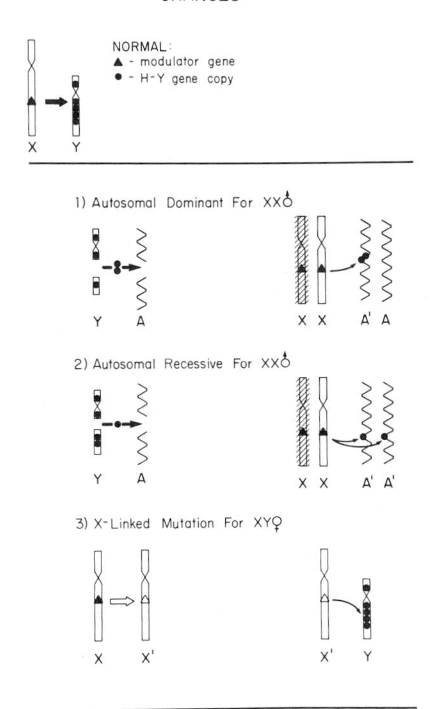

Figure 1: This scheme nicely explains the following situations. 1) An autosomal dominant gene found in the mouse and man that is transmitted from a carrier father to half of his

Figure 1 (cont'): XX progeny, thus producing XX males, represents a substantial number of H-Y gene copies tranferred from the Y. 2) When a number of H-Y gene copies transferred to an autosome is too small, the XX male condition becomes an autosomal recessive trait as in the goat and the pig. 3) When the X-linked modulator locus mutates and becomes overly repressing, the Y-linked H-Y gene copies cannot be expressed, thus, producing fertile XY wood lemming females.

NEWBORN MOUSE TESTICULAR CELLS LYSOSTRIPPED OF H-Y ANTIGEN ORGANIZE OVARIAN FOLLICLE-LIKE AGGREGATES:

The finding on the wood lemming has revealed that in the functional absence of H-Y antigen, XY gonadal cells readily organize a functional ovary. Can one experimentally force XY gonadal cells to engage in ovarian organization by depriving them of H-Y antigen?

In the presence of an excess amount of antibody, plasma membrane antigens gather over one pole of the cell: the "capping" phenomenon (13). As capped antigen-antibody complexes are subsequently consumed by the cell's lysosomes, the cell, so treated, becomes denuded of its plasma membrane antigen: the "lysostrip" phenomenon (2). Accordingly, we have exposed free testicular cells, obtained from newborn BALB/c males, to 20% concentration of H-Y antibody for 45 minutes in ice. Subsequently, they were subjected to the Moscona-type (8) rotation culture for 16 hrs. at the cellular concentration of 1,000,000 cells/0.125 ml medium 199 containing 10% Ig-free fetal calf serum and 20% mouse serum (H-Y antibody or control). Our estimate was that H-Y antibody-treated testicular cells were almost completely denuded of their plasma membrane H-Y antigen sites, at least for an initial 6 to 8 hr. period of the culture, and that unbound H-Y antibody, still present in the medium, was quite capable of dealing with regenerated H-Y antigen, at least to near the end of the culture period.

A difference between the control and the H-Y antibody treated culture was strikingly obvious at the first glance. The control primarily yielded tubular aggregates numbering 800 or so. Although most of them were still short and cylindrical, there always were 20 or so extremely long, tortuously twisting tubules that resembled testicular seminiferous tubules. In sharp contrast, the H-Y antibody treated culture yielded numerous spherical aggregates numbering some 1,000. Upon staining, large primordial germ cells, that abound in newborn testes, were found to constitute the central core of many of these spherical aggregates. The smallest class in particular consisted of a single primordial germ cell surrounded by a single layer of flattened cells. The resemblance of this class to primordial ovarian follicles that characterize newborn female gonads was particularly striking.

In a series of antibody dilution experiments, we havefurther established that an H-Y antibody concentration is the sole factor that determines the relative frequency of spherical and tubular aggregates. The results are shown in Figure 2.

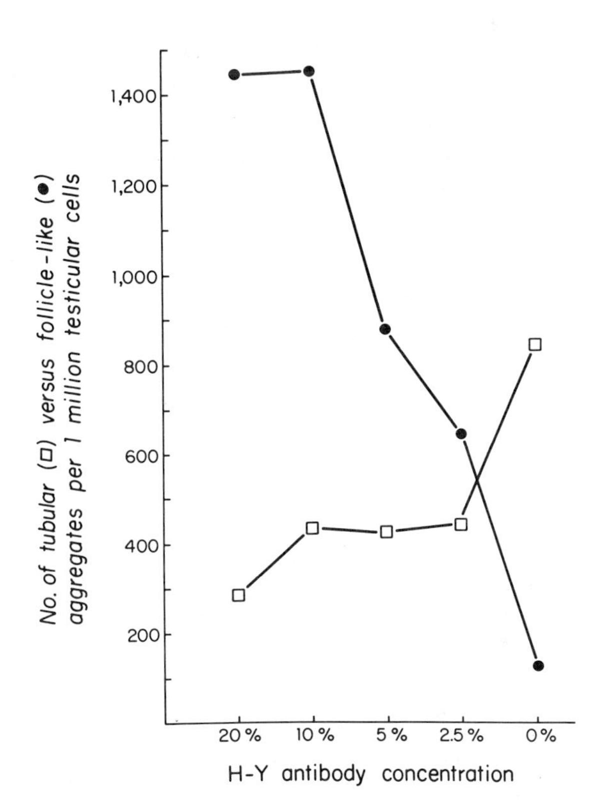

Figure 2: The effect of H-Y antibody dilution on the types of aggregates formed by free testicular cells from the BALB/c mouse testis in the Moscona-type rotation culture. Decreasing H-Y antibody concentrations, used for absorption, were accompanied by a progressive decrease in ovarian follicle-like aggregates and proportional increase in seminiferous tubule-like aggregates.

TESTIS-LIKE XX GONADS OF BOVINE FREEMARTINS ARE LOADED WITH H-Y ANTIGEN:

The above *in vitro* experiment revealed that XY gonadal cells denuded of their plasma membrane H-Y antigen sites organize ovarian, rather than testicular, structure. Such a result, of course, was anticipated by the occurence of H-Y antigen-negative, fertile XY females in the wood lemming; *Myopus schisticolor* (4, 16).

As the converse to the above, XX gonadal cells are also known to engage in testicular organization in special circumstances. For example, the very fact that a great majority of the experimentally produced XX/XY chimeric mice develop as normal males reveals that within a gonad, a majority of XX cells can be persuaded by a minority of XY cells to engage in testicular organization. Bovine freemartinism is caused by fetal vascular anastomosis between heterosexual, dizygotic twins. The XX gonad of a female twin (freemartin) originally develops along the ovarian line, but soon it is reorganized to resemble a small testis in extreme cases. Our serological study on extremely virilized testis-like gonads of 3 freemartin fetuses (150 to 175 days of gestation) revealed that, on a per cell basis, these XX gonads expressed as much H-Y antigen as the testis of normal bull fetuses (11). Thus, H-Y antigen is clearly involved in testicular organization by XX gonadal cells.

Yet the above finding created a puzzle as the origin of H-Y antigen found in the freemartin gonad. The source was thought to be a small number of blood-born XY primordial germ cells that lodged in the XX gonad (11). This, however, may not be a correct interpretation. Testicular differentiation of the fetal bull gonad begins at about 40 days of gestation, whereas the freemartin gonad noes not begin to virilize until 90 days or even later. Furthermore, freemartinism is a peculiarity of *Bovidae*, while in primates (marmoset monkeys and man), fetal vascular anastomosis with a male twin does not affect, in the slightest, the normal ovarian development of the XX gonad.

It may be that the H-Y gene does not specify a plasma membrane protein as such, rather it specifies a sugar transferase that puts particular sugar residues on a specific plasma membrane protein. A peculiarity of *Bovidae* may lie in the male fetus's tendency to disseminate a large amount of H-Y sugar transferase to blood circulation at a certain developmental stage; i.e., 90 to 150 days of gestation for cattle. Thus, the solution to one problem created a new question to be solved.

ACKNOWLEDGEMENTS

The skillful technical assistance of Mrs. Mary Romero is gratefully acknowledged. This work was supported in part by a contract, N01-CB-33907, and grants form NIH, #5 R01 CA16952 and #R01 AG00042.

REFERENCES

1. Attardi, B., Geller, L.N. and Ohno, S. (1976) Endocrinol. 98, 864.
2. Cullen, S.E., Bernoco, D., Carbonara, A.O., Jacot-Gillermod, H., Trinchieri, G. and Ceppellini, R. (1973) Transpl. Proc. 4, 1835.
3. Eichwald, E.J. and Silmser, C.R. (1955) Transpl. Bull. 2, 148.
4. Fredga, K., Gropp, A., Winking, H. and Frank, F. (1976) Nature 261, 225.
5. Goldberg, E.H., Boyse, E.A., Bennett, D., Scheid, M. and Carswell, E.A. (1971) Nature 232, 478.
6. Krco, C. and Goldberg, E.H. (1976) Science 193, 1134.
7. Jost, A. (1970) Phil. Trans. Roy. Soc. Lond. B. 259,119.
8. Moscona, A. (1957) Proc. Nat. Acad. Sci. U.S.A. 43, 184.
9. Nagai, Y. and Ohno, S. (1977) Cell 10,
10. Ohno, S. (1976) Cell 7, 315.
11. Ohno, S., Christian, L.C., Wachtel, S.S. and Koo, G.C. (1976) Nature 261, 597.
12. Scheid, M., Boyse, E.A., Carswell, E.A. and Old, J. (1972) J. Exptl. Med. 135, 928.
13. Taylor, R.B., Duffus, F.H., Raff, M.C. and Petris, de S. (1971) Nature New Biol. 233, 225.
14. Wachtel, S.S., Koo, G.C. and Boyse, E.A. (1975) Nature 254, 270.
15. Wachtel, S.S., Ohno, S., Koo, G.C. and Boyse, E.A. (1975) Nature 257, 235.
16. Wachtel, S.S., Koo, G.C., Ohno, S., Gropp, A., Dev, V.G., Tantravahi, R., Miller, D.A. and Miller, O.J. (1976) Nature 264, 638.
17. Wachtel, S.S. (1977) Transpl. Rev. 33, 33 .

Y-CHROMOSOME DNA

Louis M. Kunkel*, Kirby D. Smith and
Samuel H. Boyer*

Division of Medical Genetics, Department of Medicine,
The Johns Hopkins University School of Medicine,
Baltimore, Maryland 21205

ABSTRACT. Two techniques have been used to isolate human Y chromosome DNA sequences (1,2). The first, isolation by exclusion, yields radiolabeled reiterated DNA which reassociates to DNA isolated from human males but not to DNA from human females (1). Through analysis of Y-chromosome mutants these Y-chromosome-specific sequences have been localized to the long arm of the Y chromosome (3). Studies utilizing driver DNAs of different sizes indicate that Y-specific DNA sequences are clustered blocks of 0.6 to 1.0KB in human male DNA.

The second approach utilizes restriction endonucleases. Hae III digestion of DNA from human males yields two Y-chromosome fragments representing between 30 and 50% of the estimated Y chromosome (2). The larger of the Hae III restriction fragments ($\sim$ 3.4KB) has been purified by digestion with the restriction enzymes Hind III, Bam HI and RII (2). Reassociation data indicate that two thirds of the DNA contained within the 3.4KB fragments are homologous with highly reiterated non-Y sequences and are thus not Y-chromosome-specific. The remainder of the sequences within this fragment are Y-chromosome-specific and homologous to sequences isolated by exclusion.

INTRODUCTION

The chromosomal location, organization and functional importance of particular DNA sequences is of fundamental interest to biologists. While the overall organization of human DNA (4) has been shown to be the same as that found for most organisms (5), little is known about the organization of specific sets of sequences. We describe here two approaches for isolating DNA sequences from the human Y chromosome and report our initial studies concerning their oganization and location.

* Howard Hughes Medical Institute Laboratory of Biochemical Genetics

Reassociation of reiterated radiolabeled 47,XYY DNA with a 40,000-fold excess of 46,XX DNA, yields, after 2 challenges, tritiated Y-chromosome-specific DNA sequences (1). These reiterated Y-chromosome-specific sequences (hereafter referred to as it-Y DNA) reassociate exclusively to individuals bearing a Y chromosome (3). Based on recoveries at each step in the purification of it-Y [^{3}H] DNA, these sequences represent approximately 0.1% of a 46,XY genome or nearly 10% of the Y chromosome (1). Assuming that the $C_ot_{1/2}$ of single-copy sequences relative to *E. coli* is 3000, it-Y sequences have a reiteration frequency of 300-600. When reassociated at a size of 200 nucleotides, it-Y DNA represents between 15 and 30 different reiterated families.

Recently Cooke (2) has described the isolation and characterization of Y-chromosome fragments generated from human male DNA through endonuclease cleavage using the enzyme Hae III. We have purified the larger fragment (3.4KB) thus produced and have studied the sequences within this fragment by DNA reassociation. We have also studied the relationship between Y-chromosome sequences "isolated by exclusion" (1) and the Y-chromosome sequences generated by digestion with restriction endonuclease Hae III.

RESULTS

Characterization and localization of sequences isolated by exclusion. Having purified radiolabeled reiterated Y-chromosome-specific DNA (it-Y [^{3}H] DNA), it was of interest to determine the location of these sequences within the Y chromosome. Leukocyte DNA was isolated (3) from 10 individuals with aberrant Y chromosomes and this DNA tested by reassociation assay with it-Y [^{3}H] DNA. Each test reassociation assay was run in parallel with control assays of DNA isolated from a 46,XY individual, a 46,XX individual and *E. coli*. The results of these assays indicate that it-Y sequences are localized within the long arm of the Y chromosome (Fig. 1, Ref. 3). A 1.9-fold acceleration -- relative to assay of DNA from males -- is observed when DNA isolated from a female (Individual 1) with an isochromosome of the long arm of Y (double dose long arm, little if any short arm) is used to reassociate it-Y [^{3}H] DNA. This is the same rate increase observed when a 47,XYY (Individual 2) individual's DNA is tested (Fig. 1A). DNA isolated from 2 female individuals who have translocations of Y long arm (Individual 6, 46,X,t(X;Y); Individual 5, 46,X,t(Y;17)) reassociates it-Y [^{3}H] DNA at the same rate observed for a normal 46,XY male (Fig. 1B). Paternal male cousins who carry a deletion of

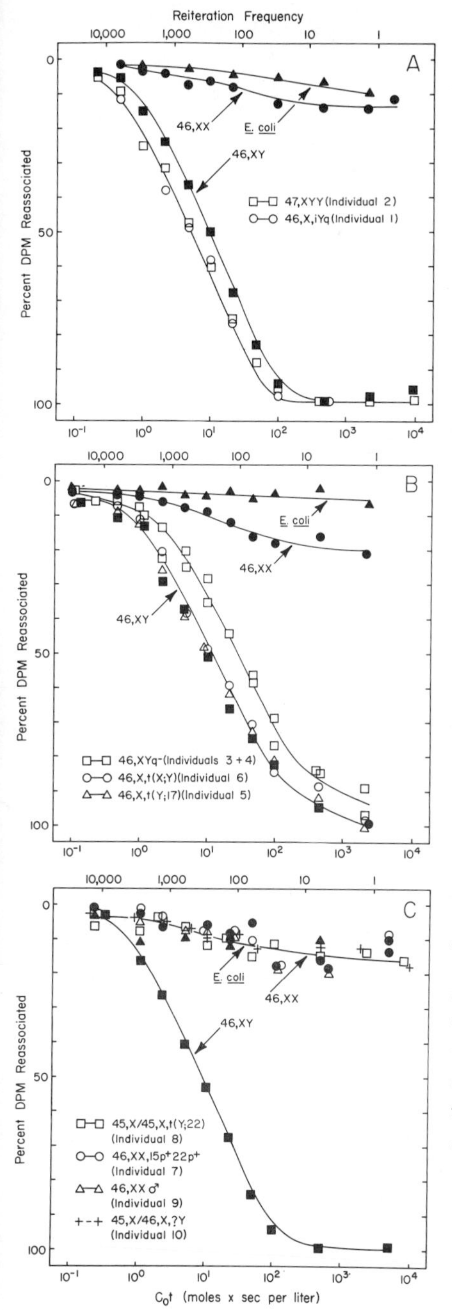

Fig. 1 Reassociation of trace levels of radiolabeled reiterated Y-chromosome-specific DNA (it-Y [^{3}H] DNA) with various excess driver DNAs. Control assays in panels A, B and C are 46,XY (■); 46,XX (●); E. coli (▲). Experimental assays: panel A - 47,XYY (Individual 2)(□); 46,X,i(Yq)(Individual 1)(O); panel B - 46,XYq- (Individuals 3 and 4)(□); 46,X,t(X;Y) (Individual 6)(O); 46,X,t(Y; 17)(Individual 5)(Δ); panel C - 45,X/45,X,t(Y;22)(Individual 8)(□), 46,XX,15p+,22p+ (Individual 7)(O); 46,XX male (Individual 9)(Δ); 45,X/46,X?Y (Individual 10)(+).

Reassociations were done at 60° in 0.12M phosphate buffer, pH 6.8. Incubations were terminated by passage through a HAP column equilibrated in 0.12M PB and held at 60°. In each instance 100% reassociation was defined as the maximum reassociation to 46,XY DNA. In all instances 80-85% of the unlabeled DNA (monitored at A_{260}) had reassociated at C_ot 10,000. Reiteration frequencies given at the top of the figure were calculated on the assumption that the $C_ot_{1/2}$ of single-copy sequences from man relative to E. coli is 3×10^3.

one half the long arm of the Y (Individuals 3 and 4, 46,XYq-) have DNA which reassociated it-Y [^{3}H] DNA at 1/2 the rate observed with normal males (Fig. 1B). This result is consistent with approximate halfing of all it-Y sequences in these two individuals' genomes.

Reassociation assays (Fig. 1C) with DNA isolated from 3 male individuals (Individual 8, mos45,X/45,X,t(Y;22); Individual 9, 46,XX; Individual 10, mos45,X/46,X/?Y) who were H-Y antigen positive (6) and lacked Y-chromosome long arm showed no reassociation beyond that observed for a normal 46,XX female.

The reassociation results obtained for all the individuals are summarized in Table 1 (page 5) along with other Y-chromosome markers. These results when considered as a group are consistent with the localization of it-Y sequences to the long arm of the Y chromosome and their absence from the short arm.

Having determined the general chromosomal location of it-Y sequences, it became of interest to determine the organization of it-Y DNA within the DNA of the Y chromosome. The interspersion pattern of it-Y sequences was determined by their rate of reassociation with fractionated driver DNAs of different sizes (4,5). DNA isolated from a 46,XY male was sheared to 3 sizes (300, 600 and 2000 nucleotides in length) as measured by agarose gel electrophoresis. The DNAs were reassociated and reiterated sequences collected on hydroxylapatite (HAP). A portion of each reiterated sample was subjected to S_1 nuclease digestion (10). These fractionated 46,XY DNAs were sonicated and used in reassociation assays of it-Y [^{3}H] DNA. The proportion of the genome represented by each driver DNA as well as the $C_ot_{1/2}$ of their reassociation with it-Y [^{3}H] is presented in Table 2 (page 6). The only difference observed in $C_ot_{1/2}$ values (rate factor of 2) was at the 2000 nucleotide size. Similar reassociation rates observed for 300 and 600 nucleotide samples indicate that reiterated duplex plus single-strand tails (HAP alone, Table 2) as well as duplex alone (HAP + S_1, Table 2) showed no difference in amounts of it-Y DNA sequences. This indicates that, at these two sizes, it-Y DNA is equally likely to be in either reiterated duplex or single-strand tails. Therefore, the clustering size must be greater than 600 nucleotides. The rate change of 2 observed for it-Y [^{3}H] DNA when reassociated with DNA fractionated at 2000 nucleotides indicates that the clustering is not as great as 2000 but closer to 1000 nucleotides. This means that it-Y sequences are probably clustered in lengths of 600-1000 nucleotides (manuscript in preparation).

TABLE 1. HUMAN Y CHROMOSOME MARKERS AND QUANTITIES

OF it-Y DNA IN NORMAL AND VARIANT INDIVIDUALS

Individual No.	Phenotype	Karyotype*	Bright Q.M. Band†	5-Methyl-Cytosine§	H-Y Antigen¶	it-Y DNA
---	Female	46,XX	-	-	-	0‖
---	Male	46,XY	+	+	+	1
1	Female	46,X,i(Yq)	++	N.T.**	N.T.	1.9
2	Male	47,XYY	++	N.T.	N.T.	1.9
3	Male	46,XYq-	-	+	+	0.5
4	Male	46,XYq-	-	N.T.	N.T.	0.5
5	Female	mos45,X,t(Y;17)(Yqter→Yq11::17p13 →17qter)/46,X,t(Y;17)(Yqter→Yq11: :17p13→17qter;17pter→17p13::Yq11→Ypter)	+	N.T.	N.T.	1
6	Female	46,X,t(X;Y)(:Xq22→Xp11 or Xpter→Xq11::Yq11→Yqter)	+	N.T.	+	1
7	Female	$46,XX,15p^{+},22p^{+}$	-	-	-	0
8	Male (Aberrant)	mos45,X/45,X,t(Y;22)(Yp11→ Ypter or Yq11→?::22pter→22q13)	-	-	+	0
9	Male (Kallman's syndrome)	46,XX	-	-	+	0
10	Male (Aberrant)	mos45,X/46,X,?Y	-	N.T.	+	0

* Ref. 7. † Ref. 8, brightly quinacrine fluorescent distal segment. § Ref. 9. ¶ Ref. 6. ‖ The relative amount of it-Y DNA in each genome was estimated from the $C_ot_{1/2}$ for its reassociation of it-Y [^{3}H] DNA compared to the $C_ot_{1/2}$ for reassociation of it-Y [^{3}H] DNA with DNA from normal 46,XY men. The $C_ot_{1/2}$ values, determined as the C_ot necessary to achieve 50% of maximum reassociation (Fig. 1), were: 47,XYY and 46,X,i(Yq), $C_ot_{1/2} = 5.4$; 46,XY, $C_ot_{1/2} = 10$; 46,X,t(Y;17) and 46,X,t(X;Y), $C_ot_{1/2} = 10$; and 46,XYq-, $C_ot_{1/2} = 20$. The value 0 represents no reassociation of it-Y [^{3}H] DNA beyond that to 46,XX or E. coli. **N.T. = not tested.

TABLE 2

RELATIVE PROPORTIONS OF THE HUMAN MALE GENOME THAT VARIOUS FRACTIONATED DNAs REPRESENT AND THE RATE OF REASSOCIATION OF it-Y [^{3}H] DNA WITH EACH OF THESE DNAs

DNA Size	Treatment of DNA	Percent of Genome*	$C_ot_{1/2}$ it-Y [^{3}H] DNA
300	HAP	38.8	3.4
	HAP + S_1	17.8	3.4
600	HAP	39.8	3.4
	HAP + S_1	19.6	3.4
2000	HAP	58.0	6.7
	HAP + S_1	20.7	3.4

Human 46,XY DNA was sheared to 3 different sizes (300, 600 and 2000 nucleotides in length) and reassociated to C_ot 100 (that C_ot at which most reiterated DNAs are reassociated) and collected on HAP. A portion of each reiterated sample was treated with S_1 nuclease (10) to remove single-stranded tails. The resultant 6 DNA fractions were sonicated to give a size of 250 nucleotides. These fractionated driver DNAs were reassociated with trace amounts of it-Y [^{3}H] DNA and $C_ot_{1/2}$ values calculated.
* The percent of the genome refers to that percentage of total recovered DNA which was assayed as reiterated either by HAP or HAP plus S_1.

Characterization of Y-chromosome sequences isolated by restriction enzymes. The Hae III generated 3.4KB fragment from human male DNA was purified as described (2). The 3.4KB fragment was radiolabeled by nick translation (11), sonicated to 300 nucleotides and then assayed for Y-chromosome-specificity. Two thirds of the radiolabeled molecules reassociated to both 46,XY and 46,XX DNA with very rapid kinetics. The other third reassociated to 46,XY DNA alone with a reiteration frequency very similar to it-Y sequences.

To determine whether the Y-specific and non-Y-specific portions of the 3.4KB fragment were linked in the same molecules, the 3.4KB fragments were selected for size ($\geq$ 2KB) after radiolabeling and tested for Y-chromosome-specificity as both sonicated and unsonicated molecules. At the larger size, the 3.4KB fragment reassociates almost as well to both male and female DNA whereas the same molecules when sonicated are still 1/3 Y-chromosome-specific. This implies that non-Y-specific and Y-specific portions of the 3.4KB fragment are adjacent to one another in the same molecules. Reassociation of it-Y [^{3}H] DNA to unlabeled 3.4KB fragment indicates that approximately 65% of it-Y sequences are homologous to the Y-specific portion of the 3.4KB fragment (manuscript in preparation).

DISCUSSION

Reassociation assays utilizing DNAs isolated from individuals who have aberrant Y chromosomes provides a model system for the mapping of DNA sequences within a chromosome. The results presented here for it-Y DNA (Fig. 1, Table 1) indicate that these particular sequences are located in the long arm of the Y chromosome. The exact topography of it-Y sequences within the long arm of the Y chromosome is as yet unclear. Two individuals with deletions of one half the long arm showed a halfing of all it-Y sequences. Therefore, it-Y sequences must be distributed equally on either side of this Y chromosome breakpoint. Whether it-Y sequences are clustered near the breakpoint in these individuals (middle of long arm) or are distributed throughout will have to await further studies of individuals in whom 1/4 or 3/4 of the long arm is deleted.

The individuals studied here are also useful for determining the relationship of this particular group of reiterated sequences to male determination. Three males with H-Y antigen (6)(Table 1) and presumed Y short arm material had no it-Y sequences in leukocyte DNA. Two females with no short arm but some long arm material translocated to other chromosomes had normal amounts of it-Y DNA. A third female with twice the long arm dose and no short arm had twice the levels of it-Y DNA. It is possible in any one individual studied here that leukocyte DNA is not representative of germline DNA, but in the aggregate, there seems to be no evident relationship between it-Y DNA and male determination.

Y-chromosome-specific sequences representing one third of the 3.4KB Hae III fragment of male DNA reassociate with similar kinetics as it-Y sequences and, like them, represent approximately the same proportion of a 46,XY genome. The whole 3.4KB fragment has been reported (2) to represent ∿ 0.4% of a 46,XY genome. Since it-Y sequences are clustered in lengths of 1.0KB (∿ 1/3 the 3.4KB fragment length) and 65% are homologous to the 3.4KB fragment, it is tempting to presume that the Y-specific sequences in the 3.4KB fragment are the same sequences as those isolated by exclusion. Whatever the relationship, there is little doubt that the non-Y-specific portion of the 3.4KB fragment is <u>not</u> the same as it-Y DNA. The ability of the non-Y-specific sequences to reassociate with DNA from either males or females indicates that they are found on chromosomes other than the Y chromosome. That these non-Y-specific sequences lie within the same 3.4KB fragment as the Y-specific sequences is shown by the loss of Y specificity when large 3.4KB radiolabeled <u>non-sonicated</u> molecules are reassociated to female DNA. In other words, the two types of sequences are adjacent to each other in the same 3.4KB fragment.

A number of conclusions can be drawn concerning the 3.4KB fragment DNA and it-Y DNA. <u>First</u>, restriction sites for Hae III must occur at regular intervals within Y-chromosome DNA. <u>Second</u>, sequences within these sites are not necessarily the same, i.e., at least 2 different types of sequences are found within the borders of the Hae III sites, one sequence of which is composed of a number of different families. <u>Third</u>, it-Y sequences probably occur in blocks of ∿ 1.1KB that are adjacent to ∿ 2.2KB lengths of non-Y-specific reiterated DNA. Single-copy DNA sequences are absent from these fragments. In this respect, the arrangement of these 2 types of reiterated DNA differs from the reported overall arrangement of reiterated DNA (4,5).

The methods used in this study, in principle, can be used not only to locate particular DNA sequences within other chromosomes but also to map the fine structure of chromosomal DNA sequences.

ACKNOWLEDGEMENTS

This work was supported in part by NIH grants HL 15026, GM 19489, and GM 00145. We wish to thank Lynne Garvin for help in preparation of this manuscript.

REFERENCES

1. Kunkel, L.M., Smith, K.D., and Boyer, S.H. (1976) Science 191, 1189-1190.
2. Cooke, H. (1976) Nature 262, 182-186.
3. Kunkel, L.M., Smith, K.D., Boyer, S.H., Borgaonkar, D.S., Wachtel, S.S., Miller, O.J., Breg, W.R., Jones, H.W. Jr., and Rary, J.M. (1977) Proc. Nat. Acad. Sci. USA 74, 1242-1244.
4. Schmid, C.W., and Deininger, P.L. (1975) Cell 6, 345.
5. Davidson, E.H., Galau, G.A., Angerer, R.C., and Britten, R.J. (1975) Chromosoma (Berl.) 51, 253-259.
6. Wachtel, S.S., Koo, G.C., Zuckerman, E.E., Hamerling, U., Scheid, M.P., and Boyse, E.A. (1974) Proc. Nat. Acad. Sci. USA 71, 1215-1218.
7. Paris Conference (1971): Standardization in Human Cytogenetics (1972) VIII(7)(The National Foundation, White Plains, N.Y.).
8. Zech, L. (1969) Exp. Cell Res. 58, 463.
9. Miller, O.J., Schnedl, W., Allen, J., and Erlander, B.F. (1974) Nature 251, 636-637.
10. Sutton, W. (1971) Biochim. Biophys. Acta 240, 522.
11. Maniatis, T., Kee, S.G., Efstratiadis, A., and Kafatos, F.C. (1976) Cell 8, 163-182.

THE DETECTION AND INDUCTION OF SISTER CHROMATID EXCHANGES

Samuel A. Latt, James W. Allen, Charles Shuler, Ken S. Loveday and Stephen H. Munroe

Department of Pediatrics, Mental Health Retardation Center, Childrens Hospital Medical Center and Center for Human Genetics, Harvard Medical School, Boston, Massachusetts 02115

ABSTRACT. Sister chromatid exchanges (S.C.E.'s), events which previously had been detected by autoradiography, can now be detected by BrdU-dye techniques in which sister chromatid differentiation is reflected by characteristic fluorescence or Giemsa staining patterns. S.C.E.'s have proved to be highly sensitive indices of the interaction of clastogens with chromosomes, and abnormalities related to S.C.E. formation occur in a number of hereditary diseases known or suspected to involve a defect in DNA repair. Methods for *in vivo* analysis of S.C.E. formation in intact animals have recently been developed. These permit the detection of tissue-specific, host mediated responses to potential clastogens and have provided preliminary data on sister chromatid differentiation in meiotic cells.

The combination of 8-methoxypsoralen plus near UV light is highly effective in inducing sister chromatid exchanges. Analysis of the number and chromosomal location of S.C.E.'s induced by 8-methoxypsoralen plus light in synchronized Chinese hamster ovary cells is being used to investigate the relationship between S.C.E. formation and DNA synthesis. These and related biochemical and cell biological studies should provide insight about the biological significance of sister chromatid exchanges.

INTRODUCTION

Sister chromatid exchanges (S.C.E.'s) reflect the interchange of DNA between replication products at similarly or identically positioned loci within a chromosome. These exchanges, which are generally detected in cytological preparations of metaphase chromosomes (Figure 1), presumably involve DNA breakage and rejoining, although little is known about the actual molecular details of S.C.E. formation. S.C.E. analysis has provided information about chromosome composition, while the associated sister chromatid differentiation has been used to characterize chromatid segregation at mitosis. Recent methodological improvements have led to extensive use of S.C.E. measurements to detect the effects of clastogens and have stimulated cytological studies of chromosome fragility diseases. These studies have focused interest on both the biochemistry of S.C.E. formation and the biological significance of S.C.E.'s.

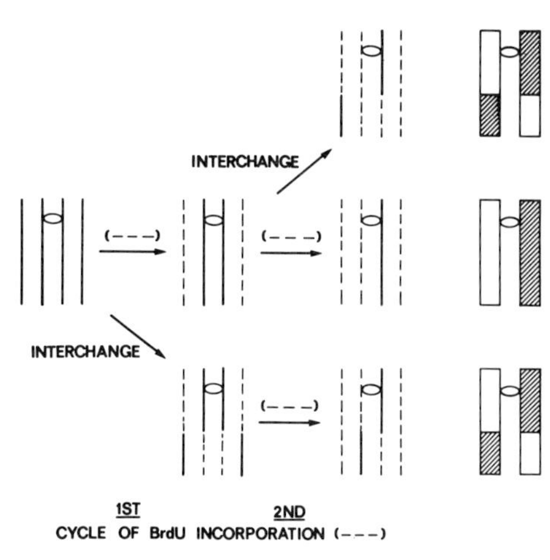

Fig. 1. Diagramatic representation of sister chromatid differentiation and S.C.E. detection. Exchanges occurring during either cycle of BrdU incorporation (---) are apparent in second division metaphases. Cross-hatched chromatid regions containing an original, unsubstituted polynucleotide chain can be differentiated by bright dye fluorescence (e.g., using 33258 Hoechst) or by intense Giemsa staining.

Sister chromatid exchanges were first described by J. Herbert Taylor and co-workers, who utilized autoradiography to detect differentially labelled sister chromatids in cells which had undergone one cycle of ^{3}HdT incorporation followed by a replication cycle in non-radioactive medium (84). S.C.E.'s were apparent as reciprocal alterations in labelling along the chromatids of metaphase chromosomes. Autoradiographic analysis of sister chromatid differentiation and exchange provided valuable information about chromosome structure and replication (83,84). However, the tedious nature of autoradiography apparently discouraged more extensive investigations, and its limited effective resolution contributed to ambiguity (90) regarding the number of DNA helices (the "nemy") in a single chromatid. Subsequent detection of S.C.E.'s by 5-bromodeoxyuridine-dye techniques (36,40,46,94) corroborated Coming's earlier hypothesis (15) that the occasional observation of photographic grains at homologous loci of both sister chromatids (isochromatid labelling) reflected not polynemy but rather the inability to resolve closely spaced exchanges.

RESULTS AND DISCUSSION

Analysis of S.C.E. formation in cytological chromosome preparations has been facilitated by techniques utilizing fluorescent dyes, Giemsa stain, or immunochemical procedures to detect BrdU incorporation and hence DNA synthesis. A few investigators had observed sister chromatid differentiation in Giemsa or orcein-stained metaphase chromosomes from cells which had incorporated BrdU (32,96), but the data were either incidental or attributed to chromosome despiralization, and the potential of BrdU-dye methodology as an alternative to autoradiography was not fully exploited. Subsequent experiments, predicated on the hypothesis that BrdU substitution of DNA might quench the fluorescence of bound dye (e.g., by a heavy atom effect (20)), led to the identification of 33258 Hoechst and related bisbenzimidazole dyes as fluorescent probes of DNA synthesis (44,51,54).

Bisbenzimidazole dyes fluoresce less efficiently when bound to DNA containing BrdU than when bound to unsubstituted DNA (44,54). Spectroscopic studies on synthetic polynucleotide-33258 Hoechst complexes are consistent with the premise that BrdU-dependent quenching of fluorescence occurs at least in part by interactions at the first excited singlet state of the dye. 33258 Hoechst has been used to detect BrdU incorporation into DNA, unfixed chromatin (54) and intact cells (51,53), as well as fixed cytological chromosome preparations (Fig. 2). BrdU is also capable of quenching the fluorescence of the dyes acridine orange (17,34) and DAPI (58), which have also been used for cytological analysis of DNA synthesis.

Routine detection of sister chromatid exchanges in cytological preparations has been assisted by the emergence of modified Giemsa techniques for BrdU detection (Fig. 2). These procedures presumably reflect selective degradation of DNA in the chromatid more highly substituted with BrdU (25,82) with (41,67) or without (43) photosensitization by bound dye. More recently, S.C.E.'s have been detected by immunological techniques which utilize antibody directed against BrdU (26).

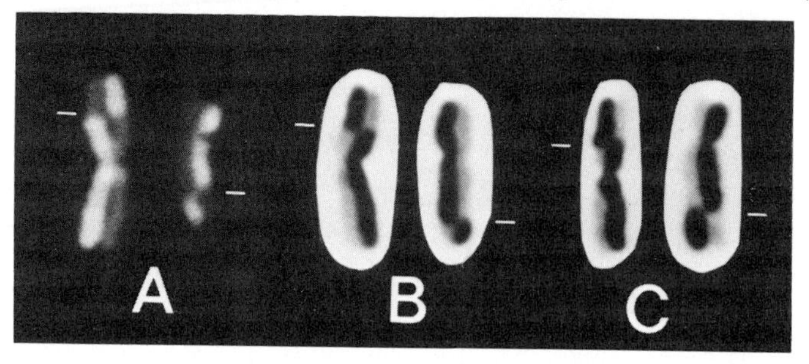

Fig. 2. S.C.E.'s in chromosomes from human lymphocytes which had replicated twice in medium containing BrdU. Chromosomes in (A) were stained with 33258 Hoechst and examined by fluorescence microscopy. Those in (B) were previously photographed to record fluorescence, as in (A), and then restained with Giemsa. Chromosomes in (C) were exposed to visible light while mounted in a solution containing 33258 Hoechst and then stained with Giemsa. S.C.E.'s are indicated by horizontal lines.

Sister chromatid differentiation has been used to study DNA polarity and chromatid segregation. Analysis of S.C.E.'s in endoreduplicated chromosomes (94), which permit assignment of the cell cycle during which the exchange occurred, corroborated Taylor's earlier conclusion that the rejoining of chromatid subunits during S.C.E. formation is restricted by DNA polarity (83). In cells which have incorporated BrdU for three replication cycles, an average of only one fourth of the chromatids will contain an unsubstituted polynucleotide chain and hence fluoresce more brightly (53) (Figure 3) or stain more darkly (60) than the remaining material. Reciprocal interchanges result from S.C.E.'s which occurred during the final replication cycle, while earlier exchanges produce non-reciprocal segments of bright fluorescence. This permitted an independent test demonstrating the DNA polarity constraint of S.C.E. formation (85). The distribution of bright chromatids in third division cells has also been used to demonstrate that segregation of sister chromatids at mitosis in pairs of homologues is random (53), consistent with previous autoradiographic studies (14). BrdU techniques have in addition provided a simple graphic method of distinguishing cells which have incorporated BrdU for different numbers of replication cycles (86).

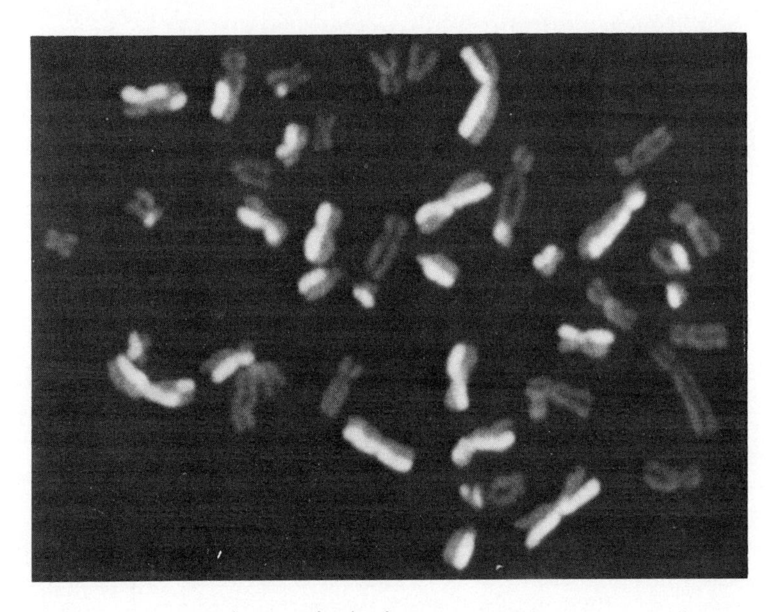

Fig. 3. Third division metaphase. A human peripheral lymphocyte incorporated BrdU for 3 replication cycles. After the chromosomes were stained with 33258 Hoechst, roughly one quarter of the chromatid material fluoresced brightly.

Incorporation of BrdU for one replication cycle has permitted differentiation of sister chromatids in chromosome regions containing DNA with an asymmetric distribution of thymine between complementary polynucleotide chains. This was initially observed in the centromeric regions of mouse chromosomes, reflecting mouse satellite DNA (59), and in the long arm of the human Y and the secondary constriction of human chromosome #16 (50). Subsequent studies have identified other regions of thymine asymmetry in human (6,22) and mouse (30) chromosomes. In metacentric mouse chromosomes and in a dicentric human Y chromosome from cells which had incorporated one round of BrdU, the thymine-rich chains had a contralateral orientation (59,50). This result was interpreted as compatible with a conservation of DNA polarity upon traversing the centromere, a result consistent with previous viscoelastic relaxation measurements on DNA from drosophila chromosomes (38). Recently, the observation of X-ray induced aberrations such as dicentric rings in the absence of S.C.E. formation has been suggested to indicate the existence of a small number of polarity switches in the DNA of Chinese hamster ovary chromosomes (23,93). Independent identification of these chromosomes by banding techniques (e.g. quinacrine) might confirm the postulated intra-chromosomal origin of these ring forms and should indicate whether the hypothesized polarity switches were confined to specific chromosomes.

A number of investigators have attempted to localize S.C.E.'s relative to chromosome banding patterns. For example, in human chromosomes, S.C.E.'s appear to occur preferentially either in between quinacrine(Q)-positive bands or at the junctions of Q-positive and Q-negative regions (45). Subsequent studies detected a clustering of S.C.E. at junctions between heterochromatic and euchromatic regions in muntjac (11), kangaroo rat (10), microtus, and hamster chromosomes (31). The significance of these "junctional" regions is as yet unknown. It has also been reported that S.C.E. occur preferentially within heterochromatic arms of microtus (63) or human chromosomes (77); any disagreement between these and the other studies may be superficial since the latter two papers did not analyze events at the level of individual bands. The recent ability (55) to detect S.C.E. in prematurely condensed chromosomes (69) may provide the resolution necessary to resolve this question.

Thus far, the most extensive use of S.C.E. analysis has been to assess the impact of clastogens on chromosomes. BrdU-dye methodology has been used to show that low doses of alkylating agents such as mitomycin C or nitrogen mustard induce large numbers of S.C.E.'s at concentrations well below those causing significant numbers of chromosome breaks (46) (Figure 4). BrdU itself was observed to induce exchanges, although this effect could be minimized by working at low BrdU concentrations. Numerous subsequent reports confirmed these observations and extended them to include other agents known to damage chromosomes either directly or after metabolic activation (29,37,39, 66,72,79). Since many of these S.C.E.-inducing agents were mutagens and/or carcinogens (66), it was suggested that S.C.E. analysis could be used as an assay for mutagenesis and carcinogenesis. The most notable exception to this correlation was irradiation with X-rays, which is well known for its genetic effects and ability to break chromosomes but is relatively ineffective at inducing S.C.E.'s.

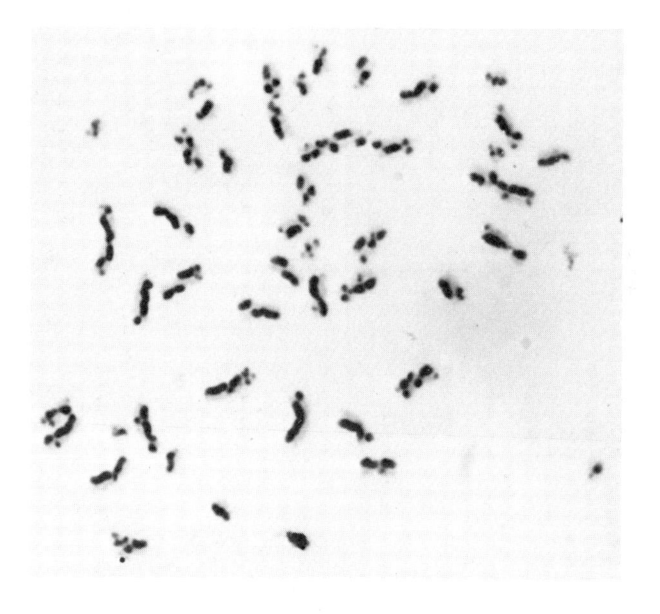

Fig. 4. Induction of sister chromatid exchanges in a human peripheral lymphocyte by mitomycin C. Mitomycin (0.075 μg/ml) was present during the third and final day of culture. Slides were stained as in Fig. 2 (C). More than 50 S.C.E.'s can be detected in this cell; untreated cells exhibit approximately 15 S.C.E.'s.

The scope of S.C.E. studies has been extended from in vitro to in vivo systems. Bloom and Hsu described the formation of S.C.E.'s in ovo in chick embryos (9). Subsequent reports described the induction by alkylating agents of S.C.E. formation in marrow or spermatogonia of mice which received repeated doses of BrdU (2,87), and extension of in vivo S.C.E. analysis to microtus (65), the rat (78) and the mud-minnow (42) has recently been accomplished. The host mediated (5,56) aspects of in vivo systems together with the obvious relevance of spermatogonial damage to germ cell formation make this approach unique for studying environmental mutagenesis.

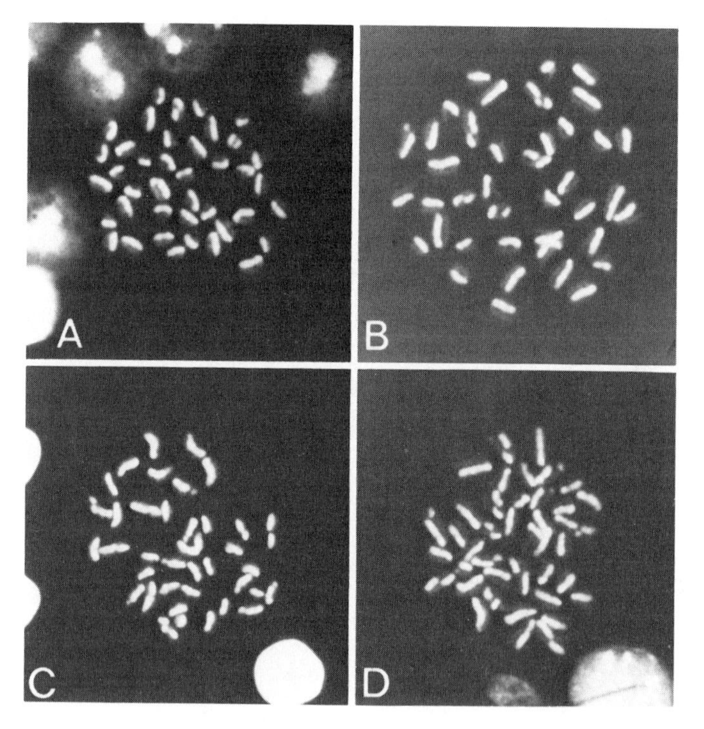

Fig. 5. Analysis of in vivo sister chromatid exchange formation in different tissues of the mouse. CBA mice were administered BrdU tablets subcutaneously, cells were harvested (4), and chromosomes stained with 33258 Hoechst were examined by fluorescence microscopy. The cells shown are from mouse spermatogonia (A), bone marrow (B), thymus (C), and spleen (D).

In contrast to "hybrid" in vivo-in vitro studies, in which lymphocytes damaged in vivo can be isolated and cultured in vitro in medium containing BrdU (80), or in which a microcosmal system capable of activating some agents is added directly to in vitro cultures (81), the in vivo systems permit examination of different processes in multiple tissues of a given organism. For example, it is possible to detect S.C.E. formation in mouse bone marrow, spermatogonia, thymus, and spleen (Figure 5), and to compare S.C.E. induction e.g., by cyclophosphamide (Figure 6) in these different tissues. The initial methodological difficulty in these in vivo studies, requiring multiple BrdU injections (2,87) or continuous BrdU infusion (65,78) because of host metabolism of BrdU, has been overcome by the use of BrdU in the form of a small tablet that can be implanted subcutaneously (4). In addition to analyses of S.C.E., in vivo studies in mice have been extended to a comparison of DNA replication kinetics in tissues such as spermatogonia and bone marrow, and an examination of these processes in meiotic cells has been initiated (3).

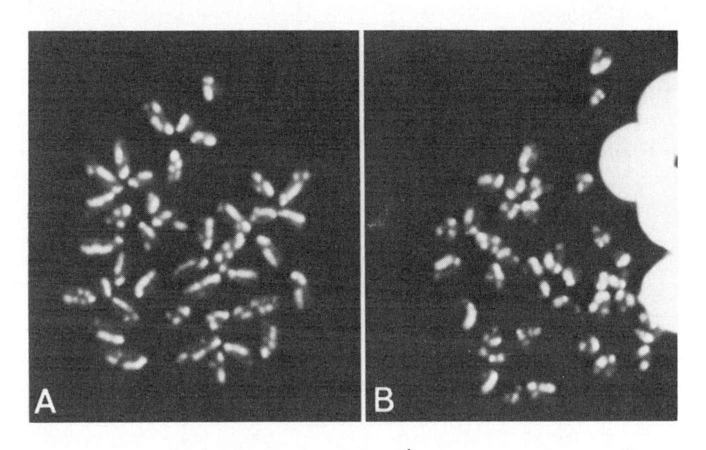

Fig. 6. Induction of sister chromatid exchanges in vivo by cyclophosphamide in mouse bone marrow and spleen cells. Animals were given a single injection of cyclophosphamide (20 mg/kg) eight hours after implantation of the BrdU tablet and seventeen hours prior to cell harvest. Marked induction of S.C.E.'s in both marrow (A) and spleen (B) cells from the same animal, relative to controls (Fig. 5), is apparent.

Analysis of S.C.E. formation has also been used to differentiate between inherited diseases characterized by chromosome fragility and a predisposition for the development of neoplasia (24). These diseases, which include Bloom's syndrome, Fanconi's anemia, and ataxia telangiectasia, presumably involve defects in DNA repair. Cells from patients with Fanconi's anemia have been shown to be highly susceptible to killing (18,19) and to chromosome breakage (7,75,76) by bifunctional alkylating agents, and they appear to exhibit reduced ability to excise UV (68) or gamma irradiation products (70). Patterson et al. (64) reported that cells from patients with ataxia telangiectasia exhibited a reduced ability to excise bases damaged by high energy radiation. While no indication of the biochemical basis of Bloom's syndrome has yet been reported, Chaganti et al. (12) detected an elevated S.C.E. frequency in cells from patients with this desease. In contrast, normal S.C.E. frequencies were detected in Fanconi's anemia, ataxia telangiectasia (21,28) and xeroderma pigmentosum (92).

An S.C.E. "stress test" was developed to probe for partial defects in S.C.E. formation. Lymphocytes from Fanconi's anemia patients, while exhibiting essentially normal S.C.E. frequencies in the presence of BrdU, did not respond to mitomycin C treatment with a normal increase in sister chromatid exchange formation (52). The reduced stimulation by mitomycin C of S.C.E. formation in Fanconi's anemia was associated with increased chromatid breakage. Interestingly, approximately half of these breaks occurred at sites of incomplete sister chromatid exchange formation (49,52), compatible with the hypothesis that the break increment and exchange deficit were causally related.

Our initial studies of lymphocytes from four patients with Fanconi's anemia have been repeated, with similar results, on two other patients with this disease, and analyses of fibroblasts obtained from cell repositories have been augmented with data on fibroblasts from some of the same patients whose lymphocytes had been examined. The results can be interpreted to suggest that Fanconi's anemia cells are defective in a form of DNA repair. More recently, it has been demonstrated that xeroderma pigmentosum cells exhibit behavior which is effectively the converse of that found in Fanconi's anemia. Lymphocytes from patients with xeroderma pigmentosum hyper-react to UV irradiation (8), or alkylating agents (95) undergoing a much greater increase in S.C.E.'s than do identically treated normal cells. Such observations of abnormalities related to S.C.E.'s in genetic diseases contain clues which might ultimately lead to a better understanding both of S.C.E. formation and of the pathogenic processes in each disease.

Implicit in many of the above studies is the assumption that S.C.E. formation bears a direct relationship to DNA repair and mutagenesis. Certain evidence lends support to this idea. For example, alkylation by psoralen plus light, a powerful inducer of S.C.E. formation (48,49), is known to stimulate DNA strand interchange in recombination proficient but not in recombination deficient (REC A) bacteria (13). This observation prompted the suggestion that S.C.E. formation in metaphase chromosomes was somehow analogous to recombinational repair (73,74) in bacteria. A feature complicating this analogy is the possible difference between post-replicational repair processes in bacterial and mammalian cells (57).

Recombinational repair in bacteria may be error-prone, e.g., UV or psoralen plus light induction of mutations in bacteria requires a functional REC A system (33,88). (The relationship of these observations to the error-prone S.O.S. repair system (89) in bacteria remains to be determined.) Also, there is a correlation between the ability of a chemical to act as a mutagen/carcinogen and its ability to induce S.C.E.'s, and S.C.E. formation defects are involved in some of the chromosome fragility diseases. However, direct demonstration of a cause and effect relationship between S.C.E. formation and mutation has not been reported, and the precise relationship of S.C.E. formation to DNA repair and carcinogenesis is still unclear.

The combination of psoralen plus ultraviolet light may prove useful to study the mechanism of S.C.E. formation in greater detail. Psoralens are furocoumarin derivatives that appear capable of intercalative binding to DNA. Upon exposure to near UV light, psoralens, e.g. 8-methoxypsoralen (Fig. 7), can alkylate and often cross-link DNA (13). 8-methoxypsoralen plus light is a powerful

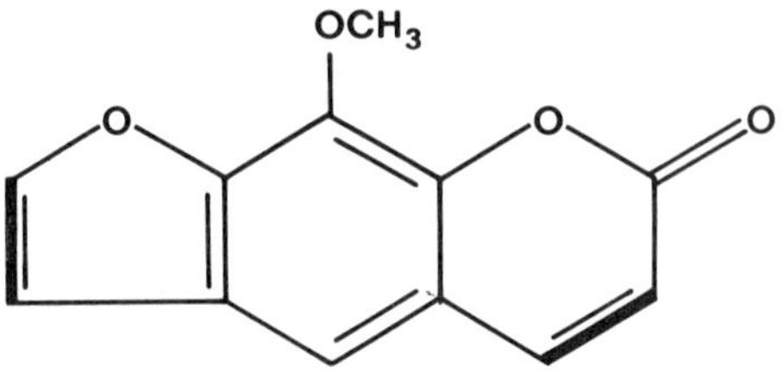

Fig. 7. The chemical structure of 8-methoxypsoralen. Reactive sites are indicated by heavy lines.

inducer of S.C.E.'s in Chinese hamster ovary cells (Fig. 8). These cells can be synchronized (27) and the timing of the DNA damage used to induce S.C.E.'s can be precisely controlled by the choice of illumination times relative to the cell cycle.

Fig. 8. The induction of S.C.E.'s in CHO cells by 8-methoxypsoralen (10^{-5}M) plus near ultraviolet light (~10^4 ergs/mm^2, predominantly at 365nm). This cell was treated prior to two cycles of BrdU incorporation. Slides were stained as described in the caption to Fig. 4. S.C.E.'s in these cells increased approximately 5-fold relative to controls.

Previous autoradiographic studies had indicated that damaged cells needed to traverse the DNA synthesis (S) phase for S.C.E. to be produced (91), and experiments utilizing direct illumination of unsynchronized BrdU-substituted CHO cells have suggested that S.C.E. induction is maximum at the middle of S (35). If BrdU-substituted CHO cells are blocked near the G1-S boundary, exposed to 8-methoxypsoralen and irradiated at fixed intervals after release from this block, the induction of S.C.E.'s (above those observed with psoralen but not light) is significant early in S but diminishes markedly after mid S (Fig. 9). This might occur if formation of S.C.E.'s following alkylation required a certain duration within the S phase, or if replication of the specific region after damage were necessary for S.C.E. formation. The latter alternative would predict that those S.C.E. induced by 8-methoxypsoralen plus light treatment of cells towards the end of S would be most heavily concentrated in late replicating chromosome regions. This indeed appears to be the case as exemplified by a comparison of S.C.E. induction in the X chromosome (the long arm of which is late replicating in these cells) and the 4z chromosome, which gen-

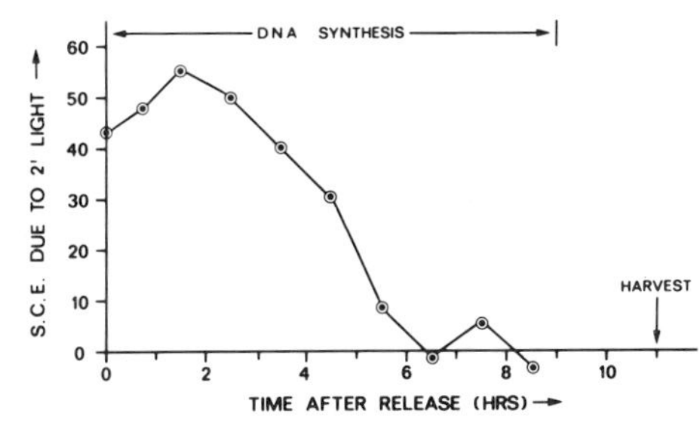

Fig. 9. Time course of S.C.E. induction during the S phase of Chinese hamster ovary cells. CHO cells were synchronized near the G1-S interface (27) after 1 cycle of BrdU incorporation, and then released to undergo a second BrdU incorporation cycle. Cells were exposed to 8-methoxypsoralen and near UV light, as described in the caption to Fig. 9, at various times after release from the hydroxyurea G1/S block. The increment of S.C.E.'s per cell relative to unirradiated controls is shown, as is the approximate interval of DNA synthesis, the latter determined by measurement of ^{3}H dT incorporation.

erally completes replication much earlier in S (16) (Figure 10). Further elucidation of the interrelationship between DNA damage, repair, and S.C.E. formation will require chemical studies on systems such as that described above.

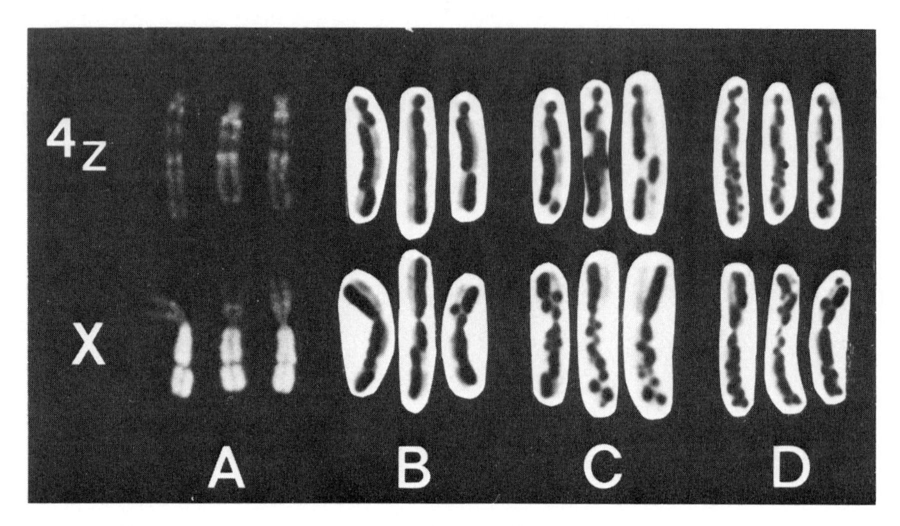

Fig. 10. The relationship between S.C.E. induction and DNA replication in CHO X and 4z chromosomes. The chromosomes in (A) are from cells which sequentially incorporated BrdU and dT so that late replication was highlighted by bright 33258 Hoechst fluorescence. This is especially prominent in the long arm of the X. The remaining chromosomes are from cells cultured as described in the caption to Fig. 9 and then stained to reveal sister chromatid exchanges. Few exchanges are apparent in the untreated control (B). In (C), 8-methoxypsoralen plus light treatment was administered 6 hours prior to cell harvest, and S.C.E. occur predominantly in late replicating regions. In (D), 8-methoxypsoralen plus light treatment was administered at the beginning of the S phase; S.C.E. induction is much more generalized.

Progress in studying DNA interchange at a physical chemical level has been reported by Rommelaere and Miller-Faures (71) and by Moore and Holliday (61). In both of these studies, cells were allowed to incorporate BrdU for nearly one cycle, so that unifilarly substituted DNA was formed. CsCl gradient analyses detected a small amount of DNA with a density intermediate between unifilarly and bifilarly substituted material, presumptively reflecting short regions bifilarly substituted with BrdU. The amount of this DNA increased with mitomycin C treatment and was interpreted in terms of Holliday's model for genetic recombination. This model predicts that there will be a small region of staggered single strand exchange bordered on either side by double strand exchange. Such a region could extend several thousand base pairs and still be too small to be detected (as isolabelling) in condensed metaphase chromosomes. More extensive characterization of this "interchange DNA" should prove interesting, since it might ultimately provide an assay both for clastogen effects and for enzymatic action related to S.C.E. formation.

Analysis of the biological significance of S.C.E. formation will be necessary in order to interpret the results of experiments testing the ability of new compounds to induce S.C.E.'s. S.C.E. induction is compatible with cell survival, as indicated by recent experiments from our laboratory using 8-methoxypsoralen-treated CHO cells, and genetic changes associated with S.C.E. formation might thus be manifest in progeny cells. Additional studies which might be performed include a comparison of the production and removal of chemical damage in the formation of S.C.E.'s and the induction of mutations in parallel cultures of the same cells. Of particular interest would be the relationship between removal of damage such as alkylation (or lack thereof) and S.C.E. formation. Specific genetic loci might be tested and the experiments performed at different points in the cycle of highly synchronized cells. In this regard, it is interesting that mutagenicity of the HGPRT locus in CHO cells is greatest early in S (1,62), as is S.C.E. inducibility, and "baseline" (perhaps BrdU-induced) S.C.E.'s occur predominantly in early replicating regions (47) between bands with bright quinacrine fluorescence (45). The correlation between S.C.E. inducibility and mutagenicity and/or carcogenicity is generally based on data from disparate systems, justifying some reservation about a cause and effect relationship between these processes. In vivo systems in which all three phenomena can be measured may prove especially valuable in investigating this relationship, and thereby increase the confidence with which results of S.C.E. analyses can be interpreted in terms of potential genetic damage.

ACKNOWLEDGMENTS

The technical assistance of Ms. Lois Juergens, Ms. Susan Brefach and Mr. Will Rogers is greatly appreciated. 33258 Hoechst and 8-methoxypsoralen were the generous gifts of Dr. H. Loewe, Hoechst AG, Frankfurt, Germany and the Paul B. Elder Co., Bryan, Ohio, respectively. CHO cells, together with a synchronization procedure, were kindly provided by Drs. Joyce Hamlin and Arthur Pardee. This research was supported by grants from the National Institute of General Medical Sciences (GM 21121), the American Cancer Society (VC-144A) and the National Foundation March of Dimes (1-353). S.L. is the recipient of a Research Career Development Award from the National Institutes of Health (GM 00122), J.W.A. is supported by postdoctoral training funds from the National Institutes of Health (GM 00156) and K.L. is the recipient of an American Cancer Society Postdoctoral Fellowship.

REFERENCES

1. Aebersold, P.M. and Burki, H.J. (1976) Mutat. Res. 40,63.
2. Allen, J.W. and Latt, S.A. (1976) Nature 260, 449.
3. Allen, J.W. and Latt, S.A. (1976) Chromosoma 58,325.
4. Allen, J.W., Shuler, C. and Latt, S.A. (1977) Unpublished Data.
5. Ames, B.N., McCann, J. and Yamasaki, E. (1975) Mutat. Res. 31,347.
6. Angell, R.R. and Jacobs, P.A. (1975) Chromosoma 51,301.
7. Auerbach, A. and Wolman, S. (1976) Nature 261, 494.
8. Bartram, C.R., Koske-Westphal, T. and Passarge, E. (1976) Ann. Hum. Genet. 40,79.
9. Bloom, S.E. and Hsu, T.C. (1975) Chromosoma 51, 261.
10. Bostock, C.J. and Christie, S. (1976) Chromosoma 56, 275.

11. Carrano, A.V. and Wolff, S. (1975) Chromosoma 53,361.
12. Chaganti, R.S.K., Schonberg, S. and German, J. (1974) Proc. Nat. Acad. Sci. USA 71, 4508.
13. Cole, R.S. (1973) Proc. Nat. Acad. Sci. USA 70,1064.
14. Comings, D.E. (1970) Chromosoma 29,428.
15. Comings, D.E. (1971) Nature New Biol. 229,24.
16. Deaven, L.L. and Petersen, D.F. (1973) Chromosoma 41,129.
17. Dutrillaux, M.B., Laurent, C., Couturier, J. and LeJeune, J. (1973) C.R. Acad. Sci. 276, 3175.
18. Finkelberg, R., Thompson, M.W. and Siminovich, L. (1974) Amer. J. Hum. Genet. 26,30a.
19. Fujiwara, Y. and Tatsumi, M. (1975) Biochem. Biophys. Res. Commun. 66, 592.
20. Galley, W.C. and Purkey, R.M. (1972) Proc. Nat. Acad. Sci. USA 69, 2198.
21. Galloway, S.M. and Evans, H.J. (1975) Cytogenet. Cell Genet. 15, 17.
22. Galloway, S.M. and Evans, H.J. (1975) Exper. Cell Res. 94, 459.
23. Geard, C.R. (1976) Chromosoma 55,209.
24. German, J. (1972) Prog. Med. Genet. 8, 61.
25. Goto, K. Akematsu, T., Shimazu, H. and Sugiyama, T. (1975) Chromosoma 53, 223.
26. Gratzner, H.G., Pollack, A., Ingram, D.J. and Leif, R.C. (1976) J. Histochem. Cytochem. 24, 34.
27. Hamlin, J. and Pardee, A.B. (1976) Exper. Cell Res. 100, 265.
28. Hatcher, N.H., Brinson, P.S. and Hook, E. (1976) Mutat. Res. 35,333.
29. Hayashi, K. and Schmid, W. (1975) Humangenetik 29, 201.
30. Holmquist, G.P. and Comings, D.E. (1975) Chromosoma 52, 245.
31. Hsu, T.C. and Pathak, S. (1976) Chromosoma 58, 269.
32. Huang, C.C. (1967) Chromosoma 23, 162.
33. Igali, S., Bridges, B.A., Ashwood-Smith, M.J. and Scott, B.R. (1970) Mutat. Res. 9, 21.
34. Kato, H. (1974) Nature 251, 70.
35. Kato, H. (1974) Nature 252, 739.
36. Kato, H. (1974) Exp. Cell Res. 89, 416.
37. Kato, H. and Shimada, H. (1975) Mutat. Res. 28,459.

38. Kavenoff, R., Klotz, L.C. and Zimm, B.H. (1973) Cold Spring Harb. Symp. Quant. Biol. 38,1.
39. Kihlman, B.A. (1975) Chromosoma 51, 11.
40. Kihlman, B.A. and Kronborg, D. (1975) Chromosoma 51, 1.
41. Kim, M.A. (1974) Humangenetik 25, 179.
42. Kligerman, A.D. and Bloom, S.E. (1976) Chromosoma 56, 101.
43. Korenberg, J. and Freedlender, E. (1974) Chromosoma 48,355.
44. Latt, S.A. (1973) Proc. Nat. Acad. Sci. USA 70, 3395.
45. Latt, S.A. (1974) Science 185, 74.
46. Latt, S.A. (1974) Proc. Nat. Acad. Sci. USA 71, 3162.
47. Latt, S.A. (1975) Somat. Cell Genet. 1, 293.
48. Latt, S.A., Allen, J.W., Rogers, W.E. and Juergens, L. Mutation Res. (in press).
49. Latt, S.A., Allen, J.W. and Stetten, G. Proceedings 1st. Int. Cong. Cell Biol. (in press).
50. Latt, S.A., Davidson, R.L., Lin, M.S. and Gerald, P.S. (1974) Exper. Cell Res. 87, 425.
51. Latt, S.A. and Stetten, G. (1976) J. Histochem. Cytochem. 24, 24.
52. Latt, S.A., Stetten, G., Juergens, L.A., Buchanan, G.R. and Gerald, P.S. (1975) Proc. Nat. Acad. Sci. USA 72, 4066.
53. Latt, S.A., Stetten, G., Juergens, L.A., Willard, H.F. and Scher, C. (1975) J. Histochem. Cytochem. 23, 493.
54. Latt, S.A. and Wohlleb, J.C. (1975) Chromosoma 52, 297.
55. Lau, Y.-F., Hittelman, W.N. and Arrighi, F.E. (1976) Experientia 32, 917.
56. Legator, M.S. and Malling, H.V. (1971) Chemical Mutagens, A. Hollaender Ed., Plenum, Vol. 2, 569.
57. Lehman, A.R. (1975) Life Sciences 15, 2005.
58. Lin, M.S. and Alfi, O.S. (1976) Chromosoma 57, 219.
59. Lin, M.S., Latt, S.A. and Davidson, R.L. (1974) Exper. Cell Res. 86, 392.
60. Miller, R.C., Aronson, M.M. and Nichols, W.W. (1976) Chromosoma 55, 1.
61. Moore, P.D. and Holliday, R. (1976) Cell 8, 573.

62. Nakazawa, N. and Klevecz, R. (1970) J. Cell Biol. 70, 181a.
63. Natarajan, A.T. and Klasterska, I. (1975) Hereditas 79, 150.
64. Paterson, M.C., Smith, B.P., Lohman, P.H., Anderson, A.K. and Fishman, L. (1976) Nature 260, 444.
65. Pera, F. and Mattias, P. (1976) Chromosoma 57, 13.
66. Perry, P. and Evans, H.J. (1975) Nature 258, 121.
67. Perry, P. and Wolff, S. (1974) Nature 251, 256.
68. Poon, P.K., O'Brien, R.L. and Parker, J.W. (1974) Nature 250, 223.
69. Rao, P.N. and Johnson, R.T. (1974) Advan. Cell. Mol. Biol. 3, 135.
70. Remsen, J.F. and Cerutti, P.A. (1976) Proc. Nat. Acad. Sci. USA 73, 2419.
71. Rommelaere, J. and Miller-Faures, A. (1975) J. Mol . Biol. 98, 195.
72. Rudiger, H.W., Kohl, F., Mangels, W., Von Wichert, P., Bartram, C.R., Wohler, W. and Passarge, E. (1976) Nature 262, 290.
73. Rupp, W.D. and Howard-Flanders, P. (1968) J. Mol. Biol. 31, 291.
74. Rupp, W.D., Wilde, C.E., III, Reno, D.L. and Howard-Flanders, P. (1971) J. Mol. Biol. 61, 25.
75. Sasaki, M. (1975) Nature 257, 501.
76. Sasaki, M.S. and Tonomura, A. (1973) Cancer Res. 33, 1829.
77. Schnedl, W., Pumberger, W., Czaker, R., Wagenbichler, P. and Schwarzacher, H.G. (1976) Human Genetics 32, 199.
78. Schneider, E.L., Chaillet, J.R. and Tice, R.R. (1976) Exper. Cell Res. 100, 396.
79. Solomon, E. and Bobrow, M. (1975) Mutat. Res. 30, 273.
80. Stetka, D.G. and Wolff, S. (1976) Mutat. Res. 41, 333.
81. Stetka, D.G. and Wolff, S. (1976) Mutat. Res. 41, 343.
82. Taichman, L. and Freedlender, E. (1970) Biochemistry 15, 447.
83. Taylor, J.H. (1958) Genetics 43, 515.
84. Taylor, J.H., Woods, P.S. and Hughes, W.L. (1957) Proc. Nat. Acad. Sci. USA 43, 122.

85. Tice, R., Chaillet, J. and Schneider, E.L. (1975) Nature 256, 642.
86. Tice, R., Schneider, E.L. and Rary, J. (1976) Exper. Cell Res. 102, 232.
87. Vogel, W. and Bauknecht, T. (1976) Nature 260, 448.
88. Witkin, E.M. (1966) Mutat. Res. 8,9.
89. Witkin, E.M. (1976) Bacteriol. Rev. 40, 869.
90. Wolff, S. (1969) Internat. Rev. Cytol. 25, 279.
91. Wolff, S., Bodycote, J., and Painter, R.B. (1974) Mutat. Res. 25, 73.
92. Wolff, S. Bodycote, J., Thomas, G.H. and Cleaver, J.E. (1975) Genetics 81, 349.
93. Wolff, S. Lindsley, D.L. and Peacock, W.J. (1976) Proc. Nat. Acad. Sci. USA 73, 877.
94. Wolff, S. and Perry, P. (1975) Exper. Cell Res. 93, 23.
95. Wolff, S., Rodin, B. and Cleaver, J.E. (1977) Nature 265, 347.
96. Zakharov, A.F. and Egolina, N.A. (1972) Chromosoma 38, 341.

THE PRODUCTION OF HARLEQUIN CHROMOSOMES BY CHEMICAL AND PHYSICAL AGENTS THAT DISRUPT PROTEIN STRUCTURE

Sheldon Wolff[1,2] and Judy Bodycote[1]

[1]Laboratory of Radiobiology and [2]Department of Anatomy, University of California, San Francisco, California 94143

ABSTRACT. If cells are grown for two rounds of DNA replication in the presence of thymidine analogs such as bromodeoxyuridine or iododeoxyuridine, the chromosomes contain sister chromatids that are chemically different from one another. These sister chromatids can be stained differentially, which allows the detection of sister chromatid exchanges without autoradiography. The earliest studies with such chromosomes indicated that one of the two chromatids was less spiralized than the other. This was attributed to an effect of chromosomal protein synthesis or to the binding of proteins to BrdUrd-containing DNA. The effect has also been attributed to a differential quenching of fluorescent dyes by DNA containing bromodeoxyuridine. The differential spiralization has been confirmed in electron micrographs.

Studies have now been carried out with chemical and physical agents that disrupt protein structure. These result in differential staining, or harlequinization, even in the absence of fluorescent dyes, implicating proteins and gross chromosomal structure in the effect.

INTRODUCTION

In 1972 Zakharov and Egolina (12) allowed chromosomes to replicate for two rounds of DNA synthesis in the presence of bromodeoxyuridine (BrdUrd) so that one of the two sister chromatids would be unifilarly substituted with BrdUrd and the other would be bifilarly substituted. They noted that the unifilarly substituted chromatid stained darkly with Giemsa whereas the other stained lightly. It was subsequently shown that the fluorescent dyes Hoechst 33258 (6) or acridine orange (1,3,8) showed the effect very dramatically and that combined treatments with fluorescent dyes and Giemsa staining (8) could be used to produce permanent preparations that could be observed with the ordinary light microscope.

Zakharov and Egolina believed that the differential staining was caused by a delay of mitotic spiralization in chromosome regions that were undergoing reduplication in the presence of bromodeoxyuridine. This was thought to be brought about either through a transient inhibition of DNA synthesis or through an inhibition of protein synthesis which contributes to the condensation of chormosomes (12). Ikushima and Wolff (2) who had observed that the effect could be brought about in iododeoxyuridine substituted chromosomes as well as those substituted with bromodeoxyuridine, attributed the effect to the differential binding of proteins to bromodeoxyuridine substituted DNA. On the other hand, Latt, who had observed a differential fluorescence of purified DNA containing bromodeoxyuridine, attributed the differential staining observed in chromatids treated with fluorescent dyes to an effect on DNA itself (6).

Studies on the morphology of chromosomes that had replicated twice in the presence of bromodeoxyuridine, or of chromosomes that had replicated once in its presence and then once again in its absence, both of which would produce sister chromatids that are chemically different from one another, have shown that the chromatid containing more bromodeoxyuridine is, indeed, less compact than its sister (4). The effect is manifest at the level of the packing of the 25 nm fiber into the larger chromosomal unit. Korenberg and Ris (4) found that the more heavily substituted chromatid is more open, with looser gyres, than is the other chromatid, and that there was no visible effect on the 10 nm or 25 nm fibers. They have attributed the effect to non-histone proteins affecting the condensation of the chromosomes.

The correlation between faint Giemsa staining and a looser coiling of the chromatids seemed analogous to Sumner's (9) finding that G-banding of chromosomes appears to be a consequence of a varying concentration of protein disulfides and sulfhydryl groups along the chromosomes, and that disruption of disulfide groups, with a concomitant loosening of chromosome structure, leads to faint Giemsa staining. Therefore an attempt was made to see if a similar loosening of the chromosomal structure, which would permit DNA chains to move apart so that they could no longer be crosslinked by either Giemsa or fluorescent dyes, leads to differential staining of sister chromatids in chromosomes from cells grown in the presence of BrdUrd for two rounds of replication.

METHODS AND RESULTS

Exponentially growing Chinese hamster ovary (CHO) cells were grown for 24 hrs (2 rounds of replication) in the presence of 10 μM BrdUrd to produce chromosomes containing unifilarly and bifilarly substituted sister chromatids that will stain differentially especially when treated with fluorescent dyes (1,3,6,8), Giemsa stain following treatment with fluorescent dyes (8,11) or Giemsa stain following treatment with hot salts at high pH (5). Colcemid (final concentration 2×10^{-7} M) was added for the last 2 hrs of culture, and the mitotic cells collected by shaking. After hypertonic treatment to swell the chromosomes and fixation in 3:1 methanol: acetic acid, air dried preparations were made.

The slides were immersed in 2×10^{-3} M solutions of dithiothreitol or β-mercaptoethanol in water (pH 5.0) and kept in the dark for 24 to 72 hrs. They were then rinsed briefly and stained in 3% Giemsa (Gurr's R66 M/15 Sorensen's buffer pH 6.8) for 15-30 minutes. Chromosomes treated in this way showed a marked harlequinization although chromosome morphology was not nearly as crisp as that seen after staining either with fluorescent dyes or the fluorescent plus Giemsa (FPG) technique (Fig. 1). Control slides incubated in either water, buffer, or ethanol, contained chromosomes that had dark uniformly stained chromatids or incipient G-bands.

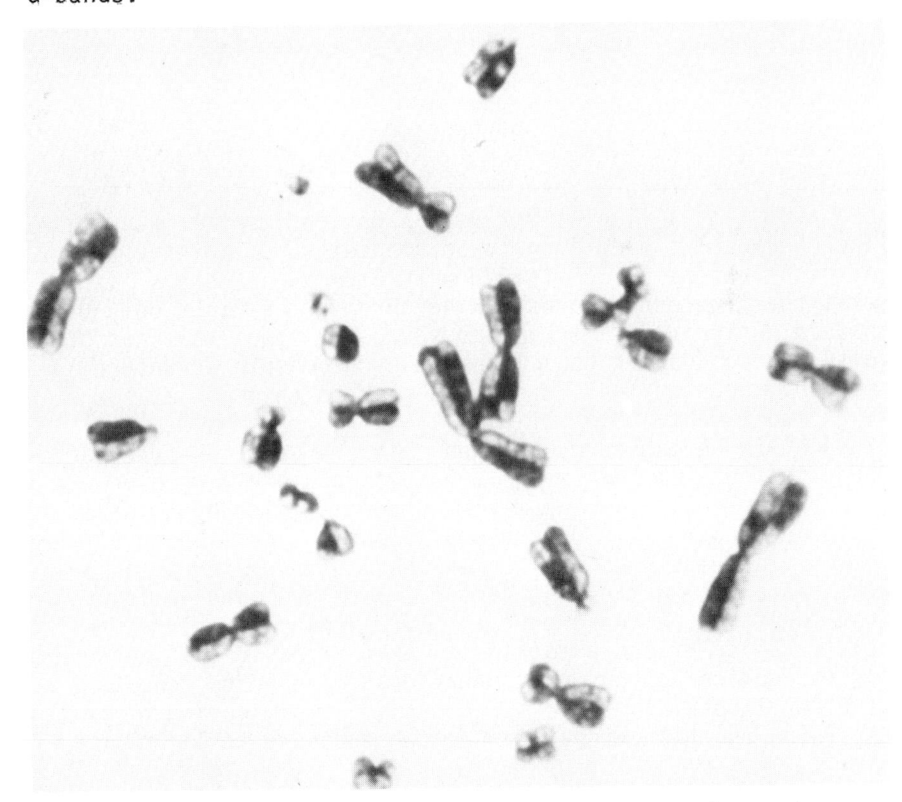

Fig. 1. Harlequin chromosomes produced by immersing slides in Cleland's reagent in the dark before staining.

These results indicated that if disulfide bonds were reduced the two chromatids would stain differentially with Giemsa, even when not treated with a fluorescent dye and light, or with hot salts. This is consistent with the idea that the differential staining usually obtained in such sister chromatids is a consequence of a loosening of the chromosomal structure which permits the DNA chains to move farther apart so that they are crosslinked by dye molecules to a lesser degree as had been proposed for G-banding (9,10).

Since disulfide bonds can be reduced by exposure to light alone (7), some slides were mounted with a cover glass and Sorensen's buffer pH 6.8 and then exposed for 3 minutes to the light from a super pressure mercury lamp (Philips SP 500 bulb, 7.5 x 10^6 J/m^2). Heating was minimized by a stream of cool air from a fan. The slides were then immediately stained in 3% Giemsa for 15-30 minutes. Control slides were simply heated.

Preparations exposed to the light and then stained with Giemsa contained differentially stained harlequin chromosomes that are indistinguishable from those obtained by treating with a fluorescent dye exposing to light and then staining with Giemsa (Fig. 2).

DISCUSSION

It thus appears that in the usual FPG technique (8,11), staining with a fluorescent dye is not in itself a necessity. The chromosomes need only to be exposed to an intense source of light before being stained with Giemsa. It seems likely that the fluorescent dyes on the chromosomes might only be providing an intense localized source of light that can break disulfide bonds and thus loosen the chromosome structure in a way analogous to that found for G-banded chromosomes.

The magenta color of Giemsa-stained chromosomes is thought to be caused by bridges that are formed between adjacent segments of DNA by two molecules of methylene blue and one of eosin (10). Presumably, when disulfide bonds of protein are reduced the chromosomal structure can loosen so that fewer dye bridges can be formed leading to weaker staining. The sister chromatid that contains more bromodeoxyuridine, has more tightly bound protein and a looser structure initially (4). Exposure to light, which breaks disulfide bonds, or to chemical agents that break disulfide bonds, causes a further loosening of the structure and a diminution of the staining in this chromatid. Longer exposure to light or to

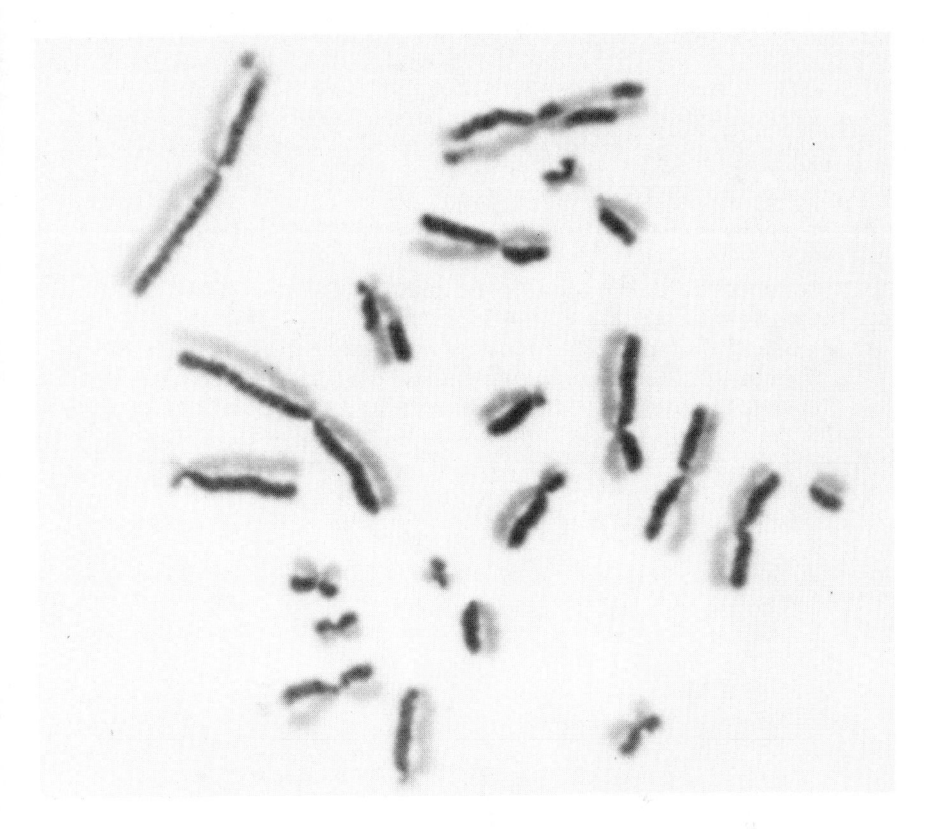

Fig. 2. Harlequin chromosomes produced by exposing slides to intense light from a high pressure mercury burner before staining with Giemsa.

the chemical reducing agents results in both chromatids becoming faintly stained, presumably because the structure of the more tightly packed chromatid now becomes sufficiently loosened to decrease the number of dye bridges.

ACKNOWLEDGEMENTS

Work performed under the auspices of the U.S. Energy Research and Development Administration.

REFERENCES

1. Dutrillaux, B., Fosse, A.M., Prieur, M., and Lejeune, J. (1974) *Chromosoma (Berl.)* 48, 327.
2. Ikushima, T., and Wolff, S. (1974) *Exptl. Cell Res.* 87, 15.
3. Kato, H. (1974) *Nature* 251, 70.
4. Korenberg, J.R., and Ris, H. (1976) Submitted to *J. Cell Biol.*.
5. Korenberg, J.R., and Freedlender, E.F. (1974) *Chromosoma (Berl.)* 48, 355.
6. Latt, S.A. (1973) *Proc. Natl. Acad. Sci. US* 70, 3395.
7. Mousseron-Canet, M., and Mani, J.-C. (1972) *Photochemistry and Molecular Reactions, Israel Program for Scientific Translations,* Keter Press, Jerusalem.
8. Perry, P., and Wolff, S. (1974) *Nature* 251, 156.
9. Sumner, A.T. (1973) *Exptl. Cell Res.* 83, 438.
10. Sumner, A.T., and Evans, H.J. (1973) *Exptl. Cell Res.* 81, 223.
11. Wolff, S., and Perry, P. (1974) *Chromosoma (Berl.)* 48, 341.
12. Zakharov, A.F., and Egolina, N.A. (1972) *Chromosoma (Berl.)* 38, 341.

DNA REPAIR MECHANISMS AND THE GENERATION OF SISTER CHROMATID EXCHANGES IN HUMAN CELL LINES FROM XERODERMA PIGMENTOSUM PATIENTS

James E. Cleaver

Laboratory of Radiobiology, University of California
San Francisco, California 94143

ABSTRACT. Sister chromatid exchanges (SCEs) are induced in xeroderma pigmentosum (XP) cells by radiation and chemicals approximately in proportion to the amount of damage that is not mended by excision repair. The perturbations of DNA replication that are observed when damaged DNA attempts to replicate and which are described as post-replication repair are not involved in SCE formation because (a) these perturbations are early, transient responses in replication whereas SCEs can be generated long after lesions have been introduced into DNA, and because (b) in the xeroderma pigmentosum variant these perturbations are most severe whereas SCE frequencies are at normal levels. SCEs appear to be formed during replication of damaged DNA by double strand exchanges which have yet to be detected and analyzed at the molecular level.

INTRODUCTION

Xeroderma pigmentosum (XP) is a human inherited skin disease in which there is a high incidence of sunlight induced cancers and which exhibits a variety of defects in repair of damaged DNA (1-3). XP cell lines therefore provide a useful system in which to study the relationship between DNA damage, DNA repair systems, cell survival, chromosome aberrations, sister chromatid exchanges, mutagenesis and carcinogenesis. We will consider for present discussion the salient features of known DNA repair systems in human cells (i.e. photoreactivation, excision repair and post-replication repair (3,4) and the published (5,6) and preliminary (7) data available for the frequencies of sister chromatid exchanges (SCEs) induced by various agents in normal and XP cells.

DNA REPAIR SYSTEMS. Three main kinds of repair systems involved in mending damage to DNA have been identified (4). One, photoreactivation, simply reverts one kind of damage in DNA to the original normal chemical state, without removing or exchanging any material. The photoreactivating system is specific for cyclobutane pyrimidine dimers that are produced by UV light and has no versatility. Although a photoreactivating enzyme has been isolated from human cells (8,9), it is only possible to demonstrate a light activated monomerization of pyrimidine dimers in living cells under unusual experimental conditions and then only to a small extent (9-11). Under most experimental conditions employed in tissue culture experiments, therefore, the action of this repair system is of no practical concern.

Excision repair is extremely versatile and can mend an almost infinite variety of UV light, X ray, and chemically-induced forms of damage in DNA (12,13). This system excises damaged single strand regions of DNA and replaces them with a new sequences of bases according to Watson Crick base pairing required by the opposing intact strands of DNA. Excision repair has two main branches according to whether the damage is excised as nucleotides or bases (4,14). These branches may correspond to the large and small patch forms of repair (13). The majority of XP cell lines are defective to varying degrees in their ability to perform nucleotide excision repair, and five mutually complementing excision repair defective groups can be classified on the basis of *in vitro* cell hybridization (3,15) (Table 1).

There is no certainty that this set of 5 groups is exhaustive, nor that the groups are completely homogenous, and the characteristics of each group have not yet been clearly established. In general, the most severe cases with extensive central nervous system disorders that have been described as the de Sanctis Cacchione syndrome fall into group A (2). Cases with central nervous system disorders that are milder often fall into group D. Group B is composed of only one patient who suffers from symptoms of Cockayne's syndrome in addition to XP, and until other cases of group B can be identified, this group has an uncertain status. Close correlation of symptoms with complementation groups cannot yet be made and group assignments are all made by *in vitro* hybridization (2).

When cell-free extracts of XP cells were assayed on a substrate consisting of purified UV-irradiated DNA all of the excision defective XP cells investigated were able to excise at normal rates (16). This evidence led to a hypothesis that XP cells may be defective in DNA binding proteins or chromatin cofactors that dissociate UV damaged DNA from histone proteins that constitute the nucleosomes (17). Once dissociated, the DNA can be repaired and subsequently repackaged (Fig. 1).

TABLE 1

CHARACTERISTICS OF XP COMPLEMENTATION GROUPS

Group	CNS Symptoms[1]	UV Sensitivity[2]	Excision Repair	Post-Replication Repair[3]
A	+ -(XP8LO)	6.5-17%	0 30%	Slow, caffeine sensitive
B	+		<10%	Slow, caffeine sensitive
C	-	25-53%	<10%	Slow, caffeine sensitive
D	+	15%	30-50%	Slow, caffeine sensitive
E	-		30-50%	Normal
V (Variant)	-	77-100%	100%	Very slow, very caffeine sensitive

[1]Central nervous system disorders associated with characteristic skin sensitivity to sunlight. + indicates presence and — absence of CNS symptoms. Group A generally has cases showing very severe symptoms from birth, the de Sanctis Cacchione Syndrome, but a few cases are negative. Group B has symptoms of Cockayne's syndrome. Some patients in group D only develope CNS symptoms during teenage.

[2]Dose modifying factors; ratio of D_{37}s of XP/normal UV survival curves (38,39).

[3]The presence of 0.5 to 1 mM caffeine prevents sealing of gaps involved in post-replication repair (23,40).

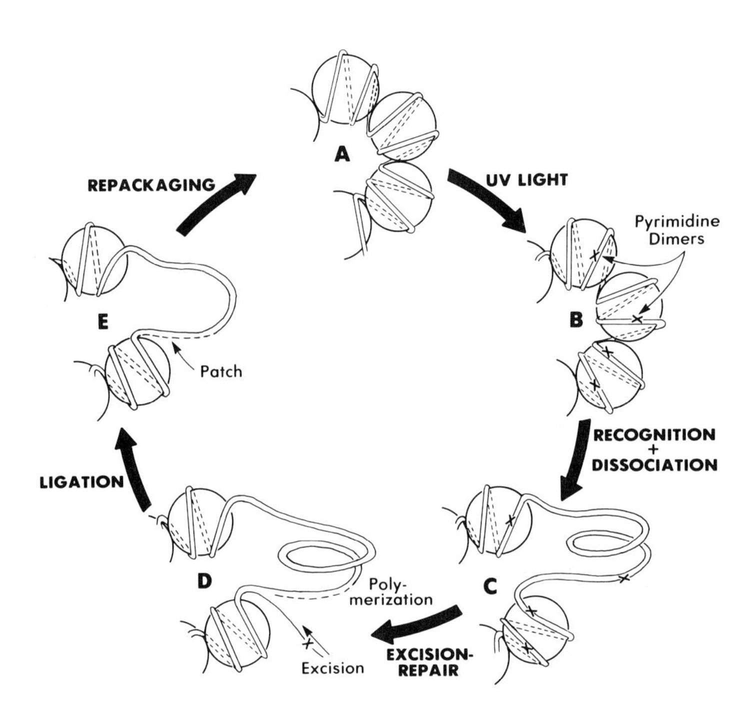

Fig. 1. Model for excision repair showing steps for disassembly and reassembly of nucleosomes in addition to the excision and repair replication of an exposed loop of protein-free DNA.

Circumstantial evidence that supports this model is that the average size of the nucleotide excision repair patch is about half or less of the length of DNA contained in one nucleosome (4). Also, if small patch repair and base excision repair are the same, this branch of excision repair may not require disassembly of nucleosomes, because XP cells perform this kind of repair normally (12,13).

The third repair system post-replication repair, is involved only during semiconservative DNA replication when this occurs within a few hours of cells being damaged. It involves a modification of normal replication in which chain growth in DNA that is replicating at the time of receiving damage is blocked (18) or interrupted (19) (Fig. 2).

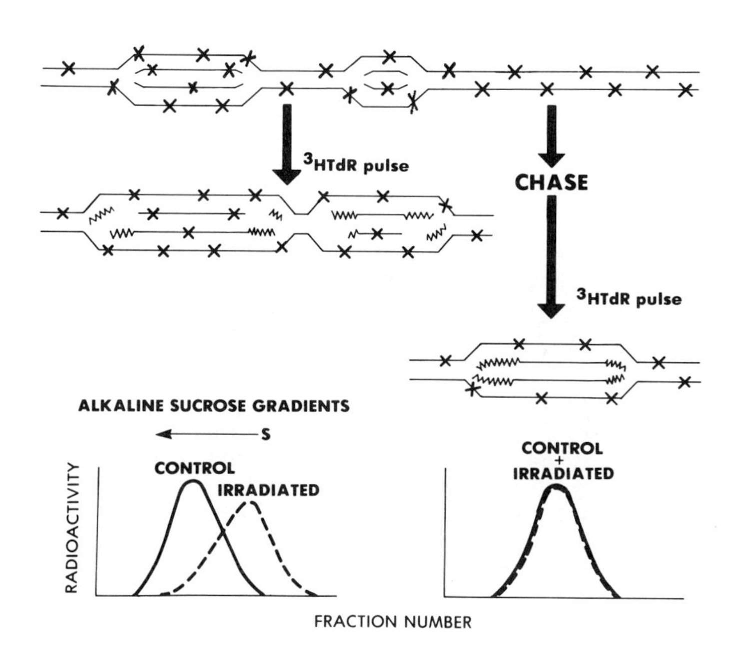

Fig. 2. Model for post replication repair showing the difference between the damage and the response of replicons active in DNA replication at the time of irradiation and those which initiate replication later. (Damage is indicated by X, newly synthesized ^{3}H labeled DNA by ⩘⩘).

Newly synthesized strands have a lower single strand molecular weight than in undamaged cells. During recovery from the block DNA replication appears to continue in the damaged replicons leaving gaps up to 1000 bases long which are filled in by de novo replication (19) and possibly a small amount of single strand exchange (20). DNA that is synthesized many hours after being damaged, possibly on DNA that had begun replication at replicon origins that were inactive when damaged, does not appear to suffer any delays or strand interruptions (21,22). The phenomena associated with post-replication repair are transient responses and confined, perhaps, to those replicons active at the time of being damaged.

The XP variant is a form of XP that appears clinically indistinguishable from most common XP patients, but one in which post-replication repair appears to be slowed much more than in other XPs and normal cells (23) (Table 1), and cells have difficulty sealing some excision repair gaps (24). The slow post-replication repair seen in excision defective XPs (23) may be a secondary consequence in replicating replicons of the defective dimer exision seen in these groups. Extending this line of argument it is possible that excision repair of replicating replicons could be different from that of inactive replicons and by this argument the very slow post-replication repair seen in the XP variants might be a secondary consequence of an excision defect that is confined to actively replicating replicons. The nature of the molecular events involved in post-replication repair are sufficiently complex and ill-understood that a great deal more information is needed about DNA replication on damaged DNA to clarify these issues. It is important, however, to attempt to make distinctions between those cellular defects that are of primary origin as compared to those that are secondary, and between those enzymatic defects that are functionally important and those that have little effect on cellular function (17).

Caffeine is known to inhibit chain elongation during post-replication repair (23, Table 1) but the mechanism is unknown. Since caffeine also can affect cyclic nucleotide levels (25) and purine pool sizes (26) any attempt to attribute an effect of caffeine in irradiated cells to one or other mechanisms is inevitably difficult.

Although all forms of XP exhibit increased UV induced mutation frequencies (27), the defect in the variant appears to have a preferential effect on lesions that are potentially mutagenic (28), whereas defects in the other XPs have equal effects on both potentially mutagenic and lethal lesions (Fig. 3).

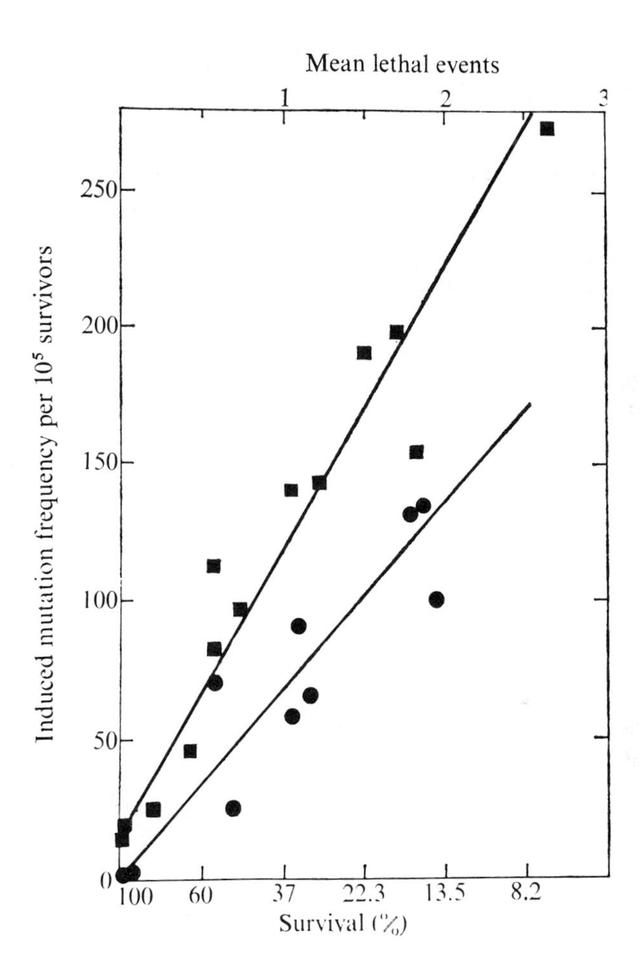

Fig. 3. The relationship between frequency of ultraviolet light-induced mutations and its cytotoxic effect in normal human cells (●) and in variant strain XP4BE (■). Results for excision defective XP cells lie along the line for normal cells (28). Reproduced with permission of the author and from Nature, Macmillan Publications Ltd.

Comparisons between SCEs in XP cells exposed to various DNA damaging agents and the DNA repair capacities and induced mutation frequencies in these cells will allow us to make some inference about what is and is not involved in SCE formation.

SPONTANEOUS AND CHEMICALLY INDUCED SCEs IN XP GROUP A CELLS. Spontaneous frequencies of SCEs in XP cells of all complementation groups are within the range obtained for normal cells (5). In contrast spontaneous levels in Bloom's syndrome are one to 2 orders of magnitude above normal, despite the absence of any known defect in DNA repair in this disease (29,30). The DNA repair defects in XP cells therefore have no effect on spontaneous SCE formation, whereas an unknown defect has a marked effect on spontaneous SCEs in Bloom's syndrome.

In radiation or chemically damaged XP cells (group A), low doses induce higher SCE frequencies than in similarly damaged normal cells (6,7,31). The increase is easily explained in UV and 4NQO damaged cells because XP cells perform low levels of excision repair of these kinds of damage (12,13,32).

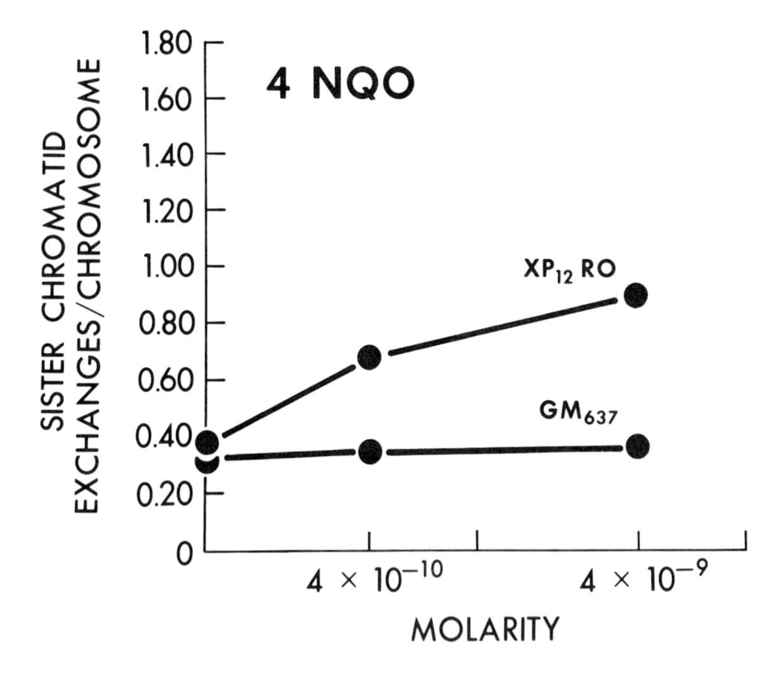

Fig. 4. Frequency of SCEs induced by 4-nitroquinoline-1-oxide in normal human (GM637) or XP12RO (group A) cells. (Figure drawn from data presented in Wolff et al. (6).

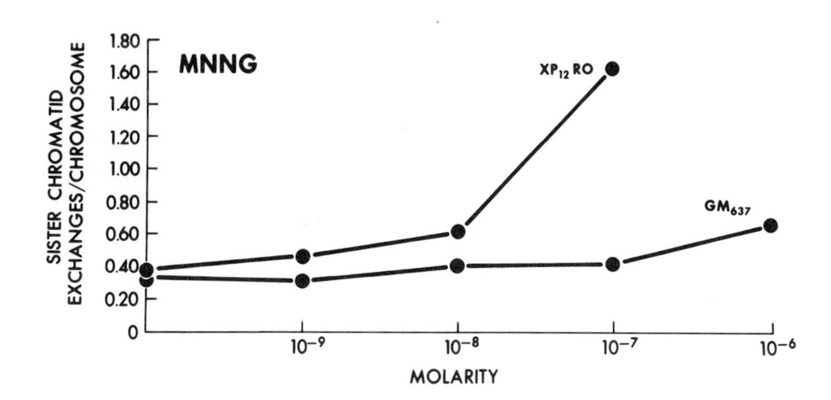

Fig. 5. Frequency of SCEs induced by N-methyl-N-nitro-N'-nitrosoguanidine in normal human (GM637) or XP12RO (group A) cells. (Figure drawn from data presented in Wolff et al. (6).

The increase caused by alkylating agents (including methyl and ethyl methane sulfonates, ethyl nitrosourea, and dimethyl sulfate, 6) requires more careful interpretation. In these cases there is a heterogeneous variety of damaged sites induced in DNA (33) and the majority of the sites are mended by normal amounts of excision repair (12,13). A minority of sites are, however, defectively repaired in XP cells, including for example 0^6 ethyl and methyl guanine (34). In consequence there is sufficient difference in residual damaged sites in XP as compared to normal cells to give rise to a difference in SCE formation. These results indicate that SCEs correlate with the amount of damage remaining in DNA unrepaired and is a very sensitive indication of this damage. Measurement of excision repair, however, indicate how much total damage is repaired and is a coarse average over many kinds of sites and models of repair (35).

ULTRAVIOLET LIGHT INDUCED SCE FORMATION IN XP COMPLEMENTATION GROUPS. A survey made of UV light induced SCE formation in various XP complementation groups by deWeerd-

Kastelein et al. (7) shows a wide range of responses in different cell lines. There is an approximate correlation between the amount of excision repair performed by a cell line and the relative SCE frequency (Fig. 6).

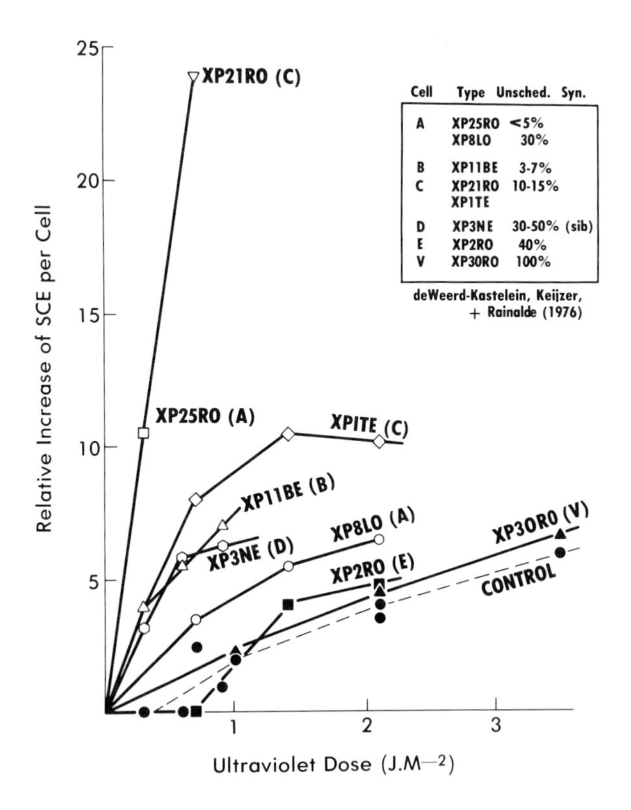

Fig. 6. Ultraviolet light induced SCE frequencies induced in normal and various XP human cells (data drawn from deWeerd-Kastelein et al. (7).

The high, near normal, excision repair levels in XP group E and XP variant cells correlate with near normal SCE frequencies; the extreme repair deficiencies of some group A & C cell lines correlate with high SCE frequency. The strength and generality of this correlation will depend on more detailed studies over a wider range of UV doses and other cell lines, but to an approximate sense these results reveal an inverse correlation between the amount of excision repair a cell line can perform and the SCE frequency. The more unrepaired damage remaining in DNA therefore, the higher the SCE frequency.

The results with the variant are especially interesting here. The abnormality this cell line has in post-replication repair has no effect on SCE frequencies and there is no correlation between the normal frequencies of UV induced SCEs and the high UV induced mutation frequency (Fig. 3).

DISCUSSION

The results described in XP cells, together with the information described previously on SCE formation (35) leads to the following general remarks about the association between repair, replication and SCEs.

(1) The more unrepaired damage that remains in DNA for relatively long periods (i.e., up to 1 or more cell cycles) the higher the frequency of SCEs (6,7,31,35).

(2) Although many kinds of damage induce SCEs, the kind that may be particularly effective is that which interferes with Watson Crick hydrogen bonding during DNA replication (6,34,38).

(3) The perturbations in DNA replication observed as post-replication repair are <u>not</u> involved in SCE formation because these are early transient perturbations, and the XP variant has normal SCE frequencies (Fig. 6).

(4) Although a correlation exists between SCE formation and mutagenesis, to the extent that both are induced by DNA damaging agents in approximate proportion to the amount of damage, the genetic defects in the XP variant increases UV induced mutation frequencies but not SCEs.

The mechanism by which SCEs are generated does not appear to be satisfactorily explained by current models for DNA repair and replication, although a small frequency of short pieces of single strand DNA has been detected after

DNA replication in damaged cells, this is the same in normal and XP cells indicating that it is not involved in SCE formation (20,39). The increased mutation frequency seen in the XP variant may indicate, however, that there are several mechanisms of mutagenesis: one of these may be associated with post-replication repair and involved in events occurring when a gene is damaged at the time of its replication and this mechanism may be altered in the variant; others may be associated with small inequalities in the amount of DNA exchanged during SCE formation and occur independently of the time a gene is damaged.

The molecular mechanisms involved in SCE formation therefore are unclear at present, other than to say that they represent double strand exchange events involving large regions of the genome. It will be necessary to search for mechanisms based on an anticipation of the expected molecular changes, but without being hidebound by current models of replication of damaged DNA that do not offer satisfactory answers or approaches. Such mechanisms, distinct from repair systems that we can currently describe, may underly the spontaneously elevated SCE frequencies in Bloom's syndrome cells (29).

ACKNOWLEDGEMENTS

I am grateful to Dr. E.A. deWeerd-Kastelein for permission to discuss her unpublished data on UV-induced SCEs (Fig. 6), and to Drs. S. Wolff and R.B. Painter for numerous discussions on the topics of this manuscript. Work supported by the U.S. Energy Research and Development Administration.

REFERENCES

1. Cleaver, J.E. (1968) *Nature* 218, 652.
2. Robbins, J.H., Kraemer, K.H., Lutzner, M.A., Festoff, B.W., and Coon, H.G. (1974) *Ann. of Int. Med.* 80, 221.
3. Cleaver, J.E., and Bootsma, D. (1975) *Ann. Rev. Gen.* 9, 19.
4. Cleaver, J.E. (1974) *Adv. in Rad. Biol.* 4, 1, ed. J.T. Lett, H. Adler, M.R. Zelle, Academic Press, New York, N.Y.
5. Wolff, S., Bodycote, J., Thomas, G.H., and Cleaver, J.E. (1975) *Genetics* 81, 349.
6. Wolff, S., Rodin, B., and Cleaver, J.E. (1977) *Nature* 265, 347.
7. deWeerd-Kastelein, E.A., Keijzer, W., and Rainalde, P. Proceedings of DNA repair workshop, May 2-6, 1976, Noordwijkerhout, The Netherlands.
8. Sutherland, B.M. (1974) *Nature* 248, 109.
9. Sutherland, B.M., Rice, M., and Wagner, E.K. (1975) *Proc. Natl. Acad. Sci. US* 72, 103.
10. Sutherland, B.M., and Oliver, R. (1976) *Biochem. Biophys. Acta* 442, 348.
11. Mortelmans, K., Cleaver, J.E., Friedberg, E.C., Paterson, M.C., Smith, B.P., and Thomas, G.H. (in press, 1977) *Mutation Res.*
12. Cleaver, J.E. (1973) *Cancer Res.* 33, 362.
13. Regan, J.D., and Setlow, R.B. (1974) *Cancer Res.* 34, 3318.
14. Lindahl, T. (1976) *Nature* 259, 64.
15. Kraemer, K.H., deWeerd-Kastelein, E.A., Robbins, J.H., Keijzer, W., Barrett, S.F., Petinger, R.A., and Bootsma, D. (1975) *Mutation Res.* 33, 327.
16. Mortelmans, K., Friedberg, E.C., Slor, H., Thomas, G.H., and Cleaver, J.E. (1976) *Proc. Natl. Acad. Sci. US* 73, 2757.
17. Cleaver, J.E., and Friedberg, E.C. (1976 in press) *Proc. of 5th Intern. Congress of Human Genetics*, Mexico City.
18. Edenberg, H. (1976) *Biophysical J.* 16, 849.
19. Lehmann, A.R. (1972) *J. Mol. Biol.* 66, 319.
20. Menighini, R., and Hanawalt, P.C. (1976) *Biochim. Biophys. Acta* 425, 428.
21. Buhl, S.N., Setlow, R.B., and Regan, J.D. (1973) *Biophys. J.* 13, 1265.
22. Lehmann, A.R. (1972) *Europ. J. Biochem.* 31, 438.

23. Lehmann, A.R., Kirk-Bell, S., Arlett, C.F., Paterson, M.C., Lohman, P.H.M., deWeerd-Kastelein, E.A., and Bootsma, D. (1975) *Proc. Natl. Acad. Sci. US* 72, 219.
24. Fornace Jr., A.J., Kohn, K.W., and Kann Jr., H.E. (1976) *Proc. Natl. Acad. Sci. US* 73, 39.
25. Waters, R.A., Gurley, L.R., and Tobey, R.A. (1974) *Biophys. J.* 14, 99.
26. Goth, R., and Cleaver, J.E. (1976) *Mutation Res.* 36, 105.
27. Maher, V.M., and McCormick, J.J. in *Biology of Radiation Carcinogenesis*, ed. R.W. Tennant and J.D. Regan, Raven Press, New York, N.Y. (1976) 129.
28. Maher, V.M., Ouellette, L.M., Curren, R.D., and McCormick, J.J. (1976) *Nature* 261, 593.
29. Chaganti, R.S.K., Schonberg, S., and German, J. (1974) *Proc. Natl. Acad. Sci. US* 71, 4508.
30. German, J., (1976, in press) *Proc. 5th Intern. Congr. of Human Genetics*, Mexico City.
31. Bartram, C.R., Koske-Westphal, T., and Passarge, E. (1976) *Ann. Human Genetics* 40, 79.
32. Stitch, H.F., and San, R.H.C. (1971) *Mutation Res.* 13, 279.
33. Sun, L., and Singer, B. (1975) *Biochem.* 14, 1795.
34. Goth-Goldstein, R., (1977, in press) *Nature.*
35. Cleaver, J.E. (1977, in press) *J. Toxicol. Environ. Health.*
36. Latt, S. (1977, this volume) *Proceedings of ICN-UCLA Symposium on Molecular and Cellular Biology.*
37. Wolff, S. (1977, this volume) *Proceedings of ICN-UCLA Symposium on Molecular and Cellular Biology.*
38. Wolff, S., Bodycote, J., and Painter, R.B. (1974) *Mutation Res.* 25, 73.
39. Waters, R., and Regan, J.D. (1977) *Biophys. J.* 17, 141a.
40. Arlett, C.F., Harcourt, S.A., and Broughton, B.C. (1975) *Mutation Res.* 33, 341.
41. Maher, V.M., Birch, N., Otto, J.R., and McCormick, J.J. (1975) *J. Natl. Cancer Inst.* 54, 1287.
42. Lehmann, A.R., Kirk-Bell, S., Arlett, C.F., Harcourt, S.A., deWeerd-Kastelein, E.A., Keijzer, W., and Hall-Smith, P. (1977, in press) *Cancer Res.*

A POSSIBLE GENETIC MECHANISM OF AGING, REJUVENATION, AND RECOMBINATION IN GERMINAL CELLS

Rolf Martin

Departments of Biochemistry and Chemistry, Brooklyn College of the City University of New York, Brooklyn, NY 11210

ABSTRACT. There is an increasing amount of evidence which relates aging with DNA damage, meiosis with DNA repair, and rejuvenation with meiosis. This evidence suggests that DNA repair during meiosis may help to rejuvenate germinal cells. It is suggested that rejuvenation takes place partly because homologous chromosomes repair each other during synapsis. Such repair is possible at this time because the homologs are closely paired, perhaps to exchange the DNA sequences required for the removal of defects which have accumulated during aging. The available evidence tends to support this model. In the systems which have been studied, it appears that repair replication takes place just after, and only if, the homologs have paired. Many thousands of possible repair sites have been estimated to be present in each cell. Incomplete pairing due to inversion is associated with increased mutation near unpaired regions, as if repair is hindered or absent where pairing does not occur. Exchange of templates for repair at this time may contribute to the genetic recombination, gene conversion and homologous chromatid exchange that are evident after meiosis.

If rejuvenation occurs because chromosomes repair each other during synapsis, then aging is very likely related to an accumulation of defects which can be removed only by recombination and repair between homologous chromosomes. It is noted that such defects could result from incomplete repair and DNA crosslinking, that aging may be related to cell division because these defects form primarily during DNA replication, and that aging and rejuvenation in germinal cells may have evolved as a by-product of mechanisms which promote genetic recombination during synapsis.

INTRODUCTION

Many strains of Paramecium must periodically undergo autogamy or conjugation or else the cells age and then die after several hundred divisions (1, 2). Many other ciliates also require conjugation for continued growth or genetic viability (2, 3). Plasmodia of the true slime mold Physarum age in a similar manner (4). As in the ciliates, the potential for further growth can be restored by sexual activity. The life cycle of Physarum is similar in many respects to that of higher organisms, including ourselves, since meiosis, gametogenesis, and fusion of haploid cells precede the formation of new, young individuals.

Although it is not presently known how rejuvenation takes place, a number of explanations have been suggested (2, 5-10). The reader is encouraged to become familiar with each of these models which may describe several complementary mechanisms for the reversal of aging.

While rejuvenation appears to occur in certain primitive eucaryotes, it is not clear whether it also takes place in germinal cells of multicellular organisms. Instead, these cells may use special protection and repair systems to maintain their genetic integrity over many generations (11). In this context, it is noted that the repair process outlined here may contribute to germ-line stability rather than to rejuvenation.

DNA DAMAGE, AGING AND REJUVENATION

There is considerable evidence that DNA damage may accompany or contribute to aging in a variety of tissues and organisms. For instance, damage to DNA by irradiation is generally associated with decreased lifespan (in man (12), mice (13, 14), insects (15, 16), Paramecium (17) and cultured fibroblasts (18)). In several mammals, chromosome aberrations in liver cells increase more quickly in species which age rapidly (19-21). The amount of DNA repair after ultraviolet irradiation is correlated with the longevity of a number of mammals (22). Progeria and Cockayne's syndrome (diseases which in certain respects resemble premature aging) are associated with deficient or abnormal DNA repair (23-26). A correlation between aging and the formation of tightly linked DNA-protein complexes has been observed (27, 28). Large numbers of ribosomal genes are damaged as certain tissues age (7, 27-29) and in some instances DNA nicks accumulate (30-34). In tissue culture, fibroblasts exhibit continually increasing rates of mutation (35) and chromosome aberration (36, 37).

Genetic instability and disorganization during senescence have been described in ciliates (38, 39) as well as in multicellular organisms.

Although the significance of this evidence is controversial (14, 25, 40, 41) it is sufficient to stimulate intense investigation of the relationship between genetic deterioration and aging.

If germ-line aging is accompanied by appreciable genetic damage, then rejuvenation undoubtedly involves either the repair or replacement of damaged DNA. This line of reasoning must lead us to wonder: Is DNA repaired during rejuvenation? If so, what types of damage are removed? And by what mechanisms? The explanations in this manuscript were developed in response to these questions.

MODEL OF AGING: INCOMPLETE REPAIR OF DNA

During excision repair, the undamaged, complementary strand of DNA serves as a repair template- information from this strand is used to insert the correct sequence of bases in the damaged region. Excision repair is not always possible however. It is temporarily prevented if replication takes place (before repair occurs) and a gap remains opposite the damaged bases which do not participate in the replication process (42, 43). Repair is prevented in this situation because the complementary strand now lacks the sequence needed as a template for repair replication (Fig. 1). When this occurs, the undamaged sister chromatid can provide the required repair template during the post-replication (recombination) repair process (42-44).

Accurate excision repair is also prevented when both strands are damaged in the same region of the double helix (44). When this occurs, neither strand contains an intact template for repair and as a result, neither strand can help to repair the other. When both strands are damaged in this way, repair templates will not be present even after replication has taken place (Fig. 1). Thus the sister chromatids cannot recombine to restore the original sequences. It is suggested that damage of this type may not be repaired because repair templates are not available. Consequently, it may accumulate and contribute to cellular aging.

A critical requirement of this mechanism is that it be reversible, at least in germ-line cells. If the aging clock is not periodically reset to zero, subsequent generations will be progressively older until they senesce and die.

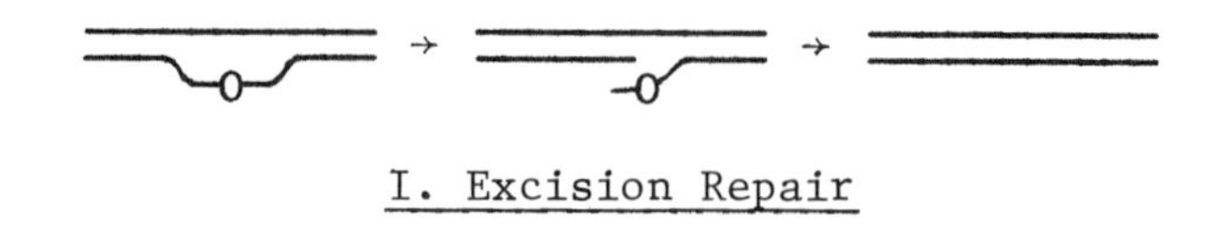

I. Excision Repair

II. Recombination Repair

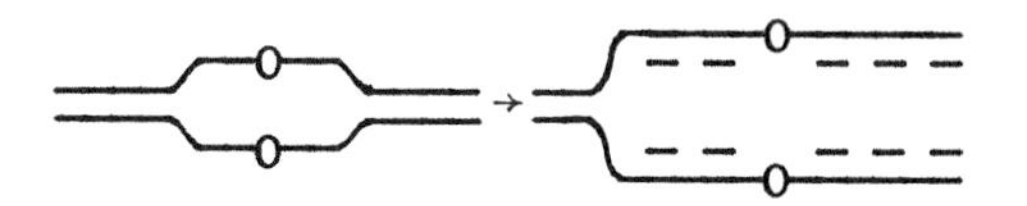

III. Damage to both complementary strands.

IV. Damage to both sister chromatids.

Fig. 1. These diagrams are not intended to be accurate representations of repair but to illustrate the principal features of information transfer after DNA has been damaged. During excision repair (I.), the undamaged complementary strand is used as a template for repair replication. During recombination repair (II.), the undamaged sister chromatid provides a repair template since none is present within the damaged chromatid. If both strands are damaged as in (III.), then neither can be used to quide repair replication. Intact templates cannot be obtained from the sister chromatid if both parent strands were damaged prior to replication (IV.). Perhaps templates contained in the homologous chromatids can be used instead.

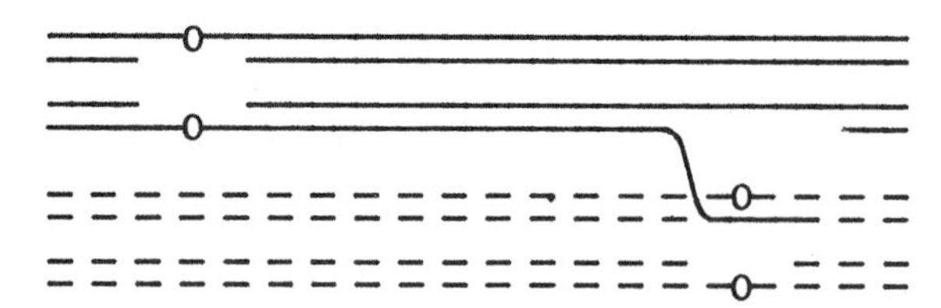

Fig. 2. Schematic representation of repair during synapsis. Homologous non-sister chromatids may be needed to provide suitable repair templates if none are available from the sister chromatids.

MODEL OF REJUVENATION: REPAIR DURING SYNAPSIS

Whenever both strands are damaged as described, the correct base sequences cannot be restored unless additional repair templates are used to provide the missing information. When "unique" sequences of DNA are involved, suitable templates can be obtained _only_ from the corresponding region of the homologous chromosome. It is therefore logically necessary that the accurate repair of accumulated damage of this kind involve repair between homologous non-sister chromosomes (Fig. 2).

But homologous chromosomes must be properly aligned and immediately adjacent to each other in order to repair each other. This requirement is satisfied at least once during the life cycle of most eucaryotes, when the homologs lie alongside each other during synapsis. Do chromosomes miss this opportunity to exchange information for repair? The association between synapsis and rejuvenation in various ciliates and in _Physarum_ is consistent with this possibility. Additional evidence of repair at this time is summarized in the next section.

EVIDENCE CONSISTENT WITH THE MODEL OF REJUVENATION

If chromosomes repair each other during synapsis, repair activity should be associated with homolog pairing. Evidence from a number of different systems suggests that repair generally does occur when pairing takes place. A rather sharp peak of radiation-induced DNA synthesis is seen in human and mouse spermatocytes at early pachytene, when the homologs have just become completely paired (45, 46). Increased DNA synthesis during pachytene is observed in undisturbed cells as well, in newt (47), mouse (45) and human (46) spermatocytes and in meiocytes of lily (47, 48) and possibly wheat (49).

Nicking and resealing of breaks are also observed during pachytene. Measurements of single-strand breakage and rejoining during pachytene in _Lilium_ suggest that more than one hundred thousand repair sites are present in each cell undergoing meiosis (48, 50, 51). Programmed nicking of DNA also takes place in _Saccharomyces_ (52). Fewer repair sites were seen, however, perhaps because resealing of gaps is more rapid in yeast than in lily.

The postulated relationship between pairing and repair is corroborated by observations that replication and nicking during pachytene appear to depend upon the prior completion of synapsis (48, 53), as if activation of repair enzymes is delayed until templates from the homolog are available. Additional evidence that pachytene DNA metabolism is "totally specialized for repair replication activity" (51) has been reviewed previously (45, 48, 51).

If repair occurs during synapsis because homologous repair templates are situated near damaged regions, then repair should be prevented or inhibited if homologous sequences are not adjacent. Thus repair should be less effective in the absence of accurate pairing and genetic instability should be apparent. Instability and incomplete pairing do seem to occur together. In irradiated Drosophila, lethal mutations are clustered near regions which have not paired due to heterozygous inversions, as if repair is incomplete in unpaired sections but not elsewhere (54). Spontaneous mutation frequency is also correlated with the extent of heterozygous inversion and mispairing during meiosis (54). Similarly, the unpaired X chromosomes of male Drosophila undergo much more spontaneous mutation than do the paired X chromosomes of female flies (55). Chromosome breaks which lead to inversion have also been linked with faulty pairing (56).

Genetic recombination and gene conversion should be observed if templates are exchanged between chromosomes. Recombination and gene conversion are, of course, associated with synapsis. And both are increased by prior damage to DNA as if related to the repair of previously acquired damage (57-60). The experiments of Hammerl and Klingmuller (60) illustrate this clearly. In these studies, a mycelial suspension of Neurospora (asco strain) was spread on agar plates and exposed to ultraviolet light, immediately or after one to several days of growth. Three days after plating, when a mycelial lawn had formed, conidia of the wild type were dusted onto the plates to initiate mating. Under these conditions, the pachytene stage of meiosis is expected after an additional three days and spores may be analyzed for gene conversion after eleven days. Two aspects of the data obtained in this manner are especially intriguing. (1.) The patterns of gene conversion in irradiated and control samples are quite similar, suggesting that irradiation does not fundamentally alter the normal mechanisms of conversion. (2.) The frequency of gene conversion is increased in all groups irradiated before plasmogamy even though an undetermined number of division cycles and up to five or six days have passed after irradiation and before pachytene, when conversion is believed to occur. Similar results were obtained by Prudhommeau and Proust (57) when they investigated the effect of ultraviolet light on recombination in Drosophila. They observed that meiotic recombination is increased by irradiation of primordial germ cells even though "un grand nombre de divisions" intervene between ultraviolet treatment and meiosis. This effect was shown to be partially photoreactivable.

Genetic recombination and unrepaired damage are further linked by biochemical evidence indicating that a nuclease apparently required for recombination in *Ustilago* will preferentially degrade DNA which contains mismatched bases and pyrimidine dimers (61).

But why do the effects of irradiation persist for so long until meiosis? And why is exchange between the homologs then increased? Is persistent damage being removed?

Gerontologists have been puzzled for decades by the delayed effects of radiation on survival. (X-irradiation of young clones of *Paramecium* or fibroblasts, or of young insects or mammals, decreases longevity by lowering viability many mitotic generations, weeks, months or years after exposure- references cited earlier.) Perhaps temporarily irreparable damage is in some way responsible for the delayed effects on both viability and genetic recombination.

Because the repair of accumulated lesions, as described in this paper, must involve the homologs rather than sister chromatids (see Figs. 1 and 2), exchange of DNA should occur primarily between homologous chromatids, and not between sister duplexes. For crossovers involving double strand exchange, this has been demonstrated (62-64). The mechanisms which have been outlined provide an additional explanation for the predominance of non-sister chromatid exchange during synapsis (but see Egel (64) who has emphasized the role of zygotene synthesis rather than pachytene repair).

Each of these observations has been reported without reference to aging or rejuvenation. This evidence indicates that further investigation is warranted to determine whether, and to what extent, homologs may repair each other.

DAMAGE TO BOTH STRANDS: A CLOCK MECHANISM?

A relationship between aging and crosslinking of DNA is suggested by evidence that bifunctional alkylating agents (which can crosslink DNA) shorten the lifespan of mice while a monofunctional alkylating agent (which does not form crosslinks) has no effect on their lifespan even when applied at doses which result in high levels of cancer (14). Crosslinking of DNA appears to be a regular occurrence within undisturbed cells since deficient repair of crosslinks is associated with high rates of spontaneous chromosome aberration and Fanconi's anemia (65). Noncovalent crosslinking of DNA may occur routinely if tightly bound chromatin proteins (which accumulate during aging- 27, 28, 66-69) prevent nucleosomes from unfolding during replication. (More detailed discussions of intrinsic crosslinking have been presented by Bjorksten (70) and Cutler (71).)

It is suggested that an accumulation and removal of crosslinking damage may take place as follows. When interstrand crosslinks form, both complementary strands of DNA are damaged within the same region and thus may not repair each other (44). As described previously (72, 73), a short segment of DNA which contains the crosslink can be excised from one of the damaged strands. If this occurs so that a gap is introduced opposite the damaged region of the unsevered strand, then accurate repair replication seems impossible without additional repair templates (44). These may be supplied by the sister chromatid or by the homologous chromatids during synapsis. It appears, however, that the homolog and not the sister chromatid must supply the missing templates whenever crosslinks occur prior to replication so that both sister chromatids are subsequently defective. Damage due to interstrand crosslinking may therefore accumulate until the homologs are aligned properly during synapsis.

During the past several decades there has been a slow increase in the amount of evidence concerning the persistence and removal of "reversible" damage. Some of this evidence is concisely described in the quotation (below) taken from Charlotte Auerbach, Annals of the New York Academy of Sciences, 68, 731-748 (1957).

> The majority of mustard gas-induced mutations or chromosome breaks occur first as labile spots on the chromosomes that subsequently, often after many cell divisions, may give rise to a mutation or chromosome break. Alternatively, chromosomes may revert to their original, normal condition. Delayed effect has also been proved for TEM, diepoxybutane and formaldehyde, and has been revealed as probable for a number of other chemical mutagens.

More recently, these results have been extended (74-76). It appears that many different types of lesions can persist through several cycles of replication.

But damage to one strand which survives over a number of cell divisions may very often result in a double-strand defect that cannot be removed until meiosis. This can occur after but one cell cycle if a gap is formed during replication (as described earlier) which is not removed prior to cell division. When this happens, one of the daughter cells will inherit the chromatid with both strands unable to repair each other.

Aging in dividing cells is very often related to cell division, rather than to metabolic or chronological time (in cultured fibroblasts (77-79), in *Paramecium* (80)). The sequence of damage that has just been described suggests an explanation for the clock mechanism: aging may be related to cell division because additional damage to both strands occurs every time DNA replicates. As a result, cumulative impairment during aging can be determined largely by the number of replication cycles. Crosslinking and coincidental damage to both strands may also contribute to the aging clock since DNA may be more susceptible to such damage when it is exposed during replication (81).

Previous explanations for the relationship between cell division and aging have been based upon incomplete DNA replication (82) or programmed modification of specific sequences (83). As noted by Holliday (84), these models fail to account for the effects of 5-fluorouracil, increased temperature, and radiation on the *in vitro* lifespan of fibroblasts. They also do not explain why cells of the same generation often have very different capacities for subsequent division (85). In contrast, explanations which include replication-dependent damage to DNA can satisfactorily account for these aspects of aging since DNA damage may be introduced by such treatments and may affect cells of the same generation quite differently.

EVOLUTION OF CLONAL AGING

Under suitable conditions, bacteria can divide indefinitely. The same is true of many unicellular eucaryotes. On the other hand, certain unicellular eucaryotes (such as *Paramecium*) age and die after a limited number of divisions unless they are periodically rejuvenated by sexual activity.

Why did clonal aging evolve? By what series of alterations did primitive eucaryotes acquire the ability to age and then be rejuvenated? What advantages were realized? Before an attempt is made to answer these questions a brief discussion of the relationship between DNA damage and recombination is necessary.

In procaryotes, repair of damaged DNA often results in genetic recombination (86). It is emphasized that DNA damage as well as repair is necessary for the operation of this recombination pathway. Because of this, damaged DNA may be considered to be raw material for recombination and formation of genetically diverse, recombinant organisms. And recombinogenic repair systems can be considered to be mechanisms by which damage can be converted to genetic diversity.

Apparently a similar recombination pathway is present in eucaryotes. Again, DNA damage leads to increased levels of recombination (57-60). But in eucaryotes, recombination takes place primarily during meiosis and between homologous chromatids rather than at random. We must wonder how mechanisms evolved to restrict the timing and nature of recombination in eucaryotes. Did systems evolve to regulate the damage and repair which constitute the recombination pathway just outlined for procaryotes? Rephrasing the model of aging and rejuvenation presented in the previous sections will enable us to see how this may have occurred.

ENRICHMENT OF RECOMBINOGENIC DAMAGE

Certain kinds of damage are repaired by recombinational mechanisms while other kinds are repaired without exchange of DNA templates between chromatids. Of those lesions whose repair requires exchange between chromatids, some can be accurately repaired with templates from the sister chromatids while others may require sequences from the homologs. It is therefore possible to have selective accumulation of those lesions which require genetic exchange between homologs for repair. This is critical since exchange between sister chromatids does not often lead to genetic recombination and since exchange between nonhomologous chromosomes very frequently results in lethal aberrations. Mechanisms for enrichment of recombinogenic damage (i.e. for those lesions which stimulate homolog exchange during repair) may therefore represent an evolutionary attempt to get the most recombination from the lowest levels of accumulated damage. Such mechanisms could at the same time serve to minimize the deleterious side-effects of damage without sacrificing a large portion of the recombination produced during repair.

The following system of enrichment is suggested: after each cycle of replication, much of the existing damage is removed by excision and recombinational repair processes but not that which requires repair templates from the homologous chromosomes. As a result, the lesions which remain after each mitotic cycle are generally (or very often) those that must be repaired during synapsis when sequences from the homologs are available. In this way the genome can be enriched during a number of division cycles just in those kinds of damage which stimulate repair and recombination between homologs at the next meiosis.

Cole has commented that crosslinks may lead to exchange at meiosis (44). Recently, Lin *et al.* (87) have reported evidence that crosslinks (and not damage to single strands) stimulate recombination between homologous sequences of phage DNA. It therefore seems that selective accumulation of crosslinks, or derivative lesions, may be particularly useful in stimulating recombination. Crosslinks, then, may represent one kind of recombinogenic damage.

REPAIR OF ACCUMULATED DAMAGE

If aging is accompanied by an accumulation of damage which is useful for recombination, then aged cells must periodically repair this damage to avoid accumulating lethal quantities. This requirement indicates the need for an additional repair system which functions at least once during the life cycle. Only with such a system can an accumulation of "enriched" damage be tolerated over a number of mitotic divisions and then used to generate recombinant progeny during repair. It is therefore suggested that the development of mechanisms by which the homologous chromosomes repair each other was a critical event in the evolution of clonal aging.

SIMULTANEOUS EVOLUTION OF CLONAL AGING AND REJUVENATION

Levels of damage cannot rise permanently if clones that accumulate damage simply age and die. Thus clonal aging may have evolved gradually- no faster than mechanisms of rejuvenation.

First, slightly elevated rates of DNA damage (or less effective repair) may have increased genetic recombination and decreased viability by only a small amount. At this point, aging may not have been severe enough to cause clonal extinction or even to slow the division rate. It is possible these organisms were similar in this respect to present day *Tetrahymena*, in which aging is too mild to slow the speed of growth or to kill all cells which fail to conjugate (3). Perhaps organisms at the next stage in the evolution of aging had levels of damage which were just barely sufficient to cause clonal senescence and extinction. It is possible they aged under certain conditions but not others. Such organisms may also have given rise to progeny with quite diverse rates of aging, as do many of the present day ciliates (3) and slime molds (4). In species of this type, many clones age and all members die if sexual activity does not accur frequently. But in other clones, aging is so mild that extinction is delayed for years. Perhaps only after the evolution of effective machinery to rejuvenate did these organisms acquire and maintain levels of damage high enough to guarantee clonal senescence. Aging may then have become sufficiently intense to ensure a steady turnover of clones and to eliminate those which do not participate frequently in sexual activity (88).

But we need not require direct evolutionary justification for clonal aging. Aging alone need not be advantageous if the mechanisms of aging and rejuvenation in tandem are beneficial. The evidence that certain types of DNA damage and repair can increase genetic recombination suggests that aging and rejuvenation, together, may represent a series of mechanisms for systematic ("senesexual") recombinogenesis.

DISCUSSION

General aspects of aging: Irradiation of paramecia, cultured fibroblasts, insects, mice and men, as described earlier, or inbreeding, e.g., of *Tetrahymena* (89), *Drosophila* (90) or mice (91), generally shorten survival, while longevous parents, as in *Paramecium* (92), *Drosophila* (93) and man (94, 95), or dietary restriction, in the ciliate, *Tokophrya* (96-98), in *Planaria* (99), dog ticks (100) and various rodents (101-104, see 104 for a recent review) are associated with greater longevity. An accumulation of pigment, e.g., in *Paramecium*, *Tokophrya*, cultured fibroblasts, insects and several mammals (105), and of DNA nicks, mutations or chromosome aberrations (in *Paramecium*, cultured fibroblasts, and a number of mammalian tissues as mentioned earlier) is very often observed. These and the many other similarities between the aging of single-celled and multicellular eucaryotes (noted in 105-108) support the possibility that certain mechanisms of aging are present throughout the animal kingdom and may even occur in all living organisms. (See 11, and 109 for discussions on this point.)

This is not to suggest there are no differences in the aging of unrelated species. Many differences, such as those outlined by Finch (110), are to be expected- particularly between organisms that differ in the extent to which aging involves intrinsic as opposed to developmentally regulated deterioration (11, 27). Differences in the aging of certain species may also arise from variations in the relative importance of aging in postmitotic compared with dividing tissues. Nevertheless, the several similarities listed above are generally evident, providing some justification for speculating that basic mechanisms for repair and rejuvenation may be equally widespread.

Unrepaired damage: To identify the basic mechanisms of aging, many investigators have focused their attention on DNA and chromatin. Hart, Trosko and Wilkins (75, 111-113) have emphasized that mammalian aging may result from accumulated, unrepaired damage of the "cyclobutane dimer form," postulating that repair is often delayed or prevented by a sheath of chromatin proteins. Many others (71, 114, 115) have noted that tightening of the chromatin sheath itself may block genic activity and contribute to aging. Cutler and coworkers (27, 28) have recently obtained evidence suggesting an accumulation of very tightly linked protein-DNA complexes, and have added to earlier observations (7, 9, 29) that ribosomal genes are especially susceptible to damage. Here it is suggested that damage to both strands within the same region of the double helix may accumulate because repair templates are not available. Each kind of damage may accompany and in some cases contribute to aging, in varying degrees in different tissues and organisms. Conversely, each kind of damage may be removed during meiosis.

Delayed effects: Inbreeding and irradiation may both be considered methods by which genetic deficiency can be introduced in young organisms. Decreased survival is very often subsequently apparent- but in many cases only after a considerable portion of the lifespan has passed. A major part of the effects of early damage is therefore delayed in its expression. If repair during meiosis reduces genetic deficiency, then by analogous mechanisms, the effects of "rejuvenation" may also be delayed. Repair between homologs may decrease mortality towards the end of the lifespan rather than rejuvenate meiocytes immediately, although both effects may be evident to a certain extent. A general increase in vigor, as often occurs after outbreeding, may be an additional consequence of repair.

Mitotic pairing: Pairing of homologous chromosomes can occur during mitosis in a variety of cells, as well as during meiosis (see 116, 117 for reviews). Mitotic pairing has been reported frequently in plant cells (118-121) which generally do not undergo clonal aging, in tumors (116, 122) which also do not age, in tissue culture (116, 117) where cells may be transformed or stressed by the unusual environment, and in germ-line cells, especially just prior to meiosis (117, 120, 123). Observations by Brown and Stack (120) that mitotic pairing during flower development is gradually enhanced in both somatic and pre-meiotic germinal cells suggest that all cells may have regulatory genes which control the length and intimacy of homolog pairing. Mitotic pairing has also been reported in myelocytes (124), in fibroblasts exposed to carcinogens in tissue culture (116, 125) and in both somatic and germ-line cells of various insects (116, 117). Attempts to demonstrate pairing in animal tissues have most often been unsuccessful or inconclusive (121, 126, 127) except when male germ cells (123, 128) or stem cells (124) which age relatively slowly (129) have been examined.

These scattered, often contradictory (127) observations should be interpreted very cautiously, since there is little evidence that mitotic chromosomes pair tightly enough, for sufficient periods of time in the presence of adequate concentrations of the required enzymes, to communicate effectively. With this in mind, it is noted that these reports tend to corroborate the suggested relationship between pairing and repair, and raise additional questions: do certain plant, tumor or stem cells control senescence partly because their chromosomes repair each other at each division? Is pairing after exposure to the carcinogen, methylcholanthrene (125) an attempt to remove excess damage? Can certain instances of paramutation (130) be explained on this basis?

In light of these considerations, the model of rejuvenation may best be qualified in several ways: (1.) Repair during synapsis may serve to remove many different kinds of damage, as noted previously (45), in addition to that which is "recombinogenic." (2.) A major part of the beneficial effects of such repair may be evident later in the lifespan, rather than immediately. (3.) Pairing of homologs may help to increase the efficiency of repair in varying degrees during mitosis as well as meiosis.

ACKNOWLEDGEMENTS

For repeated criticism and many helpful suggestions, I will always be grateful to Aaron Lukton, Costante Ceccarini, Marion Himes, Seymour Koritz, Donald Hurst and Stan Salthe. I am also thankful for helpful comments during the preparation of this manuscript from Richard Cutler, Joan Smith-Sonneborn and David Comings. This work was supported by a fellowship provided by the Graduate Center of the City University of New York.

REFERENCES

1. Sonneborn, T.M. (1954) J. Protozool. 1, 38.
2. Siegel, R.W. (1967) Symp. Soc. Exp. Biol. 21, 127.
3. Nanney, D.L. (1974) Mech. Age. Dev. 3, 81.
4. McCullough, C.H.R., Cooke, D.J., Foxon, J.L., Sudbery, P.E. and Grant, W.D. (1973) Nature NB 245, 263.
5. Sonneborn, T.M. and Schneller, M. (1960) In B.L. Strehler (Ed.) "Biology of Aging" Waverly Press, Baltimore, 283.
6. Siegel, R.W. (1970) Genetics 66, 305.
7. Johnson, R. and Strehler, B.L. (1972) Nature 240, 412.
8. Eaves, G. (1973) Mech. Age. Dev. 2, 19.
9. Cutler, R.G. (1974) Mech. Age. Dev. 2, 381.
10. Sheldrake, A.R. (1974) Nature 250, 381.
11. Cutler, R.G. (1977) In Behnke, J.A., Finch, C.E. and Moment, G.B. (Eds.) "A New Look at Biological Aging" An Amer. Inst. of Biol. Sci. publication, in press.
12. Seltser, R. and Sartwell, P.E. (1965) Amer. J. Epidemiol. 81, 2.
13. Curtis, H.J. (1967) Symp. Soc. Exp. Biol. 21, 51.
14. Alexander, P. (1967) Symp. Soc. Exp. Biol. 21, 29.
15. Baxter, R.C. and Blair, H.A. (1967) Radiat. Res. 31, 287.
16. Lamb, M.J. and Maynard-Smith, J. (1969) Radiat. Res. 40, 450.
17. Fukushima, S. (1974) Exp. Cell Res. 84, 267.
18. Macieira-Coelho, A., Lima, L. and Malaise, E. (1973) Z. Alternsforsch. 27, 255.
19. Curtis, H.J. and Miller, K. (1971) J. Gerontol. 26, 292.
20. Crowley, C. and Curtis, H.J. (1963) Proc. US Nat. Acad. Sci. 49, 626.
21. Brooks, A.L., Mead, D.K. and Peters, R.F. (1973) J. Gerontol. 28, 452.
22. Hart, R. and Setlow, R.B. (1974) Proc. US Nat. Acad. Sci. 71, 2169.
23. Epstein, J., Williams, J.R. and Little, J.B. (1973) Proc. US Nat. Acad. Sci. 70, 977.
24. Epstein, J., Williams, J.R. and Little, J.B. (1974) Biochem. Biophys. Res. Comm. 59, 850.
25. Regan, J.D. and Setlow, R.B. (1974) Biochem. Biophys. Res. Comm. 59, 858.
26. Fischman, H.K. and Joy, C. (1977) J. Supramol. Struct. suppl. 1, 117.

27. Cutler, R.G. (1977) In D. Harrison (Ed.) "Genetic Effects on Aging" Symp. at Bar Harbor, Me., White Plains, New York, National Foundation, in press.
28. Gaubatz, J.W. and Cutler, R.G., submitted for publication.
29. Johnson, R., Chrisp, C. and Strehler, B.L. (1972) Mech. Age. Dev. 1, 183.
30. Price, G.B., Modak, S.P. and Makinodan, T. (1971) Science 171, 917.
31. Massie, H.R., Baird, M.B., Nicolosi, R.J. and Samis, H.V. (1972) Arch. Biochem. Biol. 153, 736.
32. Wheeler, K.T. and Lett, J.T. (1974) Proc. US Nat. Acad. Sci. 71, 1862.
33. Chetsanga, C.J., Boyd, V., Peterson, L. and Rushlow, K. (1975) Nature 253, 130.
34. Ono, T., Okada, S. and Sugahara, T. (1976) Exp. Gerontol. 11, 127.
35. Fulder, S.J. and Holliday, R. (1975) Cell 6, 67.
36. Saksela, E. and Moorehead, P.S. (1963) Proc. US Nat. Acad. Sci. 50, 390.
37. Kato, H. and Stich, H.F. (1976) Nature 260, 447.
38. Sonneborn, T.M. and Schneller, M. (1960) In B.L. Strehler (Ed.) "Biology of Aging" Waverly Press, Baltimore 286.
39. Weindruch, R.H. and Doerder, F.P. (1975) Mech. Age. Dev. 4, 263.
40. Cleaver, J.E. and Bootsma, D. (1975) Ann. Rev. Genet. 9, 19.
41. Sheid, B., Pedrinan, L., Lu, T. and Nelson, J.H. (1975) Biochem. Biophys. Res. Comm. 66, 1131.
42. Rupp, W.D., Wilde, C., Reno, D. and Howard-Flanders, P. (1971) J. Mol. Biol. 61, 25.
43. Howard-Flanders, P., Wilkins, B.M., Rupp, W.D. and Cole, R.S. (1968) Cold Spring Harbor Symp. Quant. Biol. 33, 195.
44. Cole, R.S. (1973) Proc. US Nat. Acad. Sci. 70, 1064.
45. Kofman-Alfaro, S. and Chandley, A.C. (1971) Exp. Cell Res. 69, 33.
46. Chandley, A.C. and Kofman-Alfaro, S. (1971) Exp. Cell Res. 69, 45.
47. Hotta, Y. and Stern, H. (1971) J. Mol. Biol. 55, 337.
48. Stern, H. and Hotta, Y. (1973) Ann. Rev. Genet. 7, 37.
49. Riley, R. and Bennett, M.D. (1971) Nature 230, 182.
50. Howell, S.H. and Stern, H. (1971) J. Mol. Biol. 55, 357.
51. Hotta, Y. and Stern, H. (1974) Chromosoma 46, 279.
52. Jacobson, G.K., Pinon, R., Esposito, R.E. and Esposito, M.S. (1975) Proc. US Nat. Acad. Sci. 72, 1887.

53. Hotta, Y. and Shepard, J. (1973) Mol. Gen. Genet. 122, 243.
54. Thompson, P.E. (1960) Genetics 45, 1567.
55. Auerbach, C. (1941) J. Genet. 41, 255.
56. Novitski, E. (1946) Genetics 31, 508.
57. Prudhommeau, C. and Proust, J. (1969) Mutat. Res. 8, 317.
58. Prudhommeau, C. and Proust, J. (1973) Mutat. Res. 23, 63.
59. Williamson, J.H., Parker, D.R. and Manchester, W.G. (1970) Mutat. Res. 9, 287.
60. Hammerl, H. and Klingmuller, W. (1972) Z. Naturforsch. 27B, 68.
61. Ahmad, A., Holloman, W.K. and Holliday, R. (1975) Nature 258, 54.
62. Taylor, J.H. (1965) J. Cell Biol. 25(2), 57.
63. Craig-Cameron, T. and Jones, G.H. (1970) Heredity 25, 223.
64. Egel, R. (1973) Nature NB 241, 135.
65. Sasaki, M.S. and Tonomura, A. (1973) Cancer Res. 33, 1829.
66. Bojanovic, J.J., Jevtovic, A.D., Pantic, V.S., Dugandzic, S.M. and Jovanovic, D.S. (1970) Gerontologia 16, 304.
67. O'Meara, A.R. and Herrmann, R.L. (1972) Biochim. Biophys. Acta 269, 419.
68. Berdyshev, G.D. and Zhelabovskaya, S.M. (1972) Exp. Gerontol. 7, 321.
69. Amici, D., Gianfranceschi, G.L., Marsili, G. and Michetti, L. (1974) Experimentia 30, 633.
70. Bjorksten, J. (1968) J. Amer. Geriatr. Soc. 16, 408.
71. Cutler, R.G. (1976) In K.C. Smith (Ed.) "Aging, Carcinogenesis, and Radiation Biology" Plenum Press, New York, 443.
72. Kohn, K.W., Steigbigel, N.H. and Spears, C.L. (1965) Proc. US Nat. Acad. Sci. 53, 1154.
73. Reid, B.D. and Walker, I.G. Biochim. Biophys. Acta 179, 179.
74. Ganesan, A.K. (1974) J. Mol. Biol. 87, 103.
75. Wilkins, R.J. and Hart, R.W. (1974) Nature 247, 35.
76. Kato, H. (1974) Exp. Cell Res. 85, 239.
77. Hayflick, L. (1965) Exp. Cell Res. 37, 614.
78. Hayflick, L. (1973) Amer. J. Med. Sci. 265, 432.
79. Martin, G.M., Sprague, C.A. and Epstein, C.J. (1970) Lab. Invest. 23, 86.
80. Smith-Sonneborn, J. and Reed, J. (1976) J. Gerontol. 31, 2.

81. Drake, J.W. and Baltz, R.H. (1976) Ann. Rev. Bioch. 45, 11.
82. Olovnikov, A.M. (1971) Dokl. Akad. Nauk USSR 201, 1796- cited in (84).
83. Holliday, R. and Pugh, J.E. (1975) Science 187, 226.
84. Holliday, R. (1975) Federation Proc. 34, 51.
85. Smith, J.R. and Hayflick, L. (1974) J. Cell Biol. 62, 48.
86. Clark, A.J. (1973) Ann. Rev. Genet. 7, 67.
87. Lin, P.-F., Bardwell, E. and Howard-Flanders, P. (1977) Proc. US Nat. Acad. Sci. 74, 291.
88. Sonneborn, T.M. (1977) In Behnke, J.A., Finch, C.E. and Moment, G.B. (Eds.) "A New Look at Biological Aging" An Amer. Inst. of Biol. Sci. publication, in press.
89. Nanney, D.L. (1957) Genetics 42, 137.
90. Clarke, J.M. and Maynard-Smith, J. (1955) J. Genet. 53, 172.
91. Goodrick, C.L. (1976) J. Gerontol. 30, 257.
92. Smith-Sonneborn, J., Klass, M. and Cotton, D. (1974) J. Cell Sci. 14, 691.
93. Maynard-Smith, J. (1958) J. Genet. 56, 227.
94. Pearl, R. and Pearl, R. deWitt (1934) "The Ancestry of the Long-lived" The Johns Hopkins Press, Baltimore.
95. Kallman, F.J. (1957) Ciba Foundation Colloquia on Ageing 3, 131.
96. Rudzinska, M.A. (1951) Science 113, 10.
97. Rudzinska, M.A. (1962) Gerontologia 6, 206.
98. MacKeen, P. (1976) Ph.D. thesis, Pennsylvania State University- cited in (104).
99. Child, D.M. (1914) Biol. Bull. 26, 286.
100. Bishop, F.C. and Smith, C.N. (1938) Circ. US Dept. Agr ric. no. 478, 1- cited in (103)
101. Stuchlikova, E., Juracova-Korakova, M. and Deyl, Z. (1975) Exp. Gerontol. 10, 141.
102. Comfort, A. (1974) Mech. Age. Dev. 3, 1.
103. Comfort, A. (1964) "Ageing: the Biology of Senescence" Holt, Rinehart and Winston, Inc., New York.
104. Barrows, C.H. and Kokkonen, G.C. (1977) In H.H. Draper (Ed.) "Advances in Nutrition Research" volume 1, Plenum Press, New York, in press.
105. Sundararaman, V. and Cummings, D.J. (1976) Mech. Age. Dev. 5, 139.
106. Munkres, K.D. and Minssen, M. (1976) Mech. Age. Dev. 5, 79.
107. Munkres, K.D. (1976) Mech. Age. Dev. 5, 163.
108. Smith-Sonneborn, J. and Rodermel, S.R. (1976) J. Cell Biol. 71, 575.

109. Cutler, R.G. (1972) Adv. Gerontol. Res. 4, 219.
110. Finch, C.E. (1976) Quart. Rev. Biol. 51, 49.
111. Hart, R.W. and Trosko, J.E. (1976) Interdiscipl. Topics in Gerontol. 9, 134.
112. Trosko, J.E. and Hart, R.W. (1976) Interdiscipl. Topics in Gerontol. 9, 168.
113. Hart, R.W. (1976) In K.C. Smith (Ed.) "Aging, Carcinogenesis, and Radiation Biology" Plenum Press, New York 537.
114. Hahn, H.P. von (1970) Gerontologia 16, 116.
115. Tas, S. (1976) Exp. Gerontol. 11, 17.
116. Boss, J.M.N. (1955) Texas Rept. Biol. Med. 13, 213.
117. Brown, W.V. and Bertke, E.M. (1974) "Testbook of Cytology" (2nd edition) C.V. Mosby, Co., St. Louis, 420.
118. Watkins, G.M. (1935) Bull. Torrey Bot. Club 62, 133.
119. Therman, E. (1951) Heredity 5, 253.
120. Brown, W.V. and Stack, S.M. (1968) Bull. Torrey Bot. Club 95, 369- cited in (117).
121. Heemert, C. van (1977) Chromosoma 59, 193.
122. Ewing, J. (1940) "Neoplastic Diseases" (4th edition) W.B. Saunders, Philadelphia, 24.
123. Leblond, C.P. and Clermont, Y. (1952) Amer. J. Anat. 90, 167.
124. Kinosita, R., Ohno, S., Kaplan, W.D. and Ward, J.P. (1954) Exp. Cell Res. 6, 557.
125. Hearne, E.M. (1936) Nature 138, 291.
126. Cohen, M.M., Enis, P. and Pfeifer, C.G. (1972) Cytogenetics 11, 145.
127. Fox, D.P., Mello-Sampayo, T. and Carter, K.C. (1975) Chromosoma 53, 321.
128. McDermott, A. (1971) Can. J. Gen. Cyt. 13, 536.
129. Lajtha, L.G. and Schofield, R. (1971) Adv. Gerontol. Res. 3, 131.
130. Brink, R.A. (1973) Ann. Rev. Genet. 7, 129.

MOLECULAR EVOLUTION AND CYTOGENETIC EVOLUTION

Allan C. Wilson*, Thomas J. White*†, Steven S. Carlson*‡ and Lorraine M. Cherry*§

*Department of Biochemistry, University of California, Berkeley, California 94720
†Department of Biochemistry, University of Wisconsin, Madison, Wisconsin 53706
‡Department of Biochemistry and Biophysics, University of California, San Francisco, California 94143
§Department of Biology, San Diego State University, San Diego, California 92182

ABSTRACT. The most outstanding and surprising result of research on molecular evolution has been the discovery of the evolutionary clock. Amino acid substitutions accumulate in proteins at rather steady rates. The variation in rates within a given functional class of proteins appears to be about twice as large as that expected for a simple Poisson process such as radioactive decay. With the molecular evolutionary clock one can estimate approximately the time elapsed since a pair of present-day species had a common ancestor. When this approach is applied to primate evolution, one obtains a date of 5 ± 2 million years ago for the separation of the lineages leading to humans and African apes. Although this date is considered too recent by some paleoanthropologists, it is notable that unambiguous published evidence for the existence of the hominid lineage goes back only 3-4 million years.

With the time dimension provided by the molecular evolutionary clock, one can calculate rates of evolutionary change for any property of the species being compared. This approach and a similar approach based on fossil evidence were used to estimate rates of chromosomal evolution. Specifically, we considered rates of evolutionary change in the number of chromosomes and the number of chromosomal arms per diploid genome. These rates have been higher by an order of magnitude for most mammals than for most lower vertebrates. Likewise, herbaceous plants have experienced far faster chromosomal evolution than have the cycads or conifers. Besides having experienced rapid chromosomal evolution, mammals and herbs have had remarkably high rates of morphological evolution and speciation. However, the rates at which substitutions have accumulated in the structural genes and other unique DNA sequences appear not to have been anomalously high in mammals or herbs.

To explain these contrasting rates, a hypothesis involving the concept of effective population size (N) is considered.

The smaller N is, the more likely it is that chromosomal mutations will be fixed and that new species will arise, according to population genetics theory. By contrast, such theory indicates that small N values should have no longterm accelerating effect on the rate of accumulation of most point mutations in structural genes. Assuming the correctness of this theory, we suggest that species of mammals and herbs have unusually small effective population sizes. Factors that could reduce effective population size include clannish social behaviour and polygamy in mammals and the propensity for self-fertilization in herbs.

INTRODUCTION

Our aim here is to examine cytogenetic evolution and compare this process with evolution at other levels of biological organization. The focus is on rates of evolution and the main conclusion to be drawn is that the average rate of gross change in the chromosomes has been very high in those major taxonomic groups, such as mammals, whose rates of phenotypic evolution and rates of speciation have been especially high. We also inquire briefly into the factors that enable such groups to evolve so rapidly at the chromosomal level and beyond while evolving at the standard rate so far as the accumulation of point mutations in structural genes is concerned.

MOLECULAR EVOLUTION

We begin by discussing evolutionary change in the primary structure of proteins and nucleic acids. Comparative studies of amino acid sequences and nucleotide sequences have given insight into evolution at the gene level. By comparing sequences from present-day species whose times of evolutionary divergence were known, one could estimate approximately the rates at which mutations have accumulated over time.

The most intriguing and controversial result of these studies was that sequences have changed at nearly constant rates. This approximate rate constancy is illustrated in Figure 1 for amino acid sequence data. Although there has been much debate about the degree of rate constancy, the most reliable data indicate that the variation in rate is about twice that expected for a simple Poisson process like radioactive decay (10). To this extent, the amino acid sequences in proteins evolve in a clock-like way.

Additional evidence for a molecular evolutionary clock comes from immunological studies with the quantitative microcomplement fixation technique. As shown elsewhere (6) this method is sensitive to small differences in amino acid sequences and provides an approximate estimate of the number of amino acid sequence differences that exist between naturally occurring, globular, monomeric proteins. While the molecular

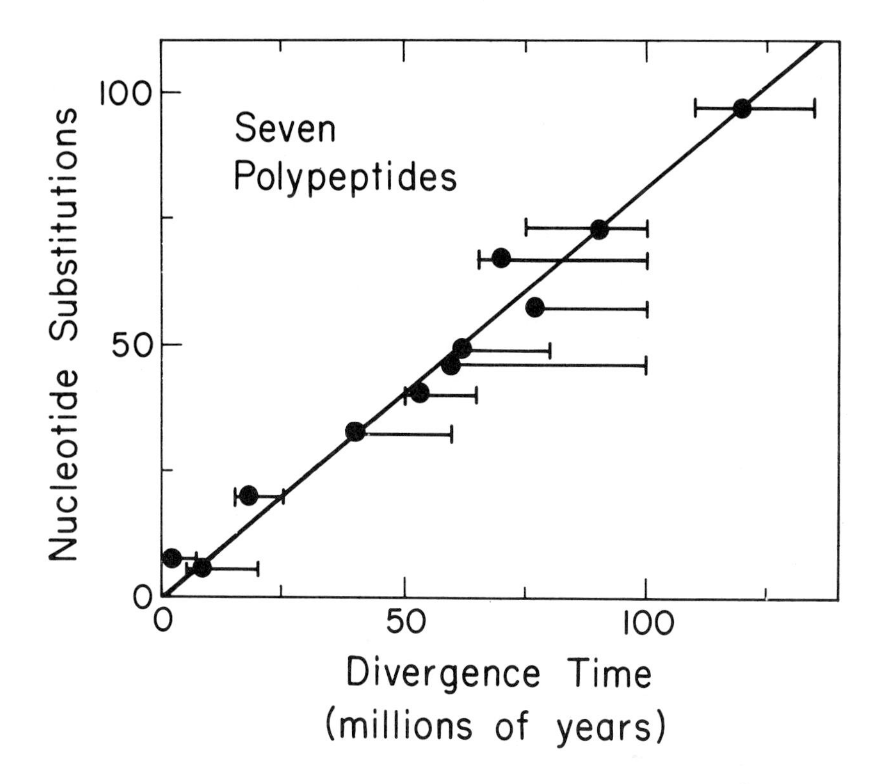

Figure 1. The evolutionary clock. The estimates of time elapsed since various mammalian species separated appear on the abscissa and are based on fossil evidence. The points represent time estimates made by L. Van Valen (personal communication to W.M. Fitch) and published by Fitch & Langley (11). To illustrate the uncertainty in these time estimates, horizontal bars have been added; they are based on additional paleontological information (51). This figure contains no intra-primate data because of the great paleontological uncertainty regarding divergence times within the higher primates. The number of nucleotide substitutions estimated by phylogenetic inference to have been fixed in seven polypeptides representing 578 codons was calculated by Fitch & Langley (11), using a maximum likelihood procedure, and is given by the ordinate on the left. The peptides are cytochrome *c*, myoglobin, hemoglobin α-chain, hemoglogin β-chain, fibrinopeptide A, fibrinopeptide B, and insulin C peptide. [From Wilson, Carlson, and White (51)].

basis for the strong correlation between the magnitude of the sequence difference and the magnitude of the antigenic difference (termed immunological distance) is uncertain, there is ample empirical evidence that the microcomplement fixation method is generally useful for comparing proteins that differ in amino acid sequence over the range from 0-30 substitutions per 100 amino acid positions. Micro-complement fixation is superior to direct determination of amino acid sequences in the sense that it is faster and more economical. With this method, albumin and transferrin evolution have been examined in more than 1000 pairs of species. In both cases the results are generally consistent with approximately clock-like evolution (25,34,51).

Annealing methods applied to the single-copy fraction of whole genome DNA have also provided information concerning rates of evolution. Most of the species compared by this method have such uncertain divergence times that a direct test of the evolutionary clock hypothesis was not possible. An indirect test, however, can be made. The ΔT values observed in mammalian DNA studies correlate strongly with albumin immunological distances (34,52) as well as with Fitch and Langley (11) estimates from amino acid sequence data. This is consistent with clock-like evolution in the single copy DNA sequences.

Although sequence evolution goes on at a nearly constant rate for proteins within a given functional class, that rate is not necessarily the same for proteins having different functions. Serum albumin, for instance, has evolved at a rate which is consistently faster than that for cytochrome *c*. Still slower in rate of evolution are histones 3 and 4. For a tabulation of rates of evolution for 50 different classes of proteins see (51).

The rate at which protein evolution occurs is thought to be proportional to the probability of fixation of mutations. The rate may depend on both P, the probability that an amino acid substitution will be compatible with the biochemical function of the protein, and on Q, the dispensability of the protein to the organism, i.e. the probability that an organism can survive and reproduce without the protein (51). Many of the most rapidly evolving proteins may be more dispensable than those proteins which evolve slowly. Serum albumin, for example, is probably much more dispensable to a vertebrate than is cytochrome *c*. This is shown by the existence of analbuminemic humans who are homozygous for a defect in albumin synthesis. Although they have less than 1% of the normal concentration of serum albumin, their health is nearly normal and they are fertile (12). Albumin has many important functions in humans and other vertebrates (31). Nevertheless, there are other proteins in serum that can carry out the albumin functions but evidently not as well as albumin does. For this reason, albumin is more dispensable to an organism than is a protein such as cytochrome *c*. Without cytochrome *c*, a vertebrate could not obtain energy and would die. It is expected, therefore, that the amino acid sequence of albumin would be freer to change than would the cytochrome *c* amino acid sequence. One problem for the future will be to devise tests for assessing the relative importance of P, the functional constraint factor and Q, the dispensability factor in protein evolution.

While we have some understanding of why some functional classes of proteins evolve faster than others, it has been harder to understand why the rate is so steady within a given class. As explanations involving positive natural selection did not seem satisfactory, some workers proposed a "non-Darwinian" explanation, referred to as the neutrality hypothesis. According to this hypothesis, the random fixation of selectively neutral substitutions is responsible for much of the sequence evolution in genes and proteins (17,18,28). Recently, a theory involving positive selection was proposed to explain the evolutionary clock (57). The available evidence has not allowed a discrimination to be made between these hypotheses.

The empirical finding that genes and proteins behave as evolutionary clocks is proving useful in the fields of paleontology, anthropology, and systematic biology, as well as in cytogenetics. Since sequence evolution is mainly divergent, it is relatively easy to reconstruct phylogenetic trees from sequence data. These trees depict the approximate order of branching of the lineages leading to modern species from a common ancestor. Phylogenetic tree analysis of sequence data is making a major contribution to knowledge of evolutionary relationships among organisms (51).

Molecular phylogenetic trees also contain information about times of divergence. If the absolute time is known for one branching event in the tree, approximate estimates of the times of other branching events in the tree can be made. This approach has been used to date the branching event which separated the human lineage from that leading to apes. The molecular approach led Sarich and Wilson in 1967 to propose a date of about 5 million years ago (35). This was upsetting to many anthropologists and paleontologists who thought in 1967 that the ape-human split was 30 million years old (32,38,39).

The paleoanthropological estimates, however, had been based on uncertain inferences from fragmentary fossils. These estimates were later revised from 30 to 15 million years ago (40,44) and the possibility of further revisions in the fossil-based estimates appears likely. Unambiguous published evidence for the existence of the human lineage goes back only 3-4 million years (15,27). In the meantime, more molecular evidence from additional proteins and nucleic acids has accumulated and this is consistent with a time of 5 ± 2 million years for the time of origin of the human lineage.

While a majority of those biochemists who study macromolecular evolution agree that sequence evolution is strongly dependent on elapsed time, there is less agreement whether the clock is geared to years or generations. Using the data for 12 peptides, Wilson, Carlson and White (51) compared the number of sequence changes that have accumulated along a lineage leading to a species characterized by unusually short generations (e.g. mouse) with the corresponding number for a lineage leading to a species having long generations (e.g. human, whale or elephant). This was done for 26 such pairs of protein lineages (51). The analysis showed that the lineages characterized by short generations exhibited no more sequence evolution than did the lineages giving rise to long generation species. For mammals, years appear to be more important than generations for sequence evolution.

ORGANISMAL *vs.* MOLECULAR EVOLUTION

Molecular evolutionists were slow to recognise that there is a contrast between rates of sequence evolution and rates of evolution at the organismal level. They expected that phenotypically conservative creatures should have experienced slower sequence evolution than had organisms whose rates of phenotypic change had been abnormally fast. To date, however, there is no convincing evidence that the proteins or nucleic acids of conservative creatures are conservative as regards their amino acid or nucleotide sequences.

Other factors contribute to the misunderstanding. Molecular evolutionists were justifiably impressed by the observation that the phylogenetic trees constructed from macromolecular sequence data usually resemble those based on morphological evidence, with regard to branching order (51). This congruence in branching order is explained simply by supposing that when a species splits into two noninterbreeding species, divergence between the two species can then begin to take place at both the morphological and the sequence level. Congruence in branching order does not require, however, that the rate of morphological change following speciation be geared to the rate of sequence change.

Another line of evidence reinforced the impression that macromolecular sequence evolution was related in a simple way to organismal evolution. A correlation exists between the degree of sequence difference, estimated by electrophoretic comparison of many proteins, and the taxonomic distance between organisms (1,2,37). This correlation probably results from the tendency of both organismal and sequence evolution to proceed generally at fairly steady rates. If sequence change and organismal change are each correlated with time, they will seem correlated with each other. The correlation does not necessarily imply a cause and effect relationship. To probe more deeply the relationship between organismal and sequence evolution, one must choose for study two or more groups of organisms whose rates of organismal change contrast greatly.

The vertebrates are well suited for such a study. Some vertebrate lineages have experienced faster rates of phenotypic evolution than others. Placental mammals, for instance, have experienced rapid organismal evolution compared with lower vertebrates, of which frogs are a typical example. Although there are thousands of frog species living today, they are so uniform phenotypically that zoologists put them all in a single order (Anura), whereas placental mammals are divided into at least 16 orders. The anatomical diversity represented by bats, whales, cats, and people is unparalleled among frogs, but frogs are a much older group than placental mammals. The present day frogs are not easily distinguishable morphologically from those living 90 million years ago. In contrast, during this same period, mammals have become extremely different in morphology from their progenitors. By way of illustration, the frog genus *Xenopus* was already in existence 90 million years ago, when the common ancestor of all placental mammals lived (9). Clearly, organismal evolution has been slow in frogs relative to mammals. Yet, at the molecular sequence level, evolution has been just as rapid in frogs as in mammals. This is apparent from immunological comparison of albumins, electrophoretic comparison of many enzymes, sequencing of hemoglobins and annealing studies with DNA (51).

This contrast between rates of evolution at the two levels accounts for the fact that frogs that are similar in phenotype can differ greatly at the sequence level. Indeed frogs that are similar enough to be included within a single genus can differ as much at the sequence level as does a bat from a whale. While frog genes have been evolving at standard rates, their phenotypes have changing slowly compared to mammalian phenotypes.

Other examples of such contrasts are known in a variety of taxonomic groups ranging from bacteria to primates (51). Most noteworthy, perhaps, is the case of ape and human evolution. Since humans and apes had a common ancestor, much more phenotypic change has occurred in the human lineage than in the chimpanzee lineage (41). No such tendency is evident at the sequence level, in either the proteins or the nucleic acid fractions examined to date (19). In spite of having evolved at an extremely high organismal rate, the human lineage does not appear to have accelerated sequence evolution.(see Fig. 2).

We are left with the strong impression that sequence evolution is primarily a function of time and proceeds as rapidly in phenotypically conservative creatures as in those which have changed radically in phenotype. Admittedly, only a few conservative groups have been examined thoroughly from the standpoint of sequence evolution. Moreover, one cannot deny that there is a basic problem involved when one tries to examine quantitatively the relationship between phenotypic

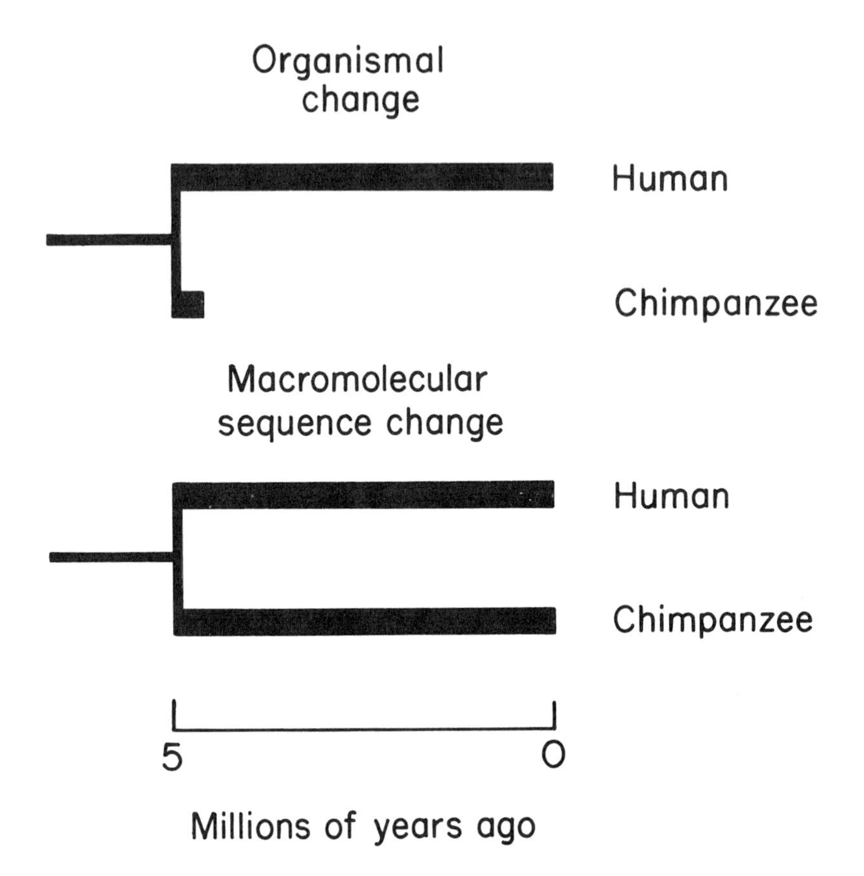

Figure 2. The contrast between biological evolution and molecular evolution since the divergence of the human and chimpanzee lineages from a common ancestor. As shown in the upper diagram, zoological evidence indicates that far more biological change has taken place in the human lineage than in the chimpanzee lineage; this illustration is adapted from that of Simpson (41). By contrast, as shown in the lower diagram, both protein and nucleic acid evidence indicate that as much change has occurred in chimpanzee genes as in human genes.

and molecular evolution. This is the non-molecular problem of estimating quantitatively and objectively the degree of phenotypic difference between organisms (36). We are optimistic that this problem can be overcome by use of appropriate numerical methods (42).

To account for the impression that sequence evolution proceeds at a rate which is independent of the rate of phenotypic evolution, two sorts of hypotheses may be entertained. First, it is possible that only a tiny fraction (less than one percent) of all evolutionary substitutions in genes (both regulatory and structural genes) are at the basis of major phenotypic changes in most kinds of organisms. While this fraction might be larger (perhaps 10 percent) in the most rapidly evolving organisms (e.g. mammals) and smaller in archconservatives (e.g. blue-green algae), the contribution of this fraction to the total rate of substitution would be too small to make the total rate measurably greater in the former organisms. The second possibility is that the most adaptively significant mutations, i.e. those producing big phenotypic effects, are regulatory. As these two hypotheses are not mutually exclusive, both may turn out to be correct. The second hypothesis may be tested more easily and evidence for it will now be reviewed briefly.

REGULATORY EVOLUTION

Major phenotypic changes, such as the acquisition of a novel metabolic ability, usually depend initially on increases in the amount of a rate-limiting protein (24,48,49,51). These quantitative effects can result from point mutations in control genes (i.e. classical regulatory genes and genes that exert control at levels other than transcription) as well as from chromosomal mutations which alter the arrangement of genes. The best examples of such quantitative mutations are from laboratory studies of bacterial evolution (24,48). It appears from these studies that such mutations are more adaptively significant than structural gene mutations. Quantitative mutations affecting enzyme levels may also have had a major role in the adaptive metabolic evolution of multicellular organisms (51).

A second line of evidence consistent with the regulatory hypothesis has been described (33,52). This evidence requires consideration of interspecific hybridization. Although there are many natural barriers to fertilization of an egg by sperm of another species, these are usually not absolute. Supposing an interspecific zygote forms, one can ask what chance it has of developing into a viable adult. Embryonic development involves an orderly program of expression of many genes that were inactive in the zygote (7). If the two genomes in an interspecific zygote are similarly regulated, so that a given block of genes will be turned on at the same time in one genome as in the other, orderly development of a hybrid organism can be expected. However, should the patterns of gene expression differ, the probability of an interspecific zygote developing successfully would be low. In accordance with this view, organismal hybrids derived from extremely different parental species often show signs of breakdowns in gene regulation (29,47).

As pointed out before, frogs have evolved much more slowly in morphology than have mammals. If regulatory evolution has also been slower in frogs than in mammals, one would expect that frog species should retain the potential to hybridize with one another much longer than mammals do. Since the rate of albumin evolution in frogs has been approximately equal to that in mammals, one expects to find small albumin immunological distances among mammals capable of hybridizing, whereas among hybridizable frogs one should encounter large immunological distances. The albumin distances are in accordance with this expectation. The mean albumin distance between hybridizable mammalian species is 3.2 units, whereas the corresponding distance for hybridizable frogs is 36 units (52). Using albumin as a clock, one can estimate that it takes about 2 million years for a distance of 3.2 units to arise and about 21 million years for a distance of 36 units to arise. Thus mammals have lost the potential for interspecific hybridization about ten times faster than frogs have (Table 1). This is consistent with the possibility that regulatory mutations affecting embryonic development have been fixed more often in mammalian evolution than in frog evolution.

TABLE 1

COMPARISON OF MEAN RATES OF EVOLUTION AT VARIOUS LEVELS IN MAMMALS AND FROGS*

Property	Mammal rate/frog rate
Amino acid sequences	1
Hybrid inviability	10
Chromosome number	11
Chromosome arm number	14
Speciation	5
Anatomy	3-20

*Taken from Wilson, Carlson and White (51).

CYTOGENETIC EVOLUTION

The molecular evolutionary clock allows the properties of organisms, even those organisms with a poor fossil record, to be viewed from a time perspective. With the time dimension provided by the clock, one can calculate rates of evolutionary change for any such property. In 1974 Wilson, Sarich and Maxson (53) used this approach to estimate rates of evolutionary change in karyotype. Until that time evolutionary cytogenetics had not been explicitly concerned with rates of chromosomal evolution (46).

Rates of chromosomal evolution can be estimated by comparing microscopically the karyotype of one species with that of another species whose time of divergence is known (from protein or fossil evidence). Two aspects of the karyotype were considered: the number of chromosomes and the number of chromosomal arms per diploid genome. Evolutionary changes in the number of chromosomes or number of arms may be brought about by mutations affecting the arrangement of large blocks of genes. These rearrangements include fusions, fissions, translocations and inversions, as well as gains or losses of heterochromatin; we refer to them collectively as chromosomal mutations.

Figure 3 illustrates the results of using this approach to compare the rates of chromosome number change in two major groups, frogs and mammals. More than three hundred pairs of frog species whose karyotypes are known were subjected to albumin comparisons with the micro-complement fixation method. The fraction of pairs having identical chromosome number is plotted against the albumin immunological distance. The same was done for more than 300 pairs of mammalian species. The albumin immunological distance scale is, in essence, an approximate time scale, where 100 units of immunological distance are equivalent to about 60 million years of divergence. The two histograms differ greatly. According to the frog histogram, one must compare frogs whose

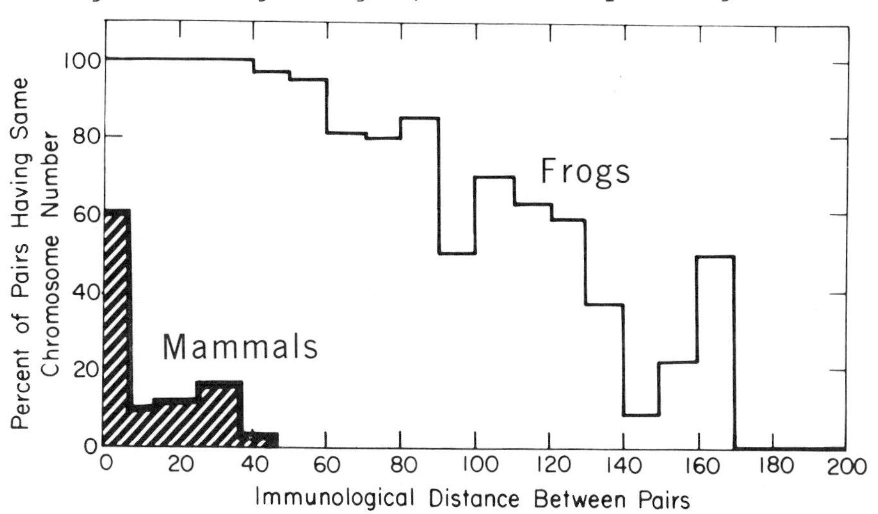

Figure 3. Proportion of species pairs having identical chromosome number as a function of the immunological distance between the albumins of the pairs. The light histogram summarises the results for 373 different pairs of frog species. The hatched histogram summarises the results for 318 different pairs of placental mammal species. [From Wilson, Sarich and Maxson (53)].

lineages separated more than 60 million years ago to have a good chance (>50%) of encountering species that differ in chromosome number. By contrast, the corresponding figure for mammals is less than 4 million years, on the average. It was inferred that the average rate of evolutionary change in chromosome number for mammals exceeds that in frogs by an order of magnitude (53). This approach showed also that arm number changes had accumulated an order of magnitude more frequently in mammals than in the frogs examined (53).

Comparable results were later obtained with another method in which the time dimension was based solely on fossil evidence (5,50). More recently, Cherry and Wilson (in prep.) devised a phylogenetic method of analysing karyotypic evolution which confirmed the above findings, as illustrated in Figure 4.

The slow karyotypic evolution exhibited by frogs was intriguing because frogs were already known to be conservative with regard to both phenotype and hybrid viability, although they were not conservative at the macromolecular sequence level. This finding raised the possibility that karyotypic evolution was connected in some way with phenotypic evolution and the evolutionary loss of hybrid viability.

To test this possibility, additional groups of phenotypically conservative animals were studied. These included 97 genera of snakes, lizards, turtles, crocodilians, salamanders, fishes and molluscs. According to the fossil-based method, each of these groups had low rates of karyotypic evolution during the history of those genera (5,50).

Similar tests were made with about 9000 species of plants (21). The most phenotypically conservative groups examined, namely cycads and conifers, had rates of chromosome number change that were more than an order of magnitude lower than in the herbaceous flowering plants. The latter group has been evolving unusually rapidly at the phenotypic level.

We conclude that when one deals with average rates for large taxonomic groups, there is a correlation between rate of karyotypic evolution and rate of phenotypic evolution. Exceptions to such a correlation are apparent, however, when one makes a more detailed analysis. As illustrated in Figure 4, primates have a high average rate of karyotypic evolution compared to the frog genus *Rana* but karyotypic rates vary markedly along the primate lineages. Whereas there have been few changes in arm number or chromosome number in the lineage leading to humans, for example, there have been many in the lineages leading to gibbons. Yet the human lineage is the one which has experienced the greatest change in phenotype and life style.

RATES OF CHROMOSOME EVOLUTION

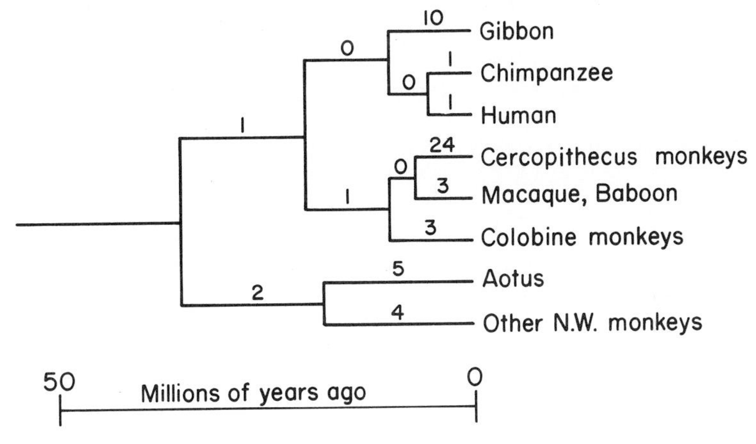

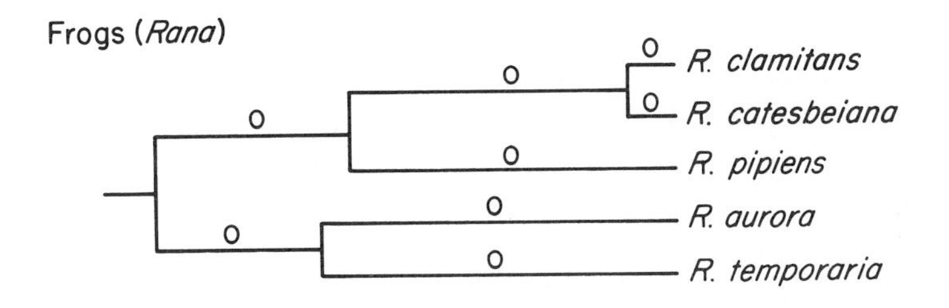

Figure 4. A phylogenetic analysis of karyotypic evolution in primates and frogs. The branching order of the evolutionary trees shown is based on biochemical evidence. The number placed on each lineage is an estimate of the minimum number of karyotypic changes which occurred along that lineage. The only events considered were those which result in a change in chromosome number or in arm number. To achieve the greatest parsimony in the tree, it was assumed that fusion and fission (i.e. events which change chromosome number) can occur with equal probability. It was further assumed that increases in arm number are as likely to occur as decreases.

A detailed presentation of this phylogenetic method of analyzing karyotypic evolution is intended for publication elsewhere (L.M. Cherry and A.C. Wilson, in preparation).

EFFECTIVE POPULATION SIZE

We now try to explain the information presented above concerning rates of evolutionary change at various levels. How can one account for the evidence that mammals and herbs have evolved remarkably rapidly in karyotype without having experienced accelerated sequence evolution?

Let us consider first the problem of the high rates of karyotypic evolution in mammals and herbs. The high rates could be due to an unusually high rate of chromosomal mutation in these two groups. Cytogeneticists could test this hypothesis. It is more likely, we suspect, that the high rates are due to an enhanced probability of fixation of chromosomal mutations in mammals and herbs. If so, the answer to our problem should be sought at the level of population structure and dynamics.

A key requirement for rapid fixation of karyotypic mutations is a small effective population size (55,56). A newly arisen chromosomal mutation is present initially in the heterozygous state in a diploid individual. During meiosis in this individual, problems of chromosome pairing often occur (22). For this and possibly additional reasons, the heterozygote often has reduced fertility. The conditions under which such a mutation can become fixed in a population are severely limited. It is unlikely that such a mutation will become fixed in a large outbreeding population. The probability (p) of random fixation is given approximately by Nei's equation

$$p = \frac{e^{-s(N-2)}}{2N}$$

where N is effective population size and s is the heterozygote disadvantage. While the fitness of the heterozygote is $1-s$, both homozygotes are assumed to have a fitness of 1. Assuming further that s will usually be small, for example $s = 0.05$, and that N is 500, there is a negligible probability ($p < 10^{-13}$) that the chromosomal mutation would ever get fixed in the population. If N is 20, however, the probability of fixation is reasonably high ($p \simeq 0.01$). For a more extensive discussion of the quantitative aspects of fixation probabilities, see References 4,5.

Within a small population ($N < 20$) there is an appreciable chance that the heterozygous individual with a new chromosomal mutation and those offspring carrying the mutation will mate with one another. Some of the offspring in the next generation will be homozygous for the mutation and have normal fertility (fitness value of 1). By a combination of inbreeding and random drift, the new mutation can become fixed in that small population.

The question then arises as to whether N is especially small for most mammals and herbs. There is evidence that many herbaceous species, especially annuals, are subdivided into populations whose effective size is less than 20 (20). Few estimates of N are available for animals, however. To obtain satisfactory estimates of N for animals will be a major undertaking for population biologists. If N is a key determinant of the rate of chromosomal evolution, a strong inverse correlation should be demonstrable between these two parameters.

SOCIAL BEHAVIOR

Wilson *et al.* (50) speculated that the effective population size may be reduced significantly in mammals by their social behavior. Thus, the mammalian type of social behavior may have an accelerating influence on chromosomal evolution. Two characteristic features of mammalian social behavior that may be relevant in this connection are the close social bonds between members of the same family or clan and the apparently high incidence of polygamy (54). The clannishness probably enhances inbreeding and reduces N. The mammalian type of polygamy may also reduce N. Dominant males characteristically prevent other males from mating. Thus the number of male mammals contributing genetically to the next generation is significantly lower, for behavioral reasons, than the number of females who contribute (30). This sexual asymmetry reduces the effective population size (N), as can be seen with the aid of the equation:

$$N = \frac{4FM}{M+F}$$

where F is the number of breeding females and M the number of breeding males in a population (54). When F is 10 and M is 1, for example, N is less than 4*.

In the case of herbs, one cannot appeal to social behavior as a factor for producing inbreeding and reducing effective population size. Perhaps the key factor here is the marked tendency in herbs for self-fertilization (20,21)--the ultimate in inbreeding.

SPECIATION

Population subdivision may be important not only for karyotypic evolution but also for speciation. Speciation is the process by which one species splits into two or more non-interbreeding species. Among evolutionary biologists there is a widespread belief that new species often arise from populations initially founded by a small group of individuals that

*It would be equally true that when M is 10 and F is 1, N is less than 4; this would be a case of polyandry.

became genetically isolated from the parental species, usually on the periphery of its range (5,26).

Stanley (43) developed a method for estimating approximate rates of speciation for taxonomic groups with a fossil record. In essence, his method considers the number of present-day species in the taxonomic group and the age of the group. From these data one can make a minimum estimate of the net rate at which new species have arisen. This estimate ignores extinction of species. To correct for extinctions one must estimate mean extinction rates from the fossil record, and this can be done only approximately. The corrected estimates of speciation rates are, therefore, likely to be approximations of the true rates of speciation.

Corrected rates of speciation have been presented for several of the major groups of vertebrates for which rates of karyotypic evolution are available (5). The speciation rates have been much higher within most mammalian genera than for genera of lower vertebrates. Moreover, the speciation rates correlate directly with estimates of the mean rates of karyotypic evolution for these groups of animals (5).

A correlation between mean rate of speciation and mean rate of karyotypic evolution has also been shown among the major groups of seed plants (21). Genera of herbs, for example, have speciated much more rapidly than have cycad or conifer genera.

The correlation between speciation rate and rate of karyotypic evolution is consistent with the hypothesis that both evolutionary processes are accelerated by subdivision of populations into demes between which there is little gene flow.

PHENOTYPIC EVOLUTION

We now consider the remarkably high rates of phenotypic evolution in mammals and herbs. Are karyotypic changes responsible for these high rates? This is a puzzling problem. One can imagine three possible ways in which karyotypic evolution could accelerate phenotypic evolution (13,45,46).

First, a karyotypic mutation that has become fixed in a given deme can act as a sterility barrier, impeding gene flow between that deme and others. Thus the mutant karyotype functions at the population level as a cytogenetic isolating mechanism. By acting as sterility barriers, and thereby reinforcing social and other barriers among demes, chromosomal mutations could facilitate not only speciation but also the fixation of recessive mutations. In small inbred populations, recessive mutations have a better chance of reaching the homozygous state, in which case their phenotypic effects can be detected by natural selection. The fixation of initially recessive mutations, we speculate, might be responsible for much of phenotypic evolution.

Second, the chromosomal mutation may link tightly two polymorphic loci that were hitherto far apart in the genome, thereby creating a particular combination of alleles, i.e. a supergene (8), which is less likely to be destroyed by recombination. This could facilitate adaptive evolution.

Third, the chromosomal mutations may be more directly involved in phenotypic change. The mutation may produce an altered pattern of gene expression that results in a new and fitter phenotype. To cytogeneticists who are impressed by the fact that balanced rearrangements often have no effects on the phenotype, this possibility will seem the least likely of the three mechanisms proposed. One recalls, however, the position effect (3) and the recent evidence, obtained with immunoglobulin genes (14), for a possible role of gene rearrangement in the process of differentiation. It would be unwise to reject the possibility that a significant fraction of karyotypic mutations has a direct role in phenotypic evolution.

CONCLUDING REMARKS

To end this analysis of rates of evolution at different levels of biological organization, it is appropriate to return to the problem of molecular evolution. The problem is to explain how mammals and herbs can have been evolving anomalously fast at the chromosomal level and beyond while evolving at standard rates so far as the accumulation of point mutations in structural genes and other unique DNA sequences is concerned. It seems reasonable to propose that small effective population size has no long-term accelerating influence on the accumulation of point mutations in these macromolecular sequences. This proposal fits well with the neutrality theory, which holds that the rate of sequence evolution is independent of population size (28). The selectionist theory may also be consistent with the proposal (23). By contrast, the rate of chromosomal evolution is expected to be inversely related to effective population size (c.f. Nei's equation). Thus according to selectionist theory as well as neutrality theory, it is not surprising that molecular evolution could follow rules that are different from those followed by evolution at the karyotypic level and beyond.

ACKNOWLEDGEMENTS

The authors thank G.L. Bush, S.M. Case and J.L. Patton for discussion, M. Nei for equation 1, as well as the National Institutes of Health and the National Science Foundation for support.

REFERENCES

1. Avise, J.C. (1976) In: *Molecular Evolution,* ed. F.J. Ayala, p. 106. Sunderland,Mass: Sinauer.
2. Ayala, F.J. (1975) *Evol. Biol.* 8, 1.
3. Bahn, E. (1971) *Hereditas* 67, 79.
4. Bengtsson, B.O. and Bodmer, W.F. (1976) *Theor. Pop. Biol.* 9, 260.
5. Bush, G.L., Case, S.M., Wilson, A.C., and Patton, J.L. (1977) *Proc. Natl. Acad. Sci. U.S.A.* (in press).
6. Champion, A.B., Soderberg, K.L., Wilson, A.C. and Ambler, R.P. (1975) *J. Mol. Evol.* 5, 291.
7. Church, R.B. and Brown, I.R. (1972) *Results Probl. Cell Differen.* 3, 11.
8. Dobzhansky, T. (1970) *Genetics of the Evolutionary Process.* New York: Columbia University Press.
9. Estes, R. (1975) *Herpetelogica* 31, 263.
10. Fitch, W.M. (1976) See Ref. 1, p. 160.
11. Fitch, W.M. and Langley, C.H. (1976) *Fed. Proc.* 35, 2092.
12. Gitlin, D. and Gitlin, J.D. (1975) In: *The Plasma Proteins,* ed. F.W. Putnam, 2, 321. New York: Academic.
13. Grant, V. (1973) *Plant Speciation.* New York: Columbia University Press.
14. Hozumi, N. and Tonegawa, S. (1976) *Proc. Natl. Acad. Sci. U.S.A.* 73, 3628.
15. Johanson, D.C. and Taieb, M. (1976) *Nature* 260, 293.
16. Jukes, T.H. (1975) *Biochim. Biophys. Res. Commun.* 66, 1.
17. Kimura, M. and Ohta, T. (1974) *Proc. Natl. Acad. Sci. U.S.A.* 71, 2848.
18. King, J.L. and Jukes, T.H. (1969) *Science* 164, 788.
19. King, M.-C. and Wilson, A.C. (1975) *Science* 188, 107.
20. Levin, D.A. and Kerster, H.W. (1974) *Evol. Biol.* 7, 139.
21. Levin, D.A. and Wilson, (1976) *Proc. Natl. Acad. Sci. U.S.A.* 73, 2086.
22. Lewis, K.R. and John, B. (1963) *Chromosome Marker.* London: Churchhill.
23. Lewontin, R.C. (1974) *The Genetic Basis of Evolutionary Change.* New York: Columbia University Press.
24. Lin, E.C.C., Hacking, A.J. and Aguilar, J. (1976) *Bioscience* 26, 548.
25. Maxson, L.R., Sarich, V.M. and Wilson, A.C. (1975) *Nature* 255, 397.
26. Mayr, E. (1970) *Population, Species and Evolution.* Cambridge, Mass.: Harvard University Press.
27. McHenry, H.M. (1975) *Science* 190, 425.
28. Nei, M. (1975) *Molecular Population Genetics and Evolution.* New York: Elsevier.
29. Ohno, S. (1969) In: *Heterospecific Genome Interaction,* ed. V. Defendi, p. 137. Philadelphia: Wistar Inst.

30. Ohno, S. (1976) *Reproduction in Mammals*, Vol. 6, eds. R.C. Austin and R.T. Short. Cambridge: Cambridge University Press.
31. Peters, T. (1975) In: *The Plasma Proteins*, ed. F.W. Putnam, 1, 133. New York: Academic Press.
32. Pilbeam, D. (1970) *The Evolution of Man*. New York: Funk and Wagnalls.
33. Prager, E.M. and Wilson, A.C. (1975) *Proc. Natl. Acad. Sci. U.S.A.* 72, 200.
34. Sarich, V.M. and Cronin, J. (1977) In: *Molecular Anthropology*, eds. M. Goodman, R.E. Tashian, p. 139. New York: Plenum.
35. Sarich, V.M. and Wilson, A.C. (1967) *Science* 158, 1200.
36. Schopf, T.J.M., Raup, D.M., Gould, S.J. and Simberloff, D.S. (1975) *Paleobiology* 1, 63.
37. Selander, R.K. and Johnson, W.E. (1973) *Ann. Rev. Ecol. Syst.* 4, 75.
38. Simons, E.L. (1964) *Sci. Am.* 211 (1), 50.
39. Simons, E.L. (1967) *Sci. Am.* 217 (6), 28.
40. Simons, E.L. (1977) In: *Molecular Anthropology*, eds. M. Goodman, R.E. Tashian, p. 35. New York: Plenum.
41. Simpson, G.G. (1963) In: *Classification and Human Evolution*, ed. S.L. Washburn, p. 1. Chicago: Aldine.
42. Sneath, P.H.A. and Sokal, R.R. (1973) *Numerical Taxonomy*. San Francisco: Freeman.
43. Stanley, S.M. (1975) *Proc. Natl. Acad. Sci. U.S.A.* 72, 646.
44. Walker, A.C. (1977) In: *Molecular Anthropology*, eds. M. Goodman, R.E. Tashian, p. 63. New York: Plenum.
45. White, M.J.D. (1968) *Science* 159, 1065.
46. White, M.J.D. (1973) *Animal Cytology and Evolution*. England: Cambridge University Press.
47. Whitt, G.S., Childers, W.F. and Cho, P.L. (1973) *J. Hered.* 64, 55.
48. Wilson, A.C. (1975) *Stadler Genet. Symp.* 7, 117.
49. Wilson, A.C. (1976) In: *Molecular Evolution*, ed. F.J. Ayala, p. 225. Sunderland, Mass.: Sinauer.
50. Wilson, A.C., Bush, G.L., Case, S.M. and King, M.-C. (1975) *Proc. Natl. Acad. Sci. U.S.A.* 72, 5061.
51. Wilson, A.C., Carlson, S.S. and White, T.J. (1977) *Ann. Rev. Biochem.* 46, 573.
52. Wilson, A.C., Maxson, L.R. and Sarich, V.M. (1974) *Proc. Natl. Acad. Sci. U.S.A.* 71, 2843.
53. Wilson, A.C., Sarich, V.M. and Maxson, L.R. (1974) *Proc. Natl. Acad. Sci. U.S.A.* 71, 3028.
54. Wilson, E.O. (1975) *Sociobiology*. Cambridge, Mass.: Harvard University Press.
55. Wright, S. (1940) *Amer. Nat.* 74, 232.
56. Wright, S. (1941) *Amer. Nat.* 75, 513.
57. Zuckerkandl, E. (1976) *J. Mol. Evol.* 7, 269.

SATELLITE DNA AND CYTOGENETIC EVOLUTION
Molecular Aspects and Implications for Man

Frederick T. Hatch and Joseph A. Mazrimas

Biomedical Sciences Division,
Lawrence Livermore Laboratory,
University of California,
Livermore, California 94550

ABSTRACT. Simple, highly reiterated DNA sequences, often observed in density gradients as satellite DNAs, exist in condensed heterochromatin. This material is predominantly located at chromosomal centromeres, occasionally at telomeres, or intercalated within arms; in a few species it occupies entire chromosome arms. Satellite DNAs are a highly variable component of the genome of most higher eukaryotes, but their functions have remained speculative.

The genus of kangaroo rats (Dipodomys) exhibits remarkable interspecies variations in content of three satellite DNAs, consisting of simple sequences 3 to 10 base pairs long, and in species karyotypes. A broad range of diploid-DNA content is correlated with satellite-DNA content. The latter is correlated positively with predominance of biarmed over uniarmed chromosomes (high fundamental number FN) and inversely with two anatomical indices (leg-bone-length ratios) of specialization for the jumping gait. Karyotypic variation is achieved via chromosomal rearrangements, e.g., Robertsonian fusion, C-band heteromorphism, and pericentric inversion. Environmental adaptation is achieved, in part, by reassortment of gene-linkage groups and regulatory controls as a result of the chromosomal rearrangements. The foregoing relationships led us to propose that highly reiterated DNA sequences play a supragenic, global role in enviromental adaptation and the evolution of new species.

We postulate that this role is played through control of the degree of chromatin condensation by repetitive DNA, and that this mechanism has analogous functions in the determination of the karyotypes of species and in the determination of the response to injury by certain clastogens and mutagens. The fundamental processes in both cases are visualized as the spontaneous or induced breakage of DNA strands and their subsequent rejoining (repair). Recent reports show slower repair-DNA synthesis in heterochromatin than in euchromatin after treatment with certain mutagens, carcinogens, and UV radiation--presumably owing to decreased accessibility of the repair enzymes to the DNA lesions in condensed chromatin. We postulate that reunion of broken DNA strands within heterochromatin occurs rarely, resulting primarily in chromosome or chromatid gaps and deletions. Delay of repair at junctions between euchromatin and heterochromatin allows time for encounters between different broken-DNA strands, forming sister chromatid exchanges, homologous and heterologous exchanges, and Robertsonian fusions. The presence of reiterated sequences in the junctions would facilitate reunions with exchange. Injury in euchromatin should usually be repaired readily, but residual errors of all types occur in proportion to the length of DNA at risk.

The misrepair phenomena may have a near-term nonevolutionary significance in the genesis of human genetic, and possibly even somatic, morbidity. Reproductive wastage, aneuploidies, and translocations are increasingly recognized as factors in our medical burden and as possible consequences of environmental pollution.

INTRODUCTION

From viruses to higher plants and animals there is a rising trend of nuclear-DNA content spanning nine orders of magnitude (1). At the lower levels most of the genome appears to consist of protein-coding sequences and their number increases proportionately to genome size. However, from the bacteria to the mammals there is perhaps only an order of magnitude of increase in number of coding sequences that is accompanied by a three-orders-of-magnitude increase in genome size (2, 3). This disparity has been attributed to the development of a "bureaucracy" of regulatory systems, in which some of the DNA may be transcribed but not translated and some may exist only as specialized forms of DNA (4, 5). A brief summary of the present status of the classification of DNA sequences follows (6, 7).

The Hierarchy of the Genome. Five levels in this hierarchy can be defined so far:

- Unique sequences, including the genes coding for the structures of proteins and other specific sequences, are present in one or two copies per cell (2). They represent about one-half of the nuclear-DNA content.
- Repeated genes (less than 5 percent of the DNA) code for several types of RNA and chromosomal proteins (8). They are present in multiple copies apparently because of transient requirements for rapid synthesis of the products.

- Moderately repetitive sequences (10 to 50 percent of the DNA) are interspersed in several patterns with the coding DNA of the first two levels and may be involved in regulation of the rates and timing of RNA and protein synthesis (4, 9, 10).
- Highly repetitive DNA (10 to 15 percent of the DNA) other than that found in satellites has received little study (11). Some of this DNA is probably analogous to satellite DNA but inseparable from the preceding classes for one reason or another; a small amount is "foldback" DNA based upon palindromic sequences.
- Satellite DNAs (5 to 50 percent of the DNA) are the most repetitive sequences, are present in 10^5 to more than 10^8 copies, and are highly variable in quantity. They are located in massive tandem arrays occupying sizeable regions of chromosomes, and are organized into condensed forms of chromatin (5, 6).

The foregoing types of DNA may account for the entire genome; however, there are large variations in the relative proportions among species. Current speculative interpretations of the significance of this hierarchy are that the moderately repetitive DNA exerts a fine tuning control over the expression of individual genes (4,9) and that the satellite DNA exerts a coarser global control over gene-linkage groups and recombination mediated by influences on chromosome structure and meiotic behavior (6). The remainder of this paper will focus on some of the evidence supporting the hypothetical role of satellite DNA.

SATELLITE DNA AND KARYOTYPIC VARIATION

As has often occurred in biological research, a particular species of animals with unique features has provided information that illuminates a subject of general significance. The kangaroo rat, genus *Dipodomys*, (Fig. 1) is the most successful and abundant mammal of the semi-arid and arid regions of North America. We serendipitously discovered that their genome was the most extraordinary yet seen in the animal kingdom (12). In the principal species studied, *D. ordii*, more than half of the nuclear DNA consists of three distinct satellite DNAs.

Nature and Distribution of Satellite DNAs. The three satellite DNAs of *D. ordii* have been designated HS-α, HS-β, and MS (12). They are well separated from main-band DNA in neutral CsCl gradients. The HS satellites have identical densities of 1.713 g/ml, but are separable in $Ag^+ - Cs_2SO_4$

Fig. 1.
A kangaroo rat species, *Dipodomys deserti*

gradients; the MS satellite bands at 1.707 g/ml and main-band DNA at 1.698 g/l. An additional component designated intermediate-density DNA is found at 1.702 g/ml in neutral CsCl, but it has not been possible to separate it completely from the main band. This component contains some repetitive DNA, but does not display the properties of a true satellite; it may represent the remainder of an ancient highly diverged satellite (6). The three satellites consist of reiterations of sequences 3, 6, and 10 to 12 base pairs in length with numerous major and minor variants (13). They are clearly visible in density gradients at DNA molecular weights at least as high as 10^8 (14, 15), which therefore is the minimum length of the tandemly repeated block in each chromosome where they occur; it is possible that the satellites occur in chromosomes in single or a few large segments. Prescott *et al.* (16) showed by *in situ* hybridization that HS-β is located in the centromeric regions of all chromosomes of *D. ordii* except three large metacentric pairs. And HS-α and probably MS are located throughout the short arms of all chromosomes except the three large pairs, in which they are found in the centromeric regions. Approximately one-half of the total length of the chromosomes contains satellite sequences, and these regions correspond accurately with C bands.

Interspecies Variations in Karyotype, DNA Content, and Satellite DNA. The 24 species and subspecies of *Dipodomys* studied with CsCl-density gradients include 19 of the approximately 22 known species of the genus (6). The interspecies variation in chromosome number 2n is unusual; there are clusters of species at 2n=52-54, 60-64, and 70-74, with no species known to have intervening values. The morphology of the autosome complement varies among the species (Table 1) from chromosomes that are entirely acrocentric or telocentric to chromosomes that are entirely with nonterminal centromeres (i.e., subtelocentric, submetacentric or metacentric); many species have mixtures of both types of chromosomes (17).

TABLE 1

KARYOTYPES OF KANGAROO RAT SPECIES

Species	2N	Autosomal Pairs		FN
		Bi-armed	Uni-armed	
spectabilis	72	0	35	70
ordii	72	35	0	140
heermanni	64	16	15	94
deserti	64	23	8	108
merriami	52	25	0	100

The average diploid-DNA content per nucleus of 11 representative species was 8.9±1.1 pg (range 6.9 to 10.9 pg). The highest 2C values were found in the species with 72-74 chromosomes, regardless of autosomal morphology. Variable lower 2C values were found in the DNA content was with the total amount of satellite DNA of densities 1.707 g/ml and 1.713 g/ml ($r=+0.79$, $p<0.01$). The subspecies D. spectabilis spectabilis was omitted from the correlation estimate because its purely acrocentric karyotype contains very little satellite DNA but a considerable amount of intermediate-density DNA. Partial linear-regression analysis was performed to obtain the standardized regression coefficients for the independent variables:

REGRESSION COEFFICIENTS

$$2C = \underset{(a)}{0.6663}\ (\text{Satellite DNA}) + \underset{(b)}{0.0269}\ (2n)$$

SIGNIFICANCE TESTS

	(a)	(b)
d.f.=7	t=2.59	0.67
	$p<0.05$	N.S.

Thus, the nuclear-DNA content is considerably expanded in some species, and the expansion consists principally of satellite DNAs.

Influence of Satellite and Intermediate-density DNAs on Karyotype. The Fundamental Number (FN) is a very useful index of karyotypic structure. Calculated as the total number of arms of the autosomal chromosomes, it takes into account both the chromosome number 2n as well as the numbers of uniarmed and biarmed chromosomes (18). Previously, we reported a correlation between the proportion of satellite DNA in the genome and the FN in 12 *Dipodomys* species (19). In Fig. 2 the species and subspecies of our more complete study are arranged from left to right in order of increasing FN. At the extreme left is *D. spectabilis spectabilis*, which has entirely acrocentric autosomes (the simplest and perhaps the most primitive karyotype according to Stock (17). Moving to the right in Fig. 2, we find that the increase in FN is accompanied by trends of increasing heavy satellite DNA (1.713 g/ml) and of decreasing intermediate-density DNA (1.702 g/ml), although variations in this pattern do occur. Alterations of 2n are of three types: minor fluctuations of one or two chromosome pairs; sharp, stepped reductions from 70-72 to 60-64 and 52-54 chromosomes; and at the right hand side of the figure, a one-step increase to 72-74 chromosomes. The final four species or subspecies have the highest possible FN for their 2n values, i.e., entirely biarmed autosomes.

Important relationships exist among satellite DNA, intermediate-density DNA, and chromosome structure. The FN increases with increasing amounts of heavy satellite DNA (HS-α plus HS-β) and decreases with increasing amounts of intermediate-density DNA. The clearest depiction of the above relationship between DNA distribution and karyotype is shown in Fig. 3. The logarithm of the ratio of heavy satellite DNA to intermediate-density DNA is highly correlated with FN. In other words, an abundance of heavy satellite DNA is associated with karyotypes consisting of mostly biarmed chromosomes, whereas a relative predominance of intermediate-density DNA is associated with more nearly acrocentric karyotypes. Several species appear to be exceptions to this relationship. Five outlying species

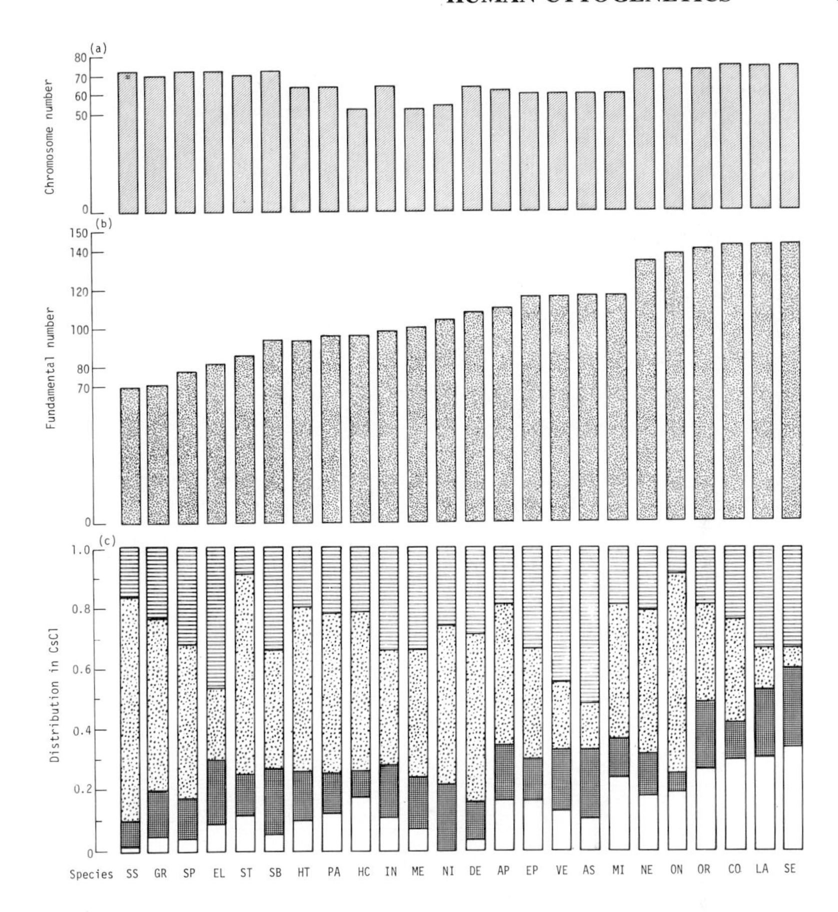

Fig. 2. Bar graph of 24 Dipodomys species and subspecies arranged in increasing order of Fundamental Number of autosomal chromosome arms. (a) Chromosome Number 2n, (b) Fundamental Number FN, (c) Distribution of DNA buoyant density fractions in neutral CsCl: white-heavy satellite (1.713 g/ml); hatched-MS (1.707 g/ml); stippled-intermediate density (1.702 g/ml); bars-main band (1.698 g/ml). Reprinted with permission from Ref. 6.

(marked as triangles) are believed to have diverged several million years earlier than the remainder, according to estimates of immunologic distance (See Fig. 4 based on (20)) and to measurements of allelic proteins by Johnson and Selander (21). Thus, the karyotypes of these divergent species may have evolved by independent mechanisms.

Common Satellite-DNA Sequences Among Rodent Genera. In nine species of kangaroo rats, the physical properties of

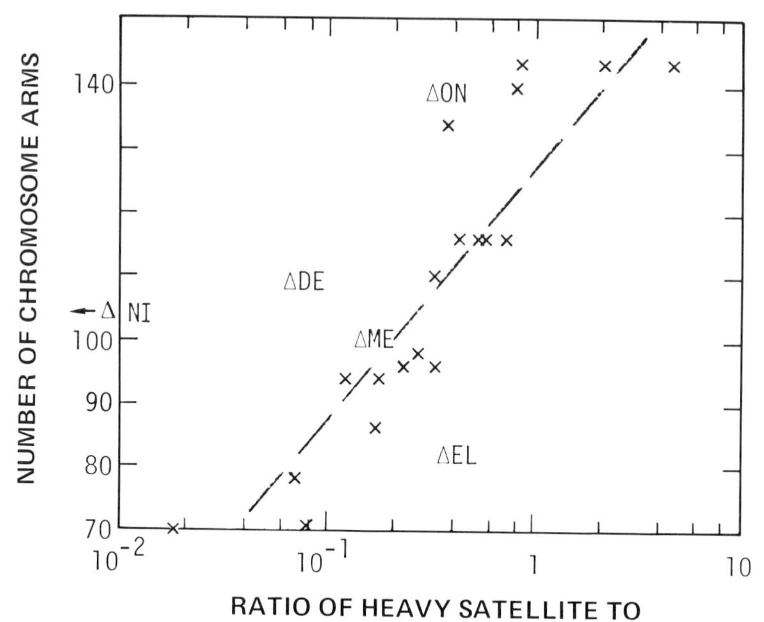

Fig. 3

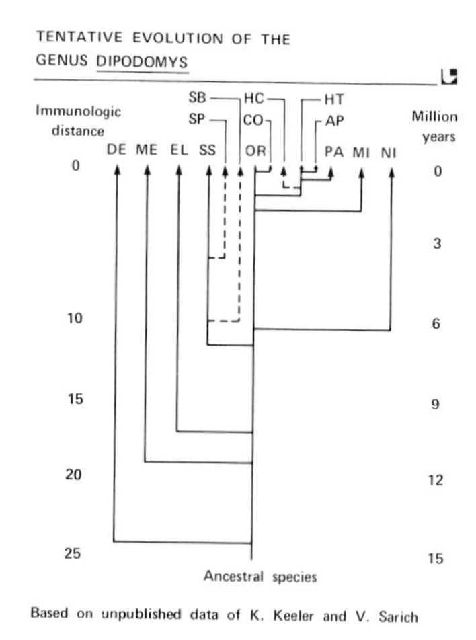

Fig. 4

the two heavy satellites show remarkable similarity (22). A satellite DNA analogous to the HS-α of the kangaroo rat has been characterized in the Pocket gopher, Guinea pig, and

Antelope ground squirrel. The physical and chemical properties of the alpha satellites from these rodents, representing four different families in two suborders, were compared. They show nearly identical Tm, nucleoside composition of single strands, and single-strand densities in alkaline CsCl gradients (22) as well as base-sequence fingerprints (23). The high degree of similarity of satellite sequences found in such a diverse group of rodents suggests a cellular function that is subject to natural selection, and implies that these sequences have been conserved over a considerable span of evolutionary time since their divergence about 50 million years ago. Among these species, an increasing proportion of alpha satellite is associated with a predominance of biarmed chromosomes, as was true within the genus Dipodomys (Table 2).

TABLE 2

PROPORTIONS OF ALPHA SATELLITE DNA AND BIARMED CHROMOSOMES IN DISTANTLY RELATED RODENTS

Species	2n	Alpha satellite	Biarmed chromosomes
Guinea pig	64	0.05	0
Pocket gopher	76	0.10	0.3-0.62
Kangaroo rat D. ordii	72	0.19	1.0
Ground squirrel	32	0.21	1.0

SATELLITE DNA AND ANATOMICAL VARIATION

One hesitates to dignify this information with a separate section because of the mind-boggling heresy involved in suggesting any relationships between the proportions of satellite DNA in the genome and the gross morphology of animals. Nevertheless, such correlations do exist within the kangaroo rat genus, and they require rational consideration. The most striking anatomical feature of kangaroo rats is their kangaroo-like appearance (Fig. 1). Their large, powerful hind legs are used for a jumping gait that can cover prodigious distances for the size of the animal; their rudimentary front legs are used for digging and transferring seeds to mouth or cheek pouches. One estimate of the extent of anatomical adaptation to this way of life is the relative lengths of their leg bones (19, 24). Amazingly, such ratios as shown in Fig. 5 are correlated with the amount of satellite DNA. The correlation indicates that the low-satellite species are the more specialized for their environment, while the high-satellite species are more generalized.

A weaker correlation has been observed between the number of subspecies that zoologists recognize on the basis of minor anatomical variations and the amount of satellite DNA (19). Similar to the previously cited anatomical correlation, the low-satellite species appear more specialized by exhibiting less variation in the form of subspecies, while the high-satellite species appear more flexible, having greater subspecies variation.

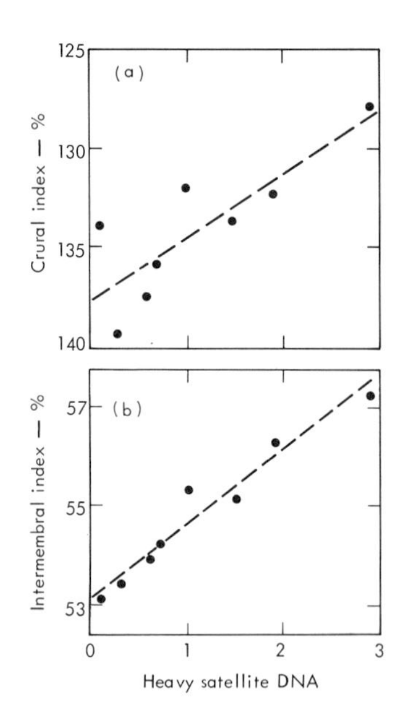

Fig. 5. Correlations of leg-bone-length ratios with the amount of heavy satellite DNA. (a) Crural index is femur/tibia; (b) Intermembral index is humerus + radius/femur + tibia.

EVOLUTION OF THE KARYOTYPE AND SPECIATION

The interspecies comparisons of quantitative DNA partition (Figs. 2 and 3) show that a high proportion of satellite DNA is associated with entirely biarmed autosomes and suggest that intermediate-density DNA was progressively eliminated in those species undergoing substantial reiteration of the major satellite DNAs. The presumed conversion of uniarmed to biarmed chromosomes could have resulted through three mechanisms:

- Species with reduced 2n show, on the average, a proportionately lower 2C value, the difference consisting predominantly of satellite DNA. The reduction of 2n probably resulted from Robertsonian translocations (fusions), although some deletions of whole chromosomes or segments may have occurred. The lower content of satellite DNA in these species probably resulted from fewer reiterations of satellite sequences than occurred in species with the full 2n of 70-74; i.e., centromeric fusion may have prevented any further reiteration of satellite DNA.
- Second chromosome arms may have arisen *de novo* by proliferation of satellite DNA from the centromere region (25).
- Conversion to biarmed chromosomes may also have arisen from pericentric inversions involving chromosome breaks through or adjacent to regions containing satellite DNA (26).

The processes of reiteration and elimination of specific DNA sequences and the rearrangement of the karyotypes are assumed to have proceeded concurrently during the evolution of the modern species of kangaroo rats (6). Thus, we propose that chromatin containing heavy satellite DNA promotes either the initiation or the fixation of chromosome rearrangements, most probably pericentric inversions, that result in the transformation of acrocentric to submetacentric or metacentric autosomes. Conversely, our evidence suggests that a relative abundance of intermediate-density DNA (1.702g/ml) or the absence of large amounts of satellite DNA does not promote, or does not permit, such inversions and thus preserves a predominantly acrocentric karyotype (Figs. 2 and 3). The mechanism by which this interaction between satellite DNA and chromosome structure arises is unknown; the structure of the chromatin containing the satellite DNAs is almost certainly involved.

By both cytologic and biochemical criteria this chromatin is highly condensed (27-29), although the mechanism by which repetitive DNA influences the state of condensation is unknown. In fact there is still no satisfactory model for the structure of heterochromatin. The significant inference is that when the amount of highly condensed chromatin exceeds that which is accommodated in the centromeres of both acrocentric and metacentric autosomes, the additional condensed chromatin achieves a more stable configuration in biarmed chromosomes in the form of second arms or telomeric and intercalary bands.

Changes in the gross structure of the genome may be more important than point mutations in the adaptive evolution and formation of new species of animals, according to a recent proposal by Wilson *et al*. (30-32, see also Wilson, this Symposium). They conclude "there may be a close parallel between chromosomal evolution and organismal evolution" (31). Our observations on the genus *Dipodomys* are highly consistent with the hypothesis of Wilson *et al*., and furthermore suggest an important role for satellite DNA and condensed chromatin in facilitating the evolution of new karyotypes and new species.

The *Rodentia* are well known to be one of the most versatile groups of animals in terms of their evolution (33). Anatomical and physiological adaptations are utilized for coping with environmental and ecological changes and for allowing expansion into new ecological niches. Such adaptations, when heritable, could have resulted from the reorganization of expression of multiple genes or "polygenes" (34). One plausible mechanism for accomplishing this is by means of structural rearrangements of the genome (35), which alter the number and size of linkage groups and may alter position effects on the expression of genes located in proximity to condensed chromatin. Such a mechanism is feasible for animals like *Dipodomys* that live in small subpopulations of limited mobility (36-38), since the resultant inbreeding is favorable for fixation of chromosomal rearrangements (34, 39). According to this viewpoint environmental adaptation, subspeciation, and speciation are progressive steps in a continuum based upon changes in the macrostructure of the genome(39-43). We propose that variability in the generation of satellite DNAs (44), and possibly in the deletion of intermediate-density DNA, is a fundamental molecular aspect of this mode of evolutionary behavior that is representative of such animals as crabs (45), *Drosophila* (46), and several rodent genera (6, 26, 35, 47).

POSSIBLE NONEVOLUTIONARY SIGNIFICANCE OF SATELLITE DNA

We may ask whether the postulated evolutionary mechanism that is based upon rearrangements of the genome has any near-term significance in, for example, determining responses to genetic injury or in causing somatic disease? Consideration of this question requires a description of the hypothetical role of satellite DNA in terms of the structures of DNA and chromatin (Fig. 6). Extensive evidence supports the close association between satellite DNA and condensed chromatin, whether this is detected by differential centrifugation of interphase chromatin (27, 28) or by C banding and *in situ* hybridization of mitotic chromosomes (16). We suggest that highly reiterated DNA sequences, through their influence on the condensation state of chromatin, play similar roles in the determination of the karyotypes of species and some of their rearrangements, and of the response to injury by certain clastogens and mutagens. The fundamental molecular processes in both cases are visualized as the spontaneous or induced breakage of DNA strands and their subsequent rejoining (repair).

Strand breakage may result from direct chemical reactions such as free radical attack or depurination followed by hydrolysis, or from endonuclease action(48). Repair requires access to the DNA of a different nuclease, a DNA polymerase, and DNA ligase together with appropriate substrates and cofactors (49). Recent reports have shown slower or less complete repair synthesis in heterochromatin than in euchromatin after treatment with certain mutagens, carcinogens, and UV radiation--presumably resulting from decreased accessibility of repair enzymes to the DNA lesions in condensed chromatin (50, 51). We postulate that reunion of broken DNA strands within heterochromatin occurs rarely, resulting primarily in chromosome or chromatid gaps and deletions (52). Such lesions are detected as aberrations in mitotic chromosomes, but are probably lethal to the cells in which they occur. Therefore these aberrations are of evolutionary or medical significance only when they cause diminished fertility or teratogenesis. Injury in euchromatin should usually be repaired rapidly, but residual errors of all types will occur with a frequency proportional to the length of the DNA at risk (i.e., the length of the euchromatic segment).

It is at the junctions between euchromatin and heterochromatin in chromosomes that some interesting and important events may occur (53). Here the chromatin is in

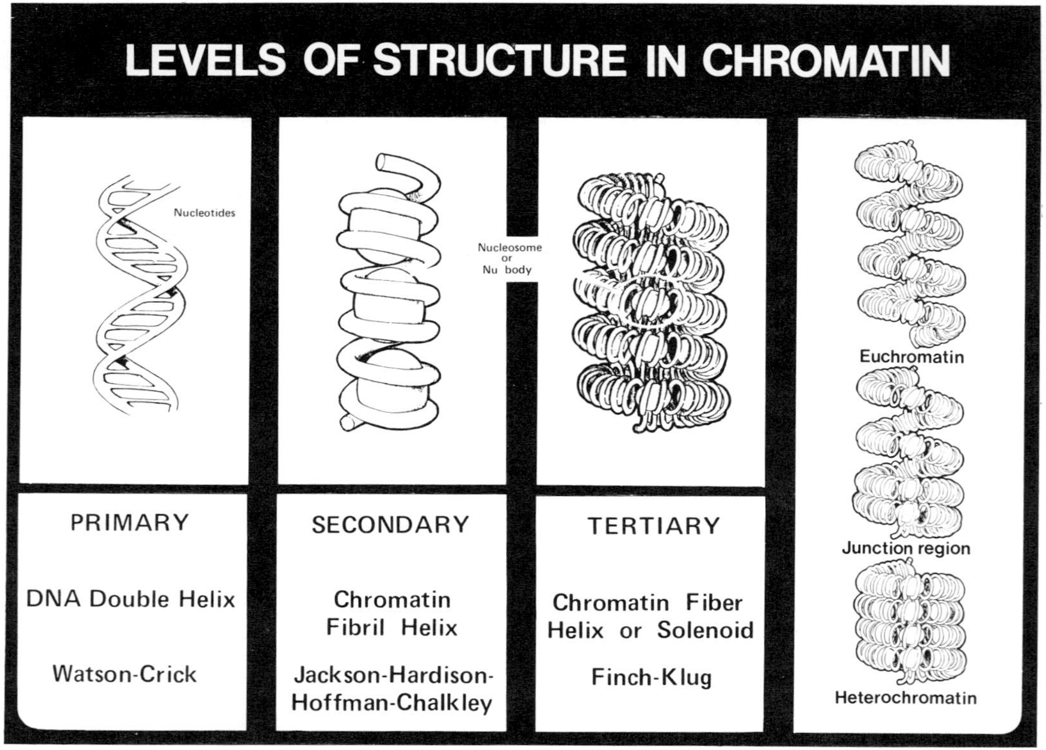

Fig. 6. Hierarchical helical model of chromatin structure (75-79). The quaternary level distinguishing extended euchromatin and condensed heterochromatin is proposed in this article.

transition between condensed and extended forms (Fig. 6). Highly reiterated DNA sequences are also present in the heterochromatic portions of each junction region; these sequences have the properties of being complementary and annealable, not only with their original partner strand at many sites, but also with nearby DNA in the same chromatid, in their sister chromatid, in homologous chromatids, and even in heterologous chromosomes possessing identical or similar reiterated sequences. Owing to the partially condensed state of chromatin in the junction regions, the access of repair enzymes might be expected to be slowed but not prevented altogether. Delay of repair at junctions between euchromatin and heterochromatin allows time for encounters between nonpartner DNA strands involved in breaks. Erroneous rejoining can lead to sister chromatid exchanges, homologous exchanges (inversions), heterologous exchanges (reciprocal translocations), and Robertsonian fusions, as well as heteromorphism of heterochromatic segments (C bands). While all of the foregoing mechanisms have been recognized in experimental systems and probably have evolutionary significance, their role in human disorders is not well established. The human genome contains 4 to 6 "cryptic" satellite DNAs, amounting to less than 5 percent of the DNA, as well as about 30 percent of additional moderately and highly repetitive DNA sequences (54, 61). It has been suggested that chromosomal rearrangements of the types discussed above may have been involved in the speciation of the primates (62). The major chromosomal disorders of man are based on trisomy, which may arise from translocation or fusion, usually between D and G group chromosomes, and from nondisjunction involving the same chromosomes (63, 69). The chromosomal satellites, or rudimentary short arms, of these particular chromosomes contain reiterated ribosomal coding and spacer sequences (70), which should be capable of participating in erroneous repair in the event they have incurred strand-breaking injury.

CONCLUSIONS

This has been a speculative essay and we wish it to serve as a stimulus to thinking about the functions of highly repetitive DNA sequences and about a possible chromosomal mechanism of evolutionary change. Experiments addressing evolutionary questions are difficult if not impossible to perform. However, some of the near-term cytogenetic and somatic implications of the suggested relationship between chromatin condensation and the repair of genetic injury are subject to investigation. We believe this is an important area for further research.

A hierarchy of types of DNA structure in the genome, which may account for the puzzlingly large amount of DNA in the higher eukaryotes, has been briefly summarized. A functional role has been proposed for the most highly reiterated class of DNA sequences (6). Through control of the condensation state of chromatin and the facilitation of DNA-strand exchange during the repair of injury, this DNA may exert a coarse control on karyotype structure and the evolution of new species. A nonevolutionary, but more immediate, significance of this mechanism may lie in the partial determination of the response to injury of the genome caused by spontaneous breakage of DNA strands or by certain clastogens and mutagens. Support for the evolutionary role comes from the remarkable interspecies variation in content of three satellite DNAs and in species karyotypes of the genus of kangaroo rats. Satellite DNA content is correlated positively with diploid DNA content and with a predominance of biarmed over uniarmed chromosomes. Support for the role in the response to injury comes from reports that many clastogens produce aberrations preferentially in heterochromatin (29, 71) and from the recent findings that sister-chromatid exchanges occur preferentially at junctions between euchromatin and heterochromatin in the chromosomes of the muntjac (72), kangaroo rat (73), and man (74).

The misrepair phenomena could have considerable significance in the genesis of human genetic, and possibly even somatic, morbidity. One may view the occurrence of these defects as a price that must be paid for possessing one of the useful mechanisms of evolutionary flexibility.

ACKNOWLEDGMENTS

This work was sponsored by the U.S. Energy Research and Development Administration (Contract No. W-7405-ENG-48). We thank Mr. John Koshiver for statistical calculations and Dr. Anthony Carrano for helpful suggestions and review of the manuscript.

REFERENCES

1. Sparrow, A.H., Price, H.J. and Underbrink, A.G.(1971) In: Brookhaven Symp. Biol. 23, 451-494.
2. Bishop, J.O. (1974) Cell 2, 81-86.
3. Lewin, B. (1975) Cell 4, 77-93.
4. Bonner, J. (1975) Ciba Found. Symp. 28, 315-335.
5. Walker, P.M.B. (1971) Prog. Biophys. and Molec. Biol. 23, 145-190.
6. Hatch, F.T., Bodner, A.J., Mazrimas, J.A. and Moore, D.H.II (1976) Chromosoma 58, 155-168.
7. Southern, E. (1974) MTP International Rev. Sci., Biochemistry of Nucleic Acids 6, 101-139.
8. Brownlee, G.G., Cartwright, E.M. and Brown, D.D. (1974) J. Molec. Biol. 89, 703-718.
9. Davidson, E.H. and Britten, R.J. (1973) Quart. Rev. Biol. 48, 565-613.
10. Davidson, E.H., Hough, B.R., Amenson, C.S. and Britten, R.J. (1973) J. Mol. Biol. 77, 1-23.
11. Cech, T.R. and Hearst, J.E. (1976) J. Molec. Biol. 100, 227-256.
12. Hatch, F.T. and Mazrimas, J.A.(1974) Nucleic Acids Res. 1, 559-575.
13. Salser, W., Bowen, S., Browne, D., El Adli, F., Fedoroff, N., Fry, K., Heindell, H., Paddock, G., Poon, R., Wallace, B. and Whitcome, P. (1976) Federation Proc. 35, 23-35.
14. Macaya, G., Thiery, J.-P. and Bernardi, G. (1976) J. Mol. Biol. 108, 237-254.
15. Mazrimas, J.A. (1977) unpublished.
16. Prescott, D.M., Bostock, C.J., Hatch, F.T. and Mazrimas, J.A.(1973) Chromosoma (Berl.) 42, 205-213.
17. Stock, A.D. (1974) J. Mammal.55, 505-526.
18. Matthey, R. (1954) Caryologia 6, 1.
19. Mazrimas, J.A. and Hatch, F.T.(1972) Nature (Lond.) New Biol. 240, 102-105.
20. Keeler, K. and Sarich, V. (1975) personal communication.
21. Johnson, W.E. and Selander, R.K.(1971) System. Zool. 20, 377-405.
22. Mazrimas, J.A. and Hatch, F.T. (1977) submitted to Cell.
23. Fry, K. and Salser, W. (1977) submitted to Cell.
24. Setzer, H.W. (1949) Univ. Kansas Publ., Mus. Nat. Hist. 1, 473.
25. Stock, A.D. and Hsu, T.C.(1973) Chromosoma (Berl.) 43, 211-224.

26. Mascarello, J.T. and Warner, J.W.(1974) Experientia (Basel) 30, 90-91.
27. Mazrimas. J.A. and Hatch, F.T. (1970) Exp. Cell Res. 63, 462-466.
28. Yunis, J.J. and Yasmineh, W.H. (1971) Science 174, 1200-1209.
29. Hsu, T.C. (1975) Genetics 79, 137-150.
30. Wilson, A.C., Maxson, L.R. and Sarich, V.M. (1974) Proc. nat. Acad. Sci. (Wash.) 71, 2843-2847.
31. Wilson, A.C., Sarich, V.M. and Maxson, L.R. (1974) Proc. nat. Acad. Sci. (Wash.) 71, 3028-3030.
32. Prager, E.M. and Wilson, A.C.(1975) Proc. nat. Acad. Sci. (Wash.) 72, 200-204.
33. Simpson, G.G. (1959) Cold Spring Harbor Symp. Quant. Biol. 24, 261.
34. Wilson, A.C. (1975) Stadler Symp. 7, 117-133.
35. Mascarello, J.T. and Hsu, T.C. (1976) Evolution 30, 152-169.
36. Lidicker, W.Z. (1960) Univ. Calif. Pubs. Zool. 67, 125-218.
37. Maza, B.G., French, N.R. and Aschwanden, A.P. (1973) J. Mammal. 54, 405-425.
38. Schroder, G.D. and Geluso, K.N. (1975) J. Mammal. 56, 363-368.
39. White, M.J.D. (1973) Animal Cytology and Evolution. Cambridge: Cambridge University Press.
40. Franklin, I. and Lewontin, R.C. (1970) Genetics 65, 707-734.
41. Stanley, S.M. (1975) Proc. nat. Acad. Sci. (U.S.) 72, 646-650.
42. Corneo, G. (1974) Accad. Naz. dei Lincei Ser. 8, 57, 458-466.
43. Miklos, G.L.G. and Nankivell, R.N. (1976) Chromosoma 56, 143-167.
44. Smith, G.P. (1976) Science 191, 528-535.
45. Beattie, W.G. and Skinner, D.M. (1972) Biochim. Biophys. Acta 281, 169-178.
46. Brutlag, D.L. and Peacock, W.J. (1975) in "The Eukaryotic Chromosome" ed. by W.J. Peacock, Australian Natl. Univ. Press, Canberra, pp 35-45.
47. Pathak, S., Hsu, T.C. and Arrighi, F.E. (1973) Cytogenet. Cell Genet. 12, 315-326.
48. Bender,M.A, Griggs,H.G. and and Bedford,J.S. (1974) Mutation Res. 23, 197-212.
49. Regan, J.D. and Setlow, R.B. (1974) Cancer Res. 34, 3318-3325.
50. Harris, C.C., Connor, R.J., Jackson, F.E. and

Lieberman, M.W. (1974) Cancer Res. 34, 3461-3468.

51. *Bodell, W.J. and Banerjee, M.R. (1976) Nucleic Acids Res. 3, 1689-1701.*
52. *Bourgeois, C.A. (1974) Chromosoma 48, 203-211.*
53. *Hatch, F.T. and Carrano, A.V. Abstracts 2nd Intl. Conf. on Environmental Mutagens, Edinburgh, July 11-15, 1977.*
54. *Gosden, J.R., Mitchell, A.R., Buckland, R.A., Clayton, R.P. and Evans, H.J. (1975) Exptl. Cell Res. 92, 148-158.*
55. *Saunders, G.F. (1974) Adv. Biol. Med. Physics 15, 19-46.*
56. *Sanchez, O. and Yunis, J.J. (1974) Chromosoma 48, 191-202.*
57. *Schmid, C.W. and Deininger, P.L. (1975) Cell 6, 345-358.*
58. *Ginelli, E. and Corneo, G. (1976) Chromosoma 56, 55-68.*
59. *Sanchez, O. and Yunis, J.J. (1976) Biochim. Biophys. Acta 435, 417-426.*
60. *Marx, K.A., Allen, J.R. and Hearst, J.E. (1976) Biochim. Biophys. Acta 425, 129-147.*
61. *Marx, K.A., Allen, J.R. and Hearst, J.E. (1976) Chromosoma (Berl.) 59, 23-42.*
62. *Turleau, C., Grouchy, J. de and Klein, M.(1972) Ann. Gênét. (Paris) 15, 225-240.*
63. *Jacobs, P.A., Buckton, K.E., Cunningham, C. and Newton, M. (1974) J. Med. Genet. 11, 50-64.*
64. *Jacobs, P.A., Melville, M., Ratcliffe, S., Keay, A.J. and Syme, J. (1974) Am. Hum. Genet. 37, 359-376.*
65. *Hamerton, J.L., Canning, N., Ray, M. and Smith, S. (1975) Clin. Genet. 8, 223-243.*
66. *Jacobs, P.A., Frackiewicz, A., Law, P., Hilditch, C.J. and Morton, N.E. (1975) Clin. Genet. 8, 169-178.*
67. *Buckton, K.E., O'Riordan, M.L., Jacobs, P.A., Robinson, J.A., Hill, R. and Evans, H.J. (1976) Ann. Hum. Genet. 40, 99-112.*
68. *Speed, R.M., Johnston, A.W. and Evans, H.J. (1976) J. Med. Genet. 13, 295-306.*
69. *Moorhead, P.S. (1976) Am. J. Human Genet. 28, 294-296.*
70. *Jones, K.W. and Purdom, I.F. (1975) Symp. Soc. for Study of Human Biology 14, 39-51.*
71. *Hsu, T.C. and Pathak, S. (1976) Chromosoma 58, 269-273.*
72. *Carrano, A.V. and Wolff, S. (1975) Chromosoma 53,*

361-369.

73. Bostock, C.J. and Christie, S. (1976) Chromosoma 56, 275-287.
74. Latt, S.A. (1974) Science 185, 74-76.
75. Watson, J.D. (1976) Molecular Biology of the Gene. 3rd ed., Menlo Park, Calif., W.A. Benjamin, Inc., pp. 208-211.
76. Jackson, V., Hardison, R., Hoffmann, P. and Chalkley, R. (1977) submitted for publication.
77. Weintraub, H., Worcel, A. and Alberts, B. (1976) Cell 9, 409-417.
78. Finch, J.T. and Klug, A. (1976) Proc. Nat. Acad. Sci. (U.S.) 73, 1897-1901.
79. Varshavsky, A.J. and Georgiev, G.P. (1975) Molec. Biol. Rep. 2, 255-262.

THE CHROMOSOMES OF NON-HUMAN PRIMATES

Peter Pearson, James Garver, Anna Estop, Truus Dijksman,
Lucy Wijnen and Meera Khan, P.

Instituut voor Anthropogenetica, Rijksuniversiteit, Leiden
Netherlands

ABSTRACT. Comparative banding studies of the chromosomes of man and the higher primates have led to the following conclusions: -
That many human chromosomes appear to have direct counterparts in the great apes and certain members of the old world monkeys. However, apart from the X-chromosome, there seem to be no similarities to either man or other primates in the gibbons and siamangs. Gene mapping studies by the use of somatic cell hybrids have been carried out to ascertain whether particular primate chromosomes, defined as being homologous to a human chromosome on the basis of banding pattern, also carry the same structural enzyme loci. The present studies, involving the chimpanzee, gorilla, orangutan and rhesus monkey, show that chromosome 1 and the X have been extremely conservative. In addition the homologues of 2p, 11, 12 in the great apes appear to carry the same gene loci as their human counterparts. Notable exceptions, however, include the homologue to chromosome 6 in all species and chromosome 15 in the chimpanzee.

The provisional conclusion is that a similarity in banding pattern is generally reflected in a similarity in gene content but that certain exceptions exist suggesting that translocations have taken place at a finer level of resolution than can be defined by current banding methods.

INTRODUCTION

The use of chromosome banding techniques has, in the course of the last five years, permitted a much greater precision in the identification of individual human and primate chromosomes and for conclusions to be drawn on the possible origin of the human karyotype. Most notable has been the hypothesis that the chromosome number of 48 found in the great apes has been reduced to the 46 in man by a process of telomeric fusion of two pairs of acrocentric chromosomes (6, 3). With this one exception, translocation differences between species are believed to be non-existent and structural differences confined to a series of pericentric inversions and heterochromatic band differences.

The nomenclature of the great ape (4) chromosomes has recently been standardised and published as a supplement to the Paris nomenclature for human chromosomes (11). In this supplement the traditional identification parameters of length and centromere position were retained for the primary ordination of the chromosomes and the banding patterns used only for identifying homologous pairs and not as a basis for determining the position of chromosomes in the karyotype. This proposal encountered a certain degree of opposition on the basis that, since the banding morphology of the hominoid apes was in many ways similar to that of man, the hominoid chromosomes should be named directly after their human counterparts as defined by chromosome banding.

Fig. 1 gives a trypsin-banded comparison from left to right of the chromosomes of man, chimpanzee, gorilla, orangutan, rhesus monkey and african green monkey respectively arranged according to their supposed human homologies. The arrangement is that chosen by the present authors and does not necessarily agree in all details with previously published studies.

However, uncertainties as to interpretating what may be considered homologous on the basis of banding criteria alone, clearly underlines the necessity for also defining chromosome homology by alternative means. We have accordingly started work on a program of somatic cell hybrid analysis, using various primate and rodent parental lines, to localise genes to individual primate chromosomes and compare this pattern to that already known for man (7).

Table 1 gives a list of the hybrid cell combinations produced to date,the number of isolated primary clones and the number of clones which have been analysed either in part or in detail. All hybrids were isolated using standard HAT selection following Sendai virus fusion and the enzymes analysed according to Meera Khan (8), and van Someren (13), and the chromosomes identified using a combination of G-, Q- and Giemsa-11 banding techniques.

Table 2 gives the preliminary results arranged according to the known syntenies. Question marks have been placed where an assignment is either completely unknown or still remains somewhat inconclusive. In the latter case the most likely chromosomal candidate for the assignment has been included.

The following general and specific conclusions can be drawn from the existing data: -

1) In general chromosomes which are identical to human chromosomes also contain the same gene loci. Thus the homologues of human chromosomes 1, 4, 8, 11, 12, 21 and X appear to carry the same enzymes as in man.
2) Exceptions to this are the homologues of chromosome 6 and SOD_2 in all great apes; the syntenic pair MPI-PK_3 with chromo-

Fig. 1a

M C G O R A

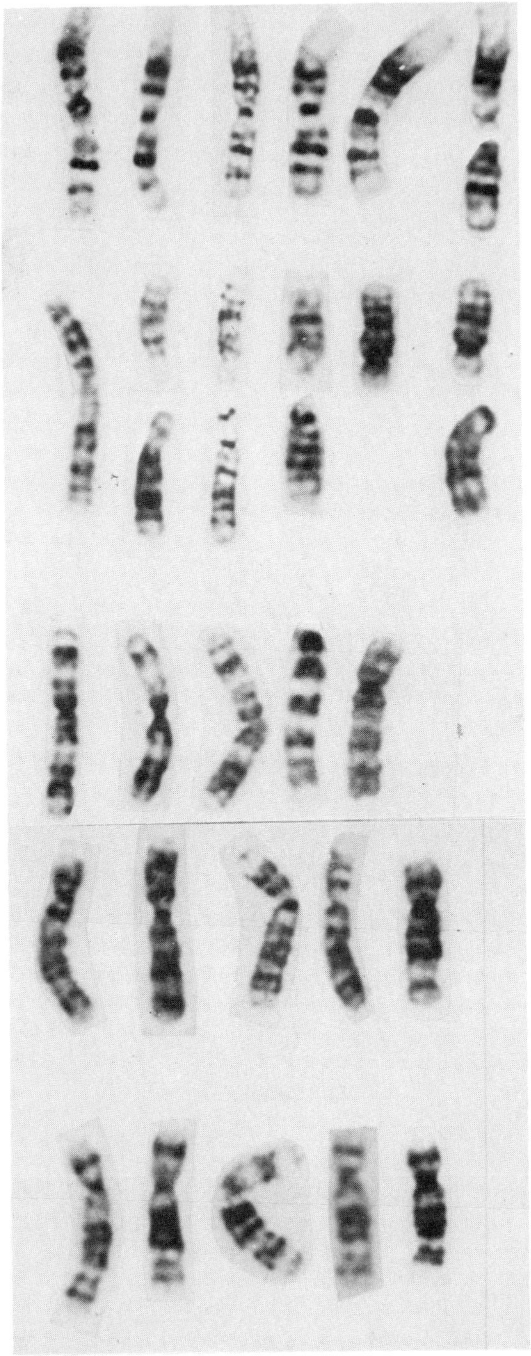

Chromosome # 1

Chromosome # 2

Chromosome # 3

Chromosome # 4

Chromosome # 5

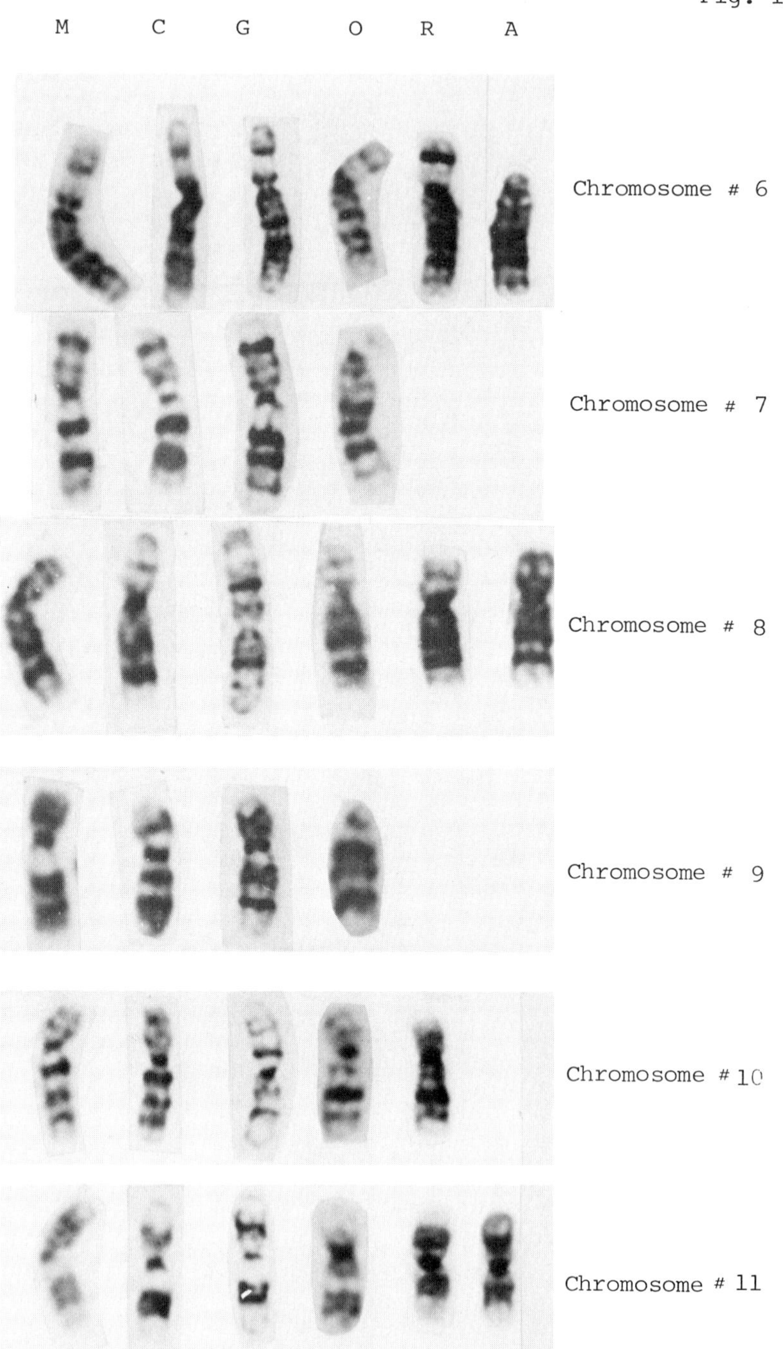
Fig. 1b
M
C
G
O
R
A
Chromosome # 6
Chromosome # 7
Chromosome # 8
Chromosome # 9
Chromosome # 10
Chromosome # 11

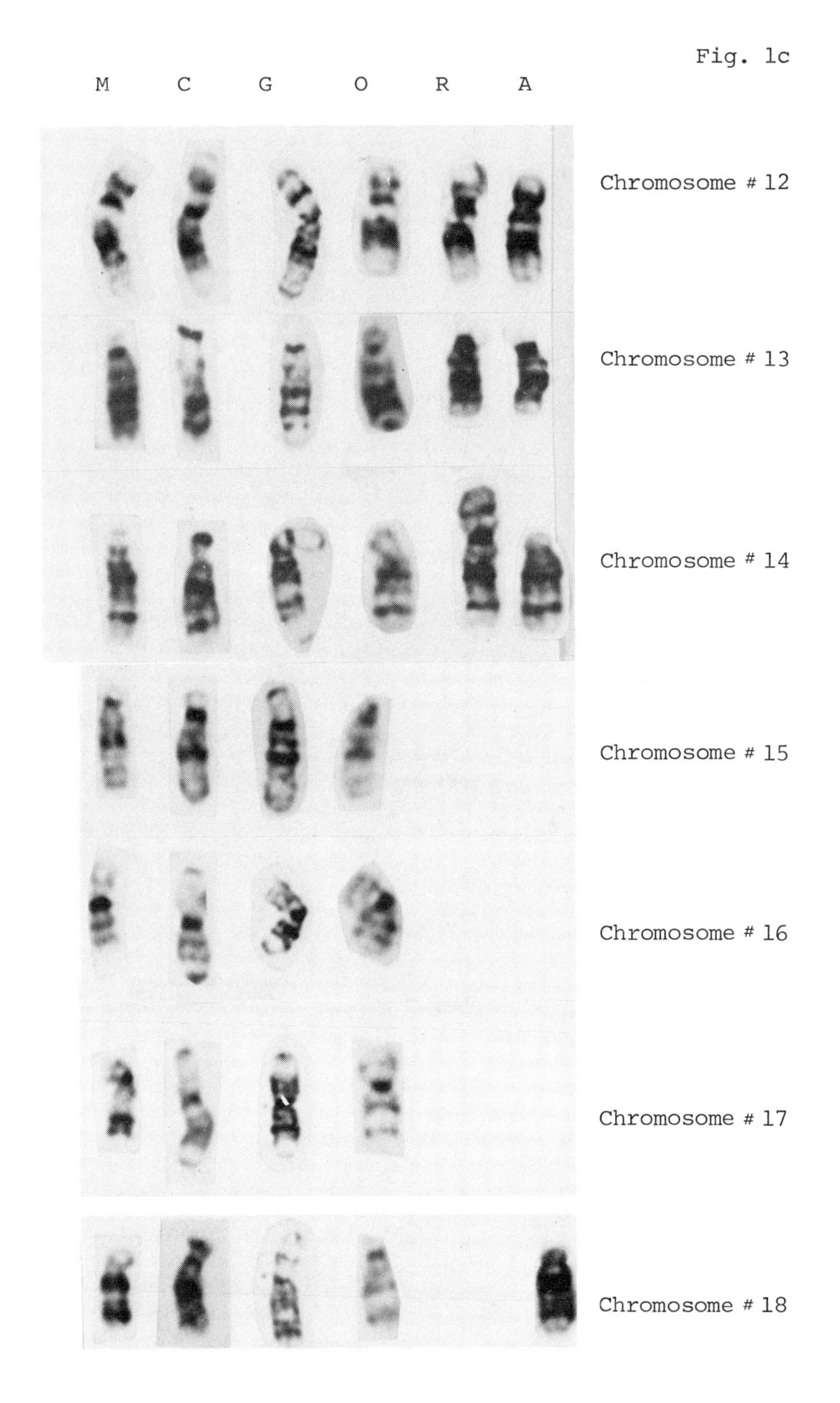
Fig. 1c
M
C
G
O
R
A
Chromosome # 12
Chromosome # 13
Chromosome # 14
Chromosome # 15
Chromosome # 16
Chromosome # 17
Chromosome # 18

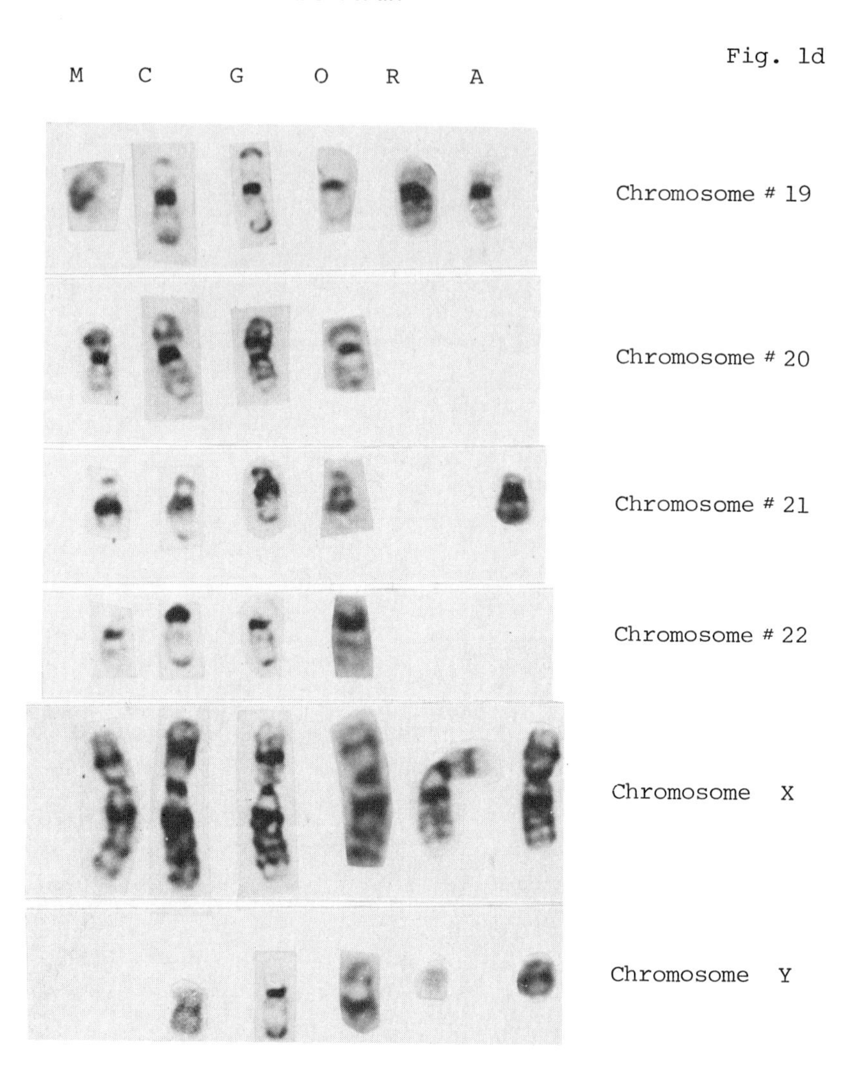

Fig. 1 The letters above each column refer to man, chimpanzee, gorilla, orangutan, rhesus and african-green monkey respectively. Spaces have been left in the rhesus and african-green monkey karyotypes where there is insufficient banding evidence for defining a homology with a particular human chromosome. Of interest is the probable homology of human chromosome 1 to two acrocentric chromosomes in the african-green monkey.

TABLE 1

PRIMATE/RODENT HYBRID CELL LINES

Parental combinations	No. fusions	No. clones isolated	No. examined
Chimpanzee - Chinese Hamster (C.H.) A_3 TK^-	1	41	10
Gorilla - C.H. A_3 TK^-	2	32 primary	32
		18 secondary	18
Orangutan - C.H. A_3 TK^-	1	45	18
Rhesus - C.H. A_3 TK^-	1	40	15
African Green - C.H.	1	23	-

TABLE 2

Chimpanzee	Gorilla	Orangutan	Rhesus
Chrom. # 1			
PGD) PGH_1) 1 PGM_1)	PGD) PPH_1) 1 PGM_1) PEP C) GUK ?	PGD) PPH_1) 1q PGM_1) GUK ?	PGD) PPH_1) 1 PGM_1)
Chrom. # 2			
MDH_1-12	MDH_1) 12? ACP_2)	MDH_1) ? ACP_2)	
Chrom. # 4			
PGM_2-3	PGM_2-3	PGM_2-?	
Chrom. # 6			
SOD_2-?	SOD_2-? PGM_3-?	SOD_2-5?	
Chrom. # 7			
β-GUR-6	β-GUR ?	β-GUR-10	
Chrom. # 8			
GR-7	-	-	
Chrom. # 9			
AK_1-11	AK_1-?	AK_1-?	
Chrom. # 10			
GOT_1-8	GOT_1-?	GOT_1-?	
Chrom. # 11			
LDH-A-9	LDH-A-9	LDH-A-8	
Chrom. # 12			
LDH-B-10	LDH-B-10	LDH-B-9	

Chrom. # 14		
NP-15	NP	NP-?
Chrom. # 15		
MPI) ? PK_3)	MPI) PK_3) 15 HEX-A)	MPI) ? PK_3)
Chrom. # 17		
TK-19	TK-19	TK-20?
Chrom. # 18		
PEP-A-17?	PEP-A-16	PEP-A-?
Chrom. # 19		
GPI-?	GPI-20	GPI-? NOT 20
Chrom. # 20		
ITP-21	ITP) ? NP)	ITP-?
Chrom. # 21		
SOD_1-22	SOD_1-22	SOD_1-? NOT 22
Chrom. # 22		
β-GAL-23	-	-
X-chrom.		
G6PD) PGK) X HPRT) α-GAL)	G6PD) PGK) X HPRT) α-GAL)	G6PD-X

A full list of enzyme abbrevations is to be found in Human Gene Mapping 3 (7).

some 15 in the chimpanzee; and SOD_1 with chromosome 22 in the orangutan.
3) Gene localisation has helped elucidate a postulated banding homology for orangutan chromosome 10 (homologous to human chromosome 7) in that both code for β-GUR. Another example has been for gorilla chromosome 18 which codes for NP and thus fits the contention of Dutrilleaux (3) that this chromosome is homologous to the human 14. The observation that NP and ITP are syntenic in the gorilla (9), although coded by different chromosomes in man thus requires verification. However, AK_1 does not seem to go with gorilla chromosome 13, thus leaving this chromosome in doubt as a possible homologue of the human 9.
4) Both MDH_1 and ACP_2 are syntenic and clearly associate with chromosome 12 in the chimpanzee and probably also with 12 in the gorilla. These assignments confirm the postulated telomeric fusion origin of chromosome 2p in man. There is as yet no hybrid cell evidence to confirm the long arm origin of human chromosome 2.
Some of the results presented here in the chimpanzee confirm those of other groups (2, 10, 5). For a summary see Warburton and Pearson (12).
5) Evidence from chromosome banding studies, in particular the Giemsa-11 technique (1), clearly shows the occurrence of structural rearrangements having arisen within the hybrid cell lines and in many cases being unidentifiable. This implies that recognition of intact chromosomes in hybrid cell lines is a prerequisite for carrying out mapping studies. This correlation of the presence of an intact chromosome with an enzyme is a much stronger evidence of the syntenic relationships of the two than absence of the chromosome and presence of the enzyme. Further, unidentified iso-enzym polymorphism may in some instances result in lack of discrimination between the rodent and primate enzymes.

At present it seems that although there are large similarities in the gene content and banding pattern of primate chromosomes, that at least some of the exceptions noted above have arisen as a result of translocations occurring at a finer lever than can be resolved using current banding techniques. This implies that theories on the evolution of the human karyotype are still an oversimplification. We are now in a phase of investigation in which theories of karyotype evolution can be rigorously tested at the molecular level and this is likely to result in significant modifications of ideas on both band and whole chromosome homology.

ACKNOWLEDGEMENTS

We wish to thank the zoological gardens in Amsterdam,

Rotterdam and the Yerkes Primate Centre for providing some of the primate specimens, and for technical assistance from Mrs A. Ebeli-Struijk and Mrs M. Monteba-van Heuvel. A.E. is a Dutch Cultural Exchange Fellow.

REFERENCES

1. Bobrow, M., Madan, K. and Pearson, P.L. (1972) *Nature (London), New Biol.* 238, 122.
2. Chen, S., McDougall, J.K., Creagan, R.P., Lewis, V. and Ruddle, F. (1976) *Orig. Artic. Series* XII, 7, 412.
3. Dutrilleaux, B., Rethoré, M.O., Prieur, M. and Lejeune, J. (1973) *Humangenetik* 20, 343.
4. Dutrilleaux, B. (1975) *Monogr. Ann. Genet.* 102.
5. Finaz, C., Cochet, C., de Grouchy, J., van Cong, N., Reboucet, R. and Frezal, J. (1975) *Ann. Genet.* 18, 169.
6. de Grouchy, J., Turleau, C., Roubin, M. and Klein, M. (1972) *Ann. Genet.*15, 79.
7. Human Gene Mapping 3. Baltimore Conference (1976) Birth Defects. *Orig. Artic. Series.* XII, 7, 3.
8. Meera Khan P.(1971) *Ph.D. Thesis, University Leiden.*
9. Meera Khan, P., Pearson, P.L., Wijnen, L.M.M., Doppert, B.A., Westerveld, A. and Bootsma, D. (1976) Birth Defects *Orig. Artic. Series,* XII, 7, 420.
10. Orkwisewski, K.G., Tedesco, T.A., Mellman, W.J. and Croce, C.M. (1976) Birth Defects: *Orig. Artic. Series,* XII, 7, 427.
11. Paris Conference (1971) Birth Defects: *Orig. Artic. Series,* 11 Suppl. 1975, no. 9.
12. Warburton, D. and Pearson, P.L. (1976) Birth Defects: *Orig. Artic. Series,* XII, 7, 75.
13. van Someren, H., Beyersbergen van Henegouwen, H., Los, W.R.T., Wurzer-Figurelli, E.M., Doppert, B.A., Vervloet, M. and Meera Khan, P. (1974) *Humangenetik* 25, 189.

OLD AND NEW SYNDROMES*

John L. Hamerton

Division of Genetics, Department of Paediatrics
University of Manitoba, Winnipeg, Manitoba

INTRODUCTION

The development of chromosome banding techniques (1,2) has resulted in a renaissance in human clinical cytogenetics and in a vast increase in our knowledge about chromosome structure and chemistry (3,4). Prior to 1971, clinical cytogenetics was restricted by the limitations of the available techniques, to the identification of 5 human chromosome pairs, Nos. 1, 2, 3, 16, and the Y chromosome by the simple morphological examination of the karyotype, while by use of autoradiography and other time consuming and not particularly satisfactory means, other chromosomes could sometimes be defined by their replication patterns. In particular, chromosomes 4 and 5, 13, 14, 15, 17, 18, and 21 and 22 could be separated from each other within their respective groups. Up to 1971, chromosome rearrangements, deletions, duplications, rings, etc., could be defined only to a very limited extent and were usually described in relation to the chromosome groups involved and only very rarely could specific chromosomes be identified. Chromosome syndromes were described, for instance, as G_1-Trisomy or Down Syndrome, D_1-Trisomy for Patau Syndrome. Eponyms, descriptive terms and generalized karyotype descriptions were common to describe syndromes of similar clinical appearance with similar and presumptively identical chromosome findings. Examples were the "Cri du Chat", 18q-, G-Monosomy, 13q-, Cat eye Syndrome and so on, an unsatisfactory state of affairs but one which, without a simple means for chromosome identification could not be resolved.

The discovery (5,6) that staining with quinacrine or its nitrogen mustard derivative produced a horizontal series of consistently occurring bright to pale fluorescent bands which allowed identification of each pair of homologous chromosomes

*Due to lack of space it has not been possible to quote individual references from which data has been taken for inclusion in the tabular summaries. The number of cases given here should therefore <u>not</u> be used for additive purposes in accumulating case totals. A full review of chromosome rearrangements in man is in preparation and detailed references will be given there.

and their differentiation from each other pair, led immediately to the development of a range of banding techniques (1) based on Giemsa as well as on other fluorochromes, antinucleoside antibodies, and more recently, BuDr incorporation for one or more S-periods followed by Hoechst 33258 or other fluorochrome staining (7). Each of these techniques results in the production of banded chromosomes and allowed the identification of whole chromosomes as well as permitting the precise location of breakpoints in chromosome rearrangements (8,9).

These technical advances make it possible to determine the chromosomal composition of derivative and recombinant chromosomes and thus correlate abnormal karyotypes to abnormal phenotypes. The present paper will be restricted to a discussion of autosomal abnormalities only, and data on new chromosomal syndromes will be discussed only when there has been a complete identification of the duplicated or deleted chromosome segment by one or more banding techniques.

CHROMOSOME ABNORMALITIES

The clinical effects of chromosome abnormalities can best be understood if the nature of the chromosome changes and their behaviour during meiosis are clearly understood (10,11). Abnormalities may be classified according to their mode of origin as follows: (i) Abnormalities resulting from nondisjunction (ii) Abnormalities resulting from chromosome rearrangement.

(i) Abnormalities resulting from nondisjunction

Nondisjunction during meiosis results in gametes with abnormal chromosome numbers. Nondisjunction during an early cleavage division results in an embryo with two or more cell lines with different chromosome numbers. The variety and range of abnormal chromosome numbers which can result from nondisjunctional events in man have been treated very fully by Hamerton (12).

(ii) Abnormalities resulting from chromosome rearrangement

Chromosome rearrangements may be 'balanced' or 'unbalanced'. Balanced rearrangements may be defined as changes in the distribution of chromosome material which do not result in any loss or gain of genetic material. Such rearrangements include, inversions, reciprocal and Robertsonian and insertional translocations. Unbalanced rearrangements are those in which loss or gain of chromosome material has occurred and include deletions, deficiencies and ring chromosomes. Balanced rearrangements can result in unbalanced segregants due to meiotic segregation either alone or in combination with crossing over (10, 11, 12).

Chromosome Abnormalities and Abnormal Phenotypes

Abnormal phenotypes seen in association with major chromosome abnormalities are generally considered to result directly from chromosome imbalance. Nondisjunction usually results in one additional (trisomy) or missing (monosomy) chromosome, although more rarely other abnormalities may occur (double trisomy, tetrasomy, etc.). Meiotic nondisjunction, when it involves the larger autosomes, usually results in full lethality and fetal wastage (13). When a live birth results after a nondisjunctional event involving one of the larger autosomes, the affected individuals are often found to have both a normal and abnormal cell line (mixoploids) so that the error must have occurred after fertilization.

When an unbalanced chromosome complement results from segregation of rearranged chromosomes at meiosis in a structural heterozygote, the chromosome number is usually normal and the chromosome imbalance results directly from duplication or deficiency of chromosome segments. More rarely, there may be an abnormal number of chromosomes resulting from 3:1 disjunction (12), in which case one additional structurally abnormal chromosome is usually present although cases with 45 chromosomes have also been reported (12,14). Unbalanced karyotypes among offspring of inversion heterozygotes usually result from crossing over within the inverted segment leading to the formation of recombinant chromosomes having both duplication and deficiency of chromosome material. The segregation of such recombinants results in unbalanced gametes. Paracentric inversions have only been reported very rarely in man and crossing over within the pachytene loop may result in 100% unbalanced or 50% balanced and 50% unbalanced gametes, depending on the number and site of the crossovers (10,11,12). Thus, in most cases, abnormal derivative or recombinant chromosomes comprise both duplicated and deficient segments of genetic material and this fact must be remembered when correlating an abnormal chromosome complement with an abnormal phenotype.

The term "partial trisomy" has been used frequently in the literature to describe the whole range of chromosome changes in which duplication of chromosome segments has been reported in individuals with abnormal phenotypes. These include segregants in which the chromosome number is normal and both duplication and deletion of chromosome segments has occurred, as well as individuals with abnormal recombinant chromosomes resulting from crossing over in an inverted segment, and subjects with an extra rearranged chromosome resulting from 3:1 disjunction. The only group to which this term legitimately might be applied is the last in which there is, at least, an extra chromosome present. This extra chromosome is, however, usually rearranged and the term "tertiary trisomic" is more appropriate (15). Trisomic subjects resulting from an extra presumptive isochromosome are termed "secondary trisomics" (15). The term 'trisomic' should be reserved for those cases in which an extra chromosome is present even if this is a derivative or recombinant chromosome, and should not be used to describe individuals with a normal chromosome number. The use of 'partial trisomy' to describe the whole range of duplication/deletion (deficiency) syndromes will lead to confusion and fails to acknowledge the possible phenotypic effect of the deleted segment. In some cases, for instance when the short arm of an acrocentric chromosome is deleted this effect may be slight; in other cases when there is a substantial autosomal segment deleted, the effect of the deleted segment is obviously much greater. Such conditions should be termed therefore "duplication/deficiency syndromes" and any description should clearly define the chromosome abnormality precisely.

CHROMOSOMAL SYNDROMES

1. Trisomics

Trisomy for chromosomes 21, 18 and 13 had been defined clinically by 1961 (16,17,18). During the sixties, a few cases of live born C-trisomics were described (12 Vol. 2). Since 1971, live born children with full or mixoploidy trisomy for chromosomes 8, 9, 10 and 20 have been described (Table 1). Between 30 and 40 patients with full or mixoploid trisomy 8 have been described (19).

Studies on spontaneous abortuses indicate that autosomal trisomy is, in most instances, fully lethal resulting in abortion early in gestation, and trisomic abortuses for each autosome have now been described (14). It has been estimated that between 80% and 95% of chromosomally abnormal conceptuses are aborted (13,10).

2. Tertiary Trisomics

True tertiary trisomics have been reported for chromosomes 9, 12, 14, 15 and 22, of which the most frequently reported have been for chromosomes 9, 14, and 15 while only one case each of tertiary trisomy for chromosomes 12 and 22 have been reported (Table 2).

TABLE 1

AUTOSOMAL TRISOMY

Chromosome	No. of Cases	No. of Mixoploids
8	30	25[1]
9	4	2
10	2	2
13	Many	≃1-2%
18	Many	≃1-2%
20	2[2]	0
21	Many	≃5-10%
22	11[3]	0

1. One mosaic with karyotype 47,XX,+8/47,XX,+21
2. The two cases reported by Opitz et al.(37) are not included as these have subsequently been found to have a rearranged and not an aneuploid chromosome complement. (Opitz pers. comm.)
3. Zellweger et al. (38) reviewed 32 cases with a non mixoploid C-trisomy of which some were trisomy 22.

TABLE 2

TERTIARY AUTOSOMAL TRISOMY

Chromosome	Number Cases	Number Parental Transloc.[1]	Breakpoints	Karyotype
9p	19	11	Cen→q13(12) q21or22(6) q33or34(1)	47,XXorXY,+(9) (q11→pter)
14q	7	6	q11→q22	47,XXorXY,+(14) (q22→pter)
15q	6	1	q13→q22	47,XXorXY,+(14) (q22→pter)
22q	6	0	q13	47,XXorXY,+(22) (q13→pter)

1. One parent carries translocation; tertiary trisomic originates from 3:1 disjunction.

Nineteen cases of tertiary trisomy 9 have been reported most of whom are trisomic for the short arm and a variable portion of the long arm of the chromosome. In most of these cases the remaining chromosomes in the set are normal. In eleven of these nineteen cases, the mother carried a reciprocal translocation between chromosome 9 and chromosomes 1, 6, 7 (1 case each), 8 (2 cases), 15 (3 cases), 20 (1 case), and 22 (2 cases) and the abnormality must have resulted from 3:1 disjunction during maternal oogensis. In six of the remaining 8 cases, both parents had normal chromosomes and the deletion or rearrangement must have arisen de novo during gametogenesis. In one case the father had the deleted chromosome 9 in 0.5% of his cells, and in one case, the father could not be studied.

Correlation of the phenotype with the chromosome abnormality is much simpler in tertiary trisomics where there is little or no deleted material and the duplicated material originates largely from one chromosome. In most cases, therefore, the phenotype can be correlated with "pure" trisomy for 9p, plus variable amounts of 9q depending on the position of the breakpoint. In 12 reports this was between the centromere and q13, in six, in q21 or q22 and in one, in q33 or 34. It is interesting to note that in over 50% of cases of tertiary trisomy 9p, the breakpoint in chromosome 9 is close to or within the variable heterochromatic regions (Fig. 1).

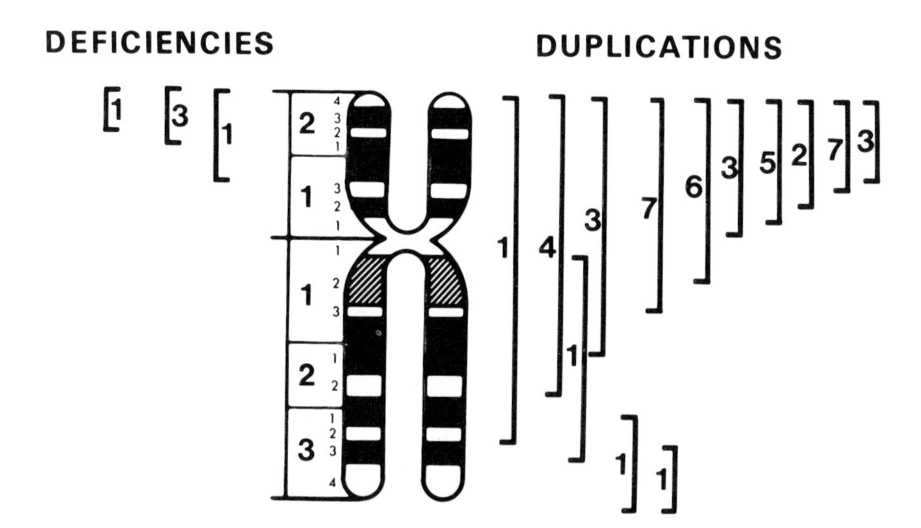

Figure 1. A diagram showing the duplicated and deficient segments in 48 rearrangements involving chromosome 9.

This suggests that this region may be more than usually subject to chromosome breakage or perhaps more able to repair DNA damage and thus form a stable telomere. Each of the eleven translocations involved in 3:1 disjunction was unequal, which would lead to an asymetrical configuration at pachytene, and this may be a factor which affects the frequency of non-disjunction (12,14). A further and more detailed analysis of the phenotype and karyotype in these cases in comparison with the duplication/deficiency syndromes will no doubt help in delimiting specific syndromes caused by specific chromosome regions.

Tertiary trisomy 14q and 15q are relatively frequent causes of duplication of the long arms of these chromosomes. Six cases of "pure" or "free" trisomy 14q are reported of which five result from 3:1 disjunction from a reciprocal translocation carried by one of the parents. On the other hand, only one out of the six cases of tertiary trisomy 15q resulted from 3:1 disjunction in a balanced translocation carried by one of the parents, while in the other five cases the parents were reported as having normal chromosomes.

3. Duplication/Deficiency Syndromes

Duplication/deficiencies have been reported for most chromosomes in the set with the possible exception of chromosome 19. Each unbalanced chromosome abnormality results in an abnormal phenotype of variable severity. Approximately 18 syndromes have been identified resulting mainly from duplication of a specific chromosome segment (Table 3) and individual cases have been reported carrying duplication/deficiencies for most of the other human chromosomes. Table 3 indicates that for each syndrome there is both variability in the amount of material duplicated and in the specificity of the deficient chromosome segment. Table 4 suggests there is non random involvement of deficient chromosomes in this series of rearrangements with the D and G groups being involved with a greater frequency than might be expected by chance (52.5%). Within the D and G groups (Table 5) chromosomes 15 and 22 are involved more frequently than chromosomes 13, 14, 21 and the short arms are deleted twice as frequently as the long arm. In addition to the variability which results from different deficiencies being present, individuals with the same or similar phenotypes may vary in the amount of duplicated genetic material. It is not surprising therefore that duplication/deficiency syndromes, in general, have more variable phenotypes than syndromes resulting from trisomy. The fact that specific duplication/deficiency syndromes can be defined despite this variability suggests that the major clinical features defining a given syndrome may be due to relatively small

TABLE 3

SYNDROMES INVOLVING MAINLY
DUPLICATION OF CHROMOSOME SEGMENTS[1]

Chromosome No. & Arm	Duplicated Segment	Chromosomes with Deficient Segments	No.of Cases
4p	cen to p14→pter	15, 21, 22, 16	16
4q	q21 to q31→qter	2, 3, 9, 13, 18, 22	10
5p	p1 to p14→pter	4, 12, 13	6
7q	q22 to q35→qter	4, 5, 12, 15, 21	8
8p	p11→pter	15, 22	2
8q	q21 to q24→qter	13, 15, 22	4
9p or p+q	q34 to p21→pter	1, 4, 6, 8, 11, 13, 14, 15, 19, 20, 21, 22	40
9q	q11 → q32	4	3
10p	cen to p14→pter	7, 14, 21, 22	6
10q	q22 to q24→qter	1, 2, 15, 17, 18, 22	9
11q	q13 to q23→qter	2, 4, 5, 6, 10, 13, 17, 21, 22	13
12p	p11→pter	6, 21, 14	3
13q	q12 to q22→qter	5, 6, 9, 13, 17, 22	13
14q	q22→pter q12 to q22→qter	10, 12, 13, 14, 19, 21	11
15q	q22 to q12→pter q13 to q25→qter	4 12, 21	7 3
18p	cen→pter	20	5[2]
20p	p11→pter	15, 18, 21, 22	5
21q	q22→qter qter→q21	10 (Non D.S.)	Uncertain
22q	q13→pter	--	6

1. Includes Tertiary Trisomics listed in Table 2.
2. Secondary trisomics.

TABLE 4

DISTRIBUTION OF CHROMOSOMES
IN 124 REARRANGEMENTS
WITH DEFICIENT SEGMENTS

GROUP	NO.	%
A	9	7.3
B	11	8.9
C	24	19.4
D	27	21.8
E	11	8.9
F	4	3.2
G	38	30.7
	124	100.0

TABLE 5

DISTRIBUTION OF D & G CHROMOSOMES
WITH DEFICIENT SEGMENTS

CHROMOSOME INVOLVEMENT

CHROMOSOME	NO.	%
13	9	13.9
14	5	7.7
15	13	20.0
21	14	21.5
22	24	36.9
	65	100.0

ARM INVOLVEMENT

ARM	NO.	%
p	42	64.6
q	23	35.4
	65	100.0

and rather specific chromosome segments. For example, it is now clear that a specific duplication of band 21q22 is sufficient to produce the Down syndrome phenotype (21,22) and the same may turn out to be true for the basic phenotype in each of these new syndromes. Phenotypic variability would then result from both variation in amount of the duplicated chromosome material as well as in the amount and constitution of the deficient chromosome. The 18q- syndrome may be taken as an example. This condition is relatively well defined clinically (19). The amount of deleted material may, however, vary from q11 to q21→qter (Fig. 2).

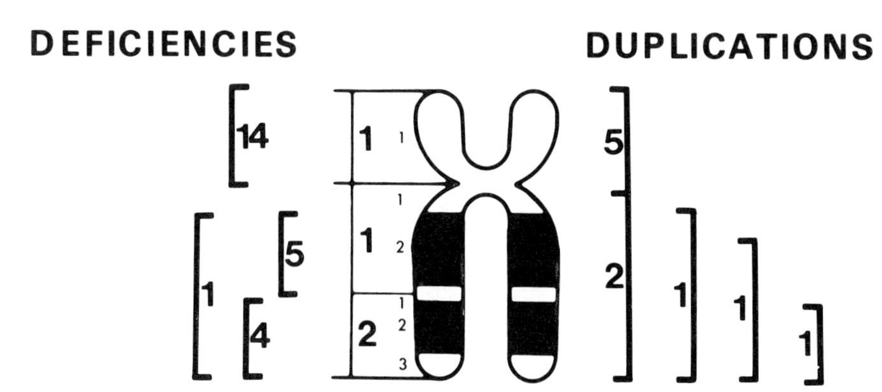

Figure 2. A diagram showing the duplicated and deficient segments in 34 rearrangements involving chromosome 18.

Five cases have now been reported with an intercalary deficiency for q12 (23,24) each of whom have the major features of the 18q- syndrome. Thus, a deficiency for this band alone may be sufficient to produce the major features of this syndrome. A precise definition of the phenotype in patients in whom overlapping chromosome duplications and deletions occur may allow the identification of specific chromosome regions which result in basic duplication or deletion syndrome for each autosomal region.

Loss of short arm material from the D and G group chromosomes might be expected to have minimal phenotypic effects, so that duplication/deficiency syndromes involving these chromosomes must be considered to be mainly caused by the duplicated segment. The fact that these chromosome regions are represented more frequently than one might expect by chance in duplication/deficiency syndromes (Table 5), tends to support this hypothesis.

4. Deletion Syndromes

Both terminal deletions and intercalary deficiencies have been reported in patients with phenotypic abnormalities. Table 6 lists 12 reasonably well defined deletion syndromes. In most, but by no means all instances, parents of children with deletion syndromes have normal chromosomes so that it must be presumed that the deletion occurred *de novo* during gametogenesis. The number of syndromes caused primarily by chromosome deletion is fewer than the number caused by duplication/deficiency. This suggests that there may be selection against major autosomal deletions so that fewer affected subjects survive to term. It is still not entirely clear whether true terminal deletions exist or whether the telomere theory originally proposed by Muller (25) is valid. True intercalary deficiencies have been reported for chromosome 7p, 7q, 8p and 18q. Such deficiencies may arise *de novo* or as the result of an insertion followed by segregation (23). In general, presumptive terminal deletions so far reported in man appear to behave normally at cell division, suggesting either that the chromosome ends have been repaired to form a new and stable telomere or that most apparent terminal deletions are, in fact, unrecognized intercalary deficiencies. It is, at present, not possible to differentiate between these two possibilities.

TABLE 6

AUTOSOMAL DELETION SYNDROMES

CHROMOSOME NO. & ARM	BREAKPOINTS	NO. OF CASES
4p	p13-p16	Numerous[1]
5p	p12-p15	Numerous[2]
7p	p15-p22	5[3]
7q	q22-q32	4[4]
8p	p21	3[5]
9p	p21-p23	5
10p	p13	3
11q	q21-q23	4
12p	p12	3
13q	q13-q32	7
18p	cen-p11	>20
18q	q11-q21	>20

1. Wolf-Hirschhorn Syndrome
2. Cri du Chat Syndrome
3. 1 int. del. (p15→p22)
4. 3 int. del. (q11→q22)
5. 1 int. del. (p21→p23)

The phenotypic effects of ring chromosomes result from the deletion of both telomeres during the formation of the ring chromosome. Ring chromosomes have been reported for chromosomes 5, 6, 12, 13 and 22. In general, rings are unstable and may vary in size and number depending upon their behaviour during replication and mitosis. Thus, the phenotypes resulting from a ring chromosome might be expected to resemble the corresponding deletion syndromes for the long and the short arm where these have been defined.

5. Inversions

Pericentric inversions leading to abnormal phenotypes have been reported for chromosomes 3, 5, 8, 13 (Table 7). The presence of recombinant chromosomes allows comparison between the phenotypic effects of duplication of one terminal segment with deletion of the other. An excellent example is a large family in which patients carrying both duplication/deletion for the terminal regions of the long and short arms of chromosome 3 are reported (26).

TABLE 7

CHROMOSOMES INVOLVED IN INVERSIONS LEADING TO DUPLICATION/DEFICIENCY

CHROMOSOME	DUPLICATION	DEFICIENCY
3	q21→qter	p25→pter
5	p14→pter	p14→pter
8	q22→q24	p23
13	q1→qter	

DISCUSSION

Chromosome banding has led to the definition of a wide range of new chromosome abnormalities. A combination of these technological advances with developments in dysmorphology has resulted in the description of a number of clinical syndromes each resulting from similar chromosome abnormalities. Variability in these new syndromes results from variation in the amount and composition of duplicated and/or deficient genetic material.

Much of this chromosome imbalance is caused by the production of unbalanced segregants from balanced chromosome rearrangements transmitted by one or other parent. Balanced reciprocal translocations occur in about 1:1000 newborn infants (27). 90% - 95% of these are inherited and it would seem that only very rarely are unbalanced segregants observed in these families (28,29,30). There is, however, some evidence to suggest that these families have a higher incidence of fetal wastage than controls (31).

The group of chromosome rearrangements under consideration in the present paper are highly selected as the result of ascertainment through a dysmorphic live born proband. A comparison between rearrangements ascertained in this way and randomly ascertained rearrangements might give some indication as to the characteristics of the rearrangement which are most likely to cause abnormalities among live born children in respect to chromosome involvement, position of breakpoints, equality or inequality of exchange and meiotic configuration.

Since 1971, an additional seven trisomic syndromes have been defined of which the best known is that caused by an additional chromosome number 8; many patients with a trisomy for a large autosome are mixoploid so that the abnormality must have arisen as the result of a post-zygotic nondisjunctional event during early cleavage, or by two events, meiotic nondisjunction followed by post -zygotic loss of the abnormal chromosome from one cell line. Monosomics are much rarer, presumably because autosomal loss is less well tolerated than gain: presumptive monosomies for chromosomes 21 and 22 (32,33,34,35) have been described although questions have been raised as to whether these are truly monosomic or result from rearrangement (36).

The finding of true 'tertiary trisomics' for chromosomes 9, 14, 15, and 22 is of interest as a detailed study of these conditions should provide a relatively good picture of the effect of trisomy for a specific chromosome region uninfluenced by deletion of other chromosome regions. A detailed comparison of full and tertiary trisomics with corresponding duplication/deletion syndromes should provide data of value in karyotype/phenotype correlation. The fact that 3 out of 4 of the chromosomes involved in tertiary trisomy are acrocentrics suggests that 3:1 disjunction may be more frequent in translocations involving an acrocentric chromosome than it is in translocations involving non acrocentrics (12,14). The finding of a relatively high frequency of tertiary trisomy for 9p with breaks adjacent to the block of paracentromeric heterochromatin may indicate that this region is more liable to breakage or more capable of repair without restitution or reunion than other regions of the set.

Examination of the duplication/deficiency syndromes indicates an excess involvement of deficiencies for 15p and 22p (Table 5) in rearrangements which result in the ascertainment of probands with a duplication for other chromosome segments. The relative genetic inertness of these chromosome regions suggests that in these instances the syndrome is most likely caused by duplication for the other chromosome segment with the deletion having little phenotypic effect. In other instances, the deleted segment is often terminal and small. The significance of this observation is not clear; it may, however, be that only very small autosomal deletions are compatible with live birth so that unequal translocations with one breakpoint close to the telomere are more likely to produce viable adjacent I and II segregants, than equal translocations in which, because of the nature of the translocation, the size of the deleted segment must be greater. Thus, it may be that a symmetrical rearrangement with proximally situated breakpoints has, first, a greater chance of the occurrence of alternate segregation or its equivalent which results in balanced gametes, or second that any adjacent I or II segregants result in non viable zygotes due to the large size of the resultant duplication/deficiency. In an unequal translocation, on the other hand, the meiotic configuration will be asymmetrical, resulting in more frequent adjacent I or II segregation. Conversely, however, the size of the duplication/deficiency may be less so that chromosomally abnormal zygotes will be viable. A careful study of families ascertained on the one hand at random and on the other, through dysmorphic probands, together with a study of early abortus material from these families would be required to test this hypothesis.

The fact that deletion syndromes are relatively less common than the duplication/deficiency syndromes in human dysmorphology provides further evidence that deletions may be less viable and less liable to come to term than duplications. It may be that the developmental disturbances caused by deletions are that much more severe than those caused by deletions so that only deletions of a small size or for less significant chromosome regions survive to term.

The variability and overlap seen in the phenotypes of the chromosome syndromes emphasizes the importance of a most careful clinical description for each case with the use of objective rather than subjective criteria. For instance, tables of normal measurements are now available for many phenotypic features so that objective measurements and percentiles are now meaningful. Only by these means will precise clinical descriptions become available which can then be correlated with chromosome findings. The cytogenetic description of a rearrangement should be as precise as possible and should include location of the breakpoints, the regions duplicated or deleted, and where necessary, more than one banding technique should be used to define rearrangement.

If this level of precision can be introduced into dysmorphology and clinical cytogenetics, then an improvement in our understanding of the effects of chromosome imbalance on the human phenotype will result. If not, then the confusion and imprecision which has plagued medical cytogenetics since 1960 will remain, and the scientific advances which have been made possible by the recent technological developments will remain unfulfilled, and an important area of human developmental biology will fail to reach its full potential.

REFERENCES

1. Dutrillaux, B. and Lejeune, J. (1975) Advances in Human Genetics 5. Eds. H. Harris and K. Hirschhorn. New York, Plenum Press. p.119.
2. Hecht, F., Wyandt, H.E. and Magenis, R.E.H. (1974). The cell nucleus, Vol. II. Ed. H. Busch., New York, Academic Press, p. 33.
3. Comings, D.E. (1972) Advances in Human Genetics 3. Eds. H. Harris and K. Hirschhorn. New York, Plenum Press. p. 237.
4. Comings, D.E. (1976). Aspects of Genetics in Paediatrics. Scientific Proceedings of the 3rd Unigate Workshop. Ed. D. Barltrop. London, Fellowship of Postgraduate Medicine. p. 17.
5. Caspersson, T., Zech, L. and Johansson, C. (1970). *Exptl. Cell Res.* 62, 490.
6. Caspersson, T., Zech, L. and Johansson, C. (1970) *Chromosoma* 30, 215.
7. Latt, S.A. (1974) *Chromosomes Today* 5, p.367.
8. Paris Conference (1971) : Standardization in Human Cytogenetics. Birth Defects: Original Article Series, VIII: 7, 1972. The National Foundation, New York
9. Paris Conference (1971), Supplement (1975): Standardization in Human Cytogenetics. Birth Defects: Original Article Series, XI, 9, 1975. The National Foundation, New York.
10. Lewis, K. R. and John, B. (1963) Chromosome marker. London, J. & A. Churchill.
11. Lewis, K. R. and John, B. (1972) The Matter of Mendelian Heredity. 2nd ed. Longman.
12. Hamerton, J. L. (1971) Human Cytogenetics. New York, Academic Press. 2 Vols.
13. Creasey, M. R., Crolla, J.A. and Alberman, E.D. (1976) *Hum. Genet.* 31, 177.
14. Lindenbaum, R. H. and Bobrow, M. (1975) *J. Med. Genet.* 12, 29.
15. Darlington, C. D. (1965) *Cytology*. London, J. & A. Churchill.

16. Lejeune, J. (1959) *Ann. Genet.*, Semaine Hop. 1, 41.
17. Edwards, J. H., Harnden, D. G., Cameron, A. H., Crosse, V. M. and Wolff, O. H. (1960) *Lancet* 1, 787.
18. Patau, K. A., Smith, D. W., Therman, E. M., Inhorn, S.L. and Wagner, H. P. (1960) *Lancet* 1, 790.
19. Smith, D.W. (1976) Recognizable Patterns of Human Malformation: Genetic Embryologic and Clinical Aspects. 2nd ed. Philadelphia, W. B. Saunders Company.
20. Polani, P. E. (1970) *Proc. Roy. Soc. Med.* 63, 50.
21. Williams, J. D., Summitt, R. L., Martens, P. R. and Kimbrell, R. A. (1975) *Am. J. Hum. Genet.*27, 478.
22. Lafourcade, J. and Rethore, M.-O. (1976) V. International Congress of Human Genetics, Mexico. Abstracts, *Excerpta Medica International Congress Series* No. 397, p.12.
23. Chudley, A. E., Bauder, F., Ray, M., McAlpine, P. J., Pena, S. D. J. and Hamerton, J. L. (1974) *J. Med. Genet.* 11, 353.
24. Waldenmaier, C., Hirsch, W., Shibata, K. and Kothe, I. (1975) *Mschr. Kinderheilk,*123, 68.
25. Muller, H. J. and Herskowitz, I. H. (1954) *Am. Nat.* 88, 177.
26. Allderdice, P. W., Browne, N. and Murphy, D. P. (1975) *Am. J. Hum. Genet.* 27, 699.
27. Hook, E. B. and Hamerton, J. L. (1977) Population Cytogenetics: studies in humans. Eds. Hook, E.B., and Porter, I. H. New York, Academic Press. 63
28. Friedrich, U. and Nielsen, J. (1974) *Humangenetik* 21, 133.
29. Jacobs, P. A., Aitken, J., Frackiewicz, A., Law, P., Newton, M.S. and Smith, P.G. (1970) *Ann. Hum. Genet.* Lond. 34, 119.
30. Hamerton, J. L., Ray, M., Canning, N., Evans, J., Martsolf, J. T. and Hunter, A. H. In preparation.
31. Jacobs, P. A., Frackiewicz, A., Law, P., Hilditch, C. J. and Morton, N. E. (1975) *Clin. Genet.* 8, 169.
32. Halloran, K. E., Breg, W. R. and Mahoney, M. J. (1974) *J. Med. Genet.* 11, 386.
33. Kaneko, Y., Ikeuchi, T., Sasaki, M., Satake, Y. and Kuwajima, S. (1975) *Humangenetik* 29, 1.
34. Dziuba, P., Dziekanowska, D. and Hubner, H. (1976) *Hum. Genet.* 31, 351.
35. Rosenthal, I. M., Bocian, M. and Krmpotic, E. (1972) Society for Pediatric Research 42nd Annual Meeting Washington, D.C. Abstract. *Ped. Res.* 6(4).
36. Schinzel, A. (1976) *Hum. Genet.* 32, 105.
37. Pallister, P. D., Herrmann, J., Meisner, L. F., Inhorn, S. L. and Opitz, J. M. (1976) *Lancet* 1, 431.
38. Zellweger, H., Ionasescu, V. and Simpson, J. (1975). *J. Genet. hum.* 23, 65.

OCCURRENCE AND SIGNIFICANCE OF CHROMOSOME VARIANTS

Herbert A. Lubs

Departments of Pediatrics and Biophysics and Genetics,
University of Colorado Medical Center,
4200 E. 9th Avenue, Denver, Colorado 80262

ABSTRACT. The majority of people have 5-10 variations in size or staining qualities of certain chromosomes with combined conventional staining, Q, C and R-banding. This high frequency contrasts with the 0.5-1% of people having definite chromosome abnormalities. These "normal" variants or polymorphisms most frequently involve chromosomes 1, 9, 16, Y and the acrocentric chromosomes and are of interest both in terms of their biological significance and their potential usefulness in clinical cytogenetics. These regions are now known to contain varying amounts of the several classes of redundant DNA. A number of studies of consecutive newborn infants and older randomly selected children, as well as several series of children referred for evaluation of mental retardation and anomalies, have shown statistically significant positive correlations between certain variants (especially large satellites or the secondary constriction region of chromosome 9) and mental retardation or congenital anomalies. In one population study, a significant correlation between a long 9qh region and low I.Q. was found in Black seven-year-old children, but not in Whites. However, in no case have the majority or even a quarter of children with a particular variant been found to be clinically abnormal. Thus, in interpreting diagnostic studies and antenatal studies these variations should still be interpreted as "normal" variants.

Further research using more sophisticated technics, such as annealing various classes of redundant DNA or RNA to these regions or combinations of several banding technics might reveal subclasses of a particular variant in which the majority of children have an associated clinical abnormality. These clinical correlations might also be explained if the variations act as one genetic factor in abnormalities that are multifactorial in origin. Better family studies would help to distinguish between these possibilities.

Other studies have shown that there are significant racial and ethnic differences in the frequencies of most variants and that there may be an increased frequency of certain variants in abortuses. The latter studies are not well controlled, however. There is also some evidence for non-random segregation of 9qh variants although most variants segregate in a Mendelian fashion and can be used as markers in linkage studies. Use of these variants to distinguish maternal cells from female fetal cells is a particularly important clinical application. Paternity determination is another. Finally, several studies suggest that the frequency of certain variants is increased in chromosomally abnormal children. In summary, these studies, to date, have revealed much that is useful to clinical medicine but the major question of the possible clinical significance remains a subject for further study by better technics.

INTRODUCTION

A number of straightforward statements can be made about variations (heteromorphisms) in the human karyotype: 1) all of us have a number of them, 2) in each of the variable regions there is generally a continuous, not discreet, variation, 3) these regions are located either in the short arms of acrocentric chromosomes, the long arm of Y chromosome, the centromere, or one of the three chromosomes with a secondary constriction adjacent to the centromere in the long arm (1, 9 or 16), 4) all contain varying amounts of different classes of highly redundant DNA, 5) in nearly all instances the heteromorphisms are inherited, 6) with each new cytogenetic technic more variation is identified; often there is no direct interrelation between a heteromorphism identified by one technic and by another, 7) since heteromorphisms are common, it is important to understand their significance. More specifically, with the advent of routine amniocentesis, it is of great importance to know whether a specific heteromorphism such as a 9 inversion, is clinically important.

The several problems inherent in assessing the importance of the heteromorphisms are listed in table 1. The most important of these are the relatively small number of children having both a specific variant and a clinical abnormality, the various biases inherent in the clinical series, and the many technical problems, some of which will be dealt with here. Nevertheless, sufficient information is available to present several tentative answers to the question of the significance of heteromorphisms.

TABLE 1

PROBLEMS IN ASSESSING IMPORTANCE OF HETEROMORPHISMS

1. Surveys of CONSECUTIVE BIRTHS:	a) long wait for IQ + behavioral data b) follow-up and family studies difficult in most countries
2. Surveys of OLDER CHILDREN OR ADULTS:	a) death or institutionalization of seriously affected prior to study b) sampling and biases

3. In both types of studies usually dealing with CORRELATIONS BETWEEN UNCOMMON EVENTS, hence N's small.

4. CLINICAL SERIES: poorly controlled, anecdotal or biased. Difficult to compare cytogenetic findings in one lab with another because of variation in technic, quality, criteria, reading and ethnic composition.

METHODS

Most of the material presented here has been drawn from two studies. The first of these (1,2) was carried out in a series of 4400 consecutively born New Haven infants in 1968 (1,2) using orcein as the principal stain. The principal collaborator in that study was Dr. Frank Ruddle. The second study also has been reported in detail (3,4) and involved G-, Q-, and C-banding in a series of more than 4000 seven-year-old infants who had been included in the perinatal collaborative study. Principal collaborators in the study were Drs. William Kimberling and S. Patil (Denver), Drs. J. Brown and M. Cohen (Buffalo), Dr. P. Gerald (Boston), Dr. F. Hecht (Portland), Dr. R. Summit (Memphis) and Dr. N. Myrianthopoulos (NINCDS, Bethesda). Both detailed descriptions of the populations under study, the logistics of the studies, and details of the cytogenetic technics have been described in these and other publications. Particular emphasis in the latter study was placed upon predetermined criteria for the size and staining criteria of the heteromorphisms, and in both studies the clinical findings were not known to the cytogeneticist. A subsample of about 200 white and 200 black children were included in the "special study subsample" reported here. These were selected using random numbers to provide equal numbers of children above and below I.Q. 85 from each center. Sequential Q-and C-banding in the same cell was carried out in 5-10 cells from each of these children. Q-banding was adequate in nearly all children, but the sequential Q-and C-banding was successful in only about half of the children.

RESULTS

Several statistically significant correlations between heteromorphisms and various clinical parameters emerged from the earlier study of New Haven infants. These are summarized in tables 2 through 6. In table 2, it can be seen that more large satellites were present in Group G chromosomes in White infants with anomalies than in normal infants and, similarly, that there was twice the frequency of large satellites present in White infants with low birth weight (under 2500 grams) (5).

TABLE 2

LARGE SATELLITES IN GROUP G AND CLINICAL PARAMETERS

(New Haven study - 10 or more cells/infant)

	Children of 3476 Caucasian Mothers		Children of 807 Negro Mothers	
	N	%	N	%
Frequency of major congenital anomalies infants with large G satellites...	(6/97)	6.2[1]	(0/30)	0
Other infants		1.8		2.5
Frequency of low birth weight in infants with large G satellites	(14/97)	14.4[1]	(5/30)	16.7
Other infants		7.2		12.6

[1] $p<.01$

Neither correlation, however, was found in the smaller number of Black infants. Similarly, major anomalies were more frequent in White children with rare heteromorphisms (table 3) than in other children. Again, this was found only in the White children.

TABLE 3

RISK OF MAJOR CONGENITAL ANOMALIES IN INFANTS WITH RARE CHROMOSOMAL VARIANTS
(New Haven study - 10 or more cells/infant)

Rare Variants	Caucasian		Negro
Increased secondary constriction region A-1	0/8[1]		0/3
9qh inversion	0/2		0/10
Y long (>18)	1/7		0/1
Y short	0/7		0/2
Y metacentric	1/3		0/0
Total	2/27	(7.4%)	0/16
Rare variants D and G groups			
Short arm of Group D chromosomes:			
Increased length (> 18 short arm)	1/10		0/7
Tandem satellite	0/3		0/2
Streaked satellite	0/1		0/1
Short arm absent	0/2		0/0
Short arm of Group G chromosomes:			
Increased length (> 18 short arm)	1/3		0/0
Tandem satellite	1/4		0/0
Streaked satellite	0/1		0/0
Total	3/24	(12.5%)	0/10
Grand total	5/51	(9/8%)[2]	0/26

[1] Number of infants with variant and major anomaly/number with variant in 3476 Caucasian and 807 Negro infants.

[2] $p<.01$ compared to 1.76% of all Caucasian newborns with major anomalies.

More recently, we reviewed the interrelationship of long Y chromosomes and the frequency of spontaneous abortions in the same family (6). Table 4 shows that the frequency of long Y chromosomes was greater in sons of mothers who had 3 or more spontaneous abortions, and more significantly, in table 5, that the overall proportion of prior abortions in mothers of long Y infants was twice as great as in all other mothers. Several anecdotal reports of long Y chromosome in spontaneous abortuses prompted this re-evaluation.

TABLE 4

FREQUENCY OF CAUCASIAN INFANTS WITH LONG Y
(New Haven Study)

$\leq$2 PRIOR SPONTANEOUS ABORTIONS IN MOTHER	33/1,736[1]	1.9%
$\geq$3 PRIOR SPONTANEOUS ABORTIONS IN MOTHER	3/29	10.3%

[1] 26 mothers had no prior spontaneous abortions, 5 had one, and 2 had two spontaneous abortions.

TABLE 5

PROPORTION OF SPONTANEOUS ABORTIONS IN
PRIOR PREGNANCIES OF CAUCASIAN MOTHERS
(New Haven Study)

	MOTHERS IN THE OVERALL STUDY	MOTHERS OF LONG Y INFANTS
NUMBER PRIOR PREGNANCIES	4,773	54
NUMBER SPONTANEOUS ABORTIONS	776	18
PROPORTION PRIOR ABORTIONS	0.16	0.33[1]

[1] χ^2, $p<0.01$

Some of the more recent data from the perinatal collaborative study is given in the remaining tables. The first point to emerge from this study was the profound effect of variation in quality on the number of heteromorphisms that were detected (table 6.) Even though this was predictable, the effect is so strong and its importance in interpreting the literature so great that it merits comment. For example, when both C-banding and Q-banding were poor in quality, fewer than 4 heteromorphisms per child were found. When Q-banding was good and C-banding excellent, more than 12 were found, a three-fold difference. On the average, there were 6.7 heteromorphisms per child when the combined sequence of Q- and C-banding was carried out. If acridine orange R-banding had also been used, the number would have been still larger. We are rapidly approaching the point where each person's karyotype will be as unique as his fingerprints.

TABLE 6

NUMBER OF Q AND C HETEROMORPHISMS COMBINED/CHILD BY QUALITY OF BANDING

Q BANDING QUALITY	C BANDING QUALITY					
	NO DATA	POOR	INTERMEDIATE	GOOD	EXCELLENT	MEAN
POOR	2.0	3.7	3.0	3.0	-	3.3
INTERMEDIATE	2.2	5.2	7.1	9.5	10.0	6.5
GOOD	4.5	5.7	7.3	8.3	12.5	7.6
EXCELLENT	-	2.0	-	-	-	2.0
MEAN	2.5	5.0	7.2	8.7	11.3	6.7

The difference in frequency of many heteromorphisms between Blacks and Whites was clearly demonstrated in both studies. The results from the New Haven study have been published previously (7) and showed that in every instance where there was a significant difference in frequency, the frequency was greater in the Black children. Although both the existance of the 9qh inversion (or inversion-like) heteromorphism and the marked racial difference in frequency were first established in this study, most of the heteromorphisms involved increased amounts of chromatin. Similar results were obtained, both by Q- and by C-banding in the more recent study of seven-year-old children (3). With few exceptions, wherever there was a racial difference, the heteromorphism again was more frequent in Blacks than in Whites. With the exception of chromosome 4, wherever there was a racial difference in the frequency of a given intensity of fluorescence by Q-banding, there was approximately a two-fold increase in the frequency of bright heteromorphisms in the Black children (table 7). For the remaining chromosomes, even though there was no significant difference, the results almost always showed a greater proportion of heteromorphisms in Black children.

TABLE 7

FREQUENCY OF Q HETEROMORPHISMS BY RACE

CHROMOSOME REGION	LEVEL(S)[2]	WHITE (N=205)		BLACK (N=210)		
		N	PROP.[1]	N	PROP.	
3c	4	57	0.28	89	0.42	χ^2 9.5,p<.01
	5	119	0.58	119	0.57	n.s.
4c	4&5	42	0.20	20	0.09	χ^2 9.8, p<.01
13p	4	87	0.42	113	0.54	χ^2 7.9, p<.01
	5	22	0.11	70	0.33	$\chi^2$30.7, p<.001
13s	4&5	9	0.04	20	0.10	χ^2 4.2, p<.01
14s	4&5	23	0.11	27	0.13	n.s.
15p	4&5	2	0.01	6	0.03	n.s.
15s	4&5	21	0.10	21	0.10	n.s.
21p	4&5	1	0.005	4	0.02	n.s.
21s	4&5	16	0.08	28	0.13	n.s.
22p	4&5	4	0.02	8	0.04	n.s.
22s	4&5	12	0.06	29	0.14	χ^2 7.1, p<.01

[1] Proportion of people, either heterozygous or homozygous.

[2] Level 1 = least fluorescence, level 5 = brightest, as in distal Y, etc.

There was no significant difference in the quality of cultures between the Black and the White children in this special study sub-sample (3), hence, the results cannot be explained on that basis. Similar results were also obtained with C-banding, although no significant racial difference was noted for the heteromorphic regions of chromosomes 1, 9, 16 and Y. Although the numbers of centromere heteromorphisms were too small to permit statistical analysis for individual chromosomes, there was a two-fold increase in the number of large centromere heteromorphisms in the Black children (table 8). Indeed, several were found only in Black children (chromosomes 18 and 19).

TABLE 8

CENTROMERE HETEROMORPHISM FREQUENCIES BY RACE
(C-BANDING)

CHROMOSOME REGION	WHITE (N=95) SMALL	LARGE	TOTAL	BLACK (N=97) SMALL	LARGE	TOTAL
2c	1	–	1	–	3	3
3c	3	4	7	3	3	6
4c	3	–	3	3	7	10
5c	–	3	3	6	3	9
6c	2	3	5	1	1	1
7c	–	4	4	1	3	4
8c	–	1	1	–	–	–
10c	1	3	4	1	3	4
11c	6	–	6	3	1	4
12c	–	1	1	–	5	5
13c	2	3	5	5	4	9
14c	1	2	3	3	–	3
15p	–	4	4	–	5	5
15c	6	4	10	9	1	10
17c	4	2	6	2	4	6
18p	1	–	1	–	1	1
18c	9	–	9	7	8	15
19c	2	–	2	4	9	13
20c	6	1	7	1	5	6
21c	–	–	–	–	4	4
22c	1	2	3	4	8	12
Xc	–	5	5	–	1	1
TOTAL	48	42[1]	90	53	79[1]	132

[1] $\chi^2 = 11.01$, $p < .005$

As found in the New Haven study by orcein staining, there was a significant difference in 9 inversions in the two races (table 9) in respect to "complete inversions". Partial inversions were similar in frequency in both races and 1qh partial inversions were more frequent in Black children.

Many types of clinical assessments were carried out. These included inspection of the 220 routinely analysed clinical parameters for all children with a given heteromorphism, correlation analysis, and wherever possible, regres-

TABLE 9

1 AND 9 qh INVERSIONS BY RACE

(Overall Study)

	WHITE (N=3084)		BLACK (N=1780)		
	N	%	N	%	
1qh "partial inversion"[1]	14	0.45	1	0.06	$\chi^2 = 5.8$, $p<0.05$
9qh "partial inversion"[1]	17	0.55	8	0.45	
9qh "complete inversion"	4	0.13	19	1.07	$\chi^2 = 21$, $p<0.001$

[1] "Partial inversions" were defined as those in which the centromere was in the middle of the h region and "complete inversions" as those in which there was a metacentric 9 with the entire h region being on the short arm.

---

sion analysis. The regression analyses were carried out using the added heteromorphism scores (1 = a small or dim heteromorphism, 5 = a large or bright heteromorphism, etc.) for both homologs for Q- or C-banding and certain continuous variables such as height, weight, I.Q., etc. In most cases, these analyses have shown little. No clinically significant associations with any Q heteromorphism have been found, for example.

The several interesting exceptions relate to C-band heteromorphisms. The most important was a significant ($p<.01$) regression analysis using I.Q. score as the dependent variable. This showed a steady drop in I.Q. score in 105 Black children as the amount of 9qh material increased, and there was a net decrease of 25 I.Q. points from a summed score of 4 to a score of 10. Similar results were found in the 107 White children; however, the 4 children having a 9qh score of 9 also had high I.Q.'s and the results were not significant statistically. If the four children just referred to were omitted from the analysis, the results for the White children were also significant. It may be that these children were "overscored" and that the effect is similar in both races. Somewhat similar findings were observed for chromosome 1 but these were not significant in either race; for chromosome 16 there was clearly no such relationship. No relationship between I.Q. and the quality of culture was found in either race and no artifactual reason for this significant result has been discerned in spite of very rigorour inspection and analysis of the data. The results were essentially the same when either the 4 year or 7 year I.Q. scores were used.

Special care was taken to analyse the information in respect to the 9 inversions, since, if indeed these are inversions, they might be especially likely to have an associated risk. However, the clinical findings in these children with either complete or partial inversions and their other family histories were not significantly different from other infants. Since the Boués (8) have recently reported a suggestively high frequency of such inversions in abortuses (but without control data), the negative reproductive data shown in table 10 is especially pertinent. It shows that the proportion of viable births is well within the expected limits for the complete inversions (41 of 42 prior births in families with 9 inversions were viable) and that the findings were essentially the same for families having partial inversions.

Finally, the analysis of the immense amount of data in this study has many pitfalls, one of which is relying too completely on the statistical analysis. Hence, although the information in table 11 is not statistically different from the average of 0.23 minor or major anomalies per child found in the overall study, the numbers are small and the occurrence of roughly twice the expected frequency of anomalies for each of the most easily readable heteromorphisms (one average plus one very large h region) for 1, 9 and 16 is at least suggestive and warrants further study. It is quite possible that technical factors may have obscured a more significant relationship.

TABLE 10

9 INVERSIONS AND REPRODUCTION

	No. of Mothers	Ratio of Viable Births to No. of Pregnancies
Complete Inv.	14	41/42
Partial Inv.	13	21/22

TABLE 11

CONGENITAL ANOMALIES AND 1, 9, 16 HETEROMORPHISMS

1 average (3) and 1 very large (5) h region
(Perinatal Collaborative Study)

	Average No. Anomalies/Child	
1qh	0.50[1]	(8/16)
9qh	0.38	(6/16)
16qh	0.43	(3/7)

[1]Average all children = 0.23 anomalies/child.

DISCUSSION

The importance of technical factors and the racial differences appear well documented. Jacobs (9) recently reviewed the literature on heteromorphisms and their possible significance and concluded that although these same two factors were clearly documented and important, little else was. Some, but not all, of the present information was available to her. Our present evidence can be summarized briefly. In no instance have we found a specific variant present only in abnormal children, and in no instance have even 50% of children with a specific heteromorphism been abnormal. Thus, even where statistically significant associations have been found, the odds are quite good that the child will be clinically normal. This does not mean, however, that there is no biological, clinical, or selective significance to these heteromorphisms. The recurrance of small, or selective but significant associations in the population studies for large satellites and for chromosome 9 as well as in several unpublished clinical series suggests that, indeed, certain variants may influence either intelligence or development and that still more needs to be known. Perhaps with the combined use of more sophisticated technics, better cytogenetics, and carefully constructed samples, closer relationships will be found.

Many explanations have been put forward to explain such possible relationships. There is clearly no simple, direct relationship between any heteromorphism and any clinical defects. The most tenable explanation is simply that certain heteromorphisms may act as one factor in the multi-factorial systems which are known to influence the clinical parameters under study here. An explanation for the racial differences is completely lacking, but may well be a fascinating area for further research once more is known of the several human satellite DNA's. It seems premature to write these differences off as random or meaningless variations.

ACKNOWLEDGEMENTS

This work has been supported by NICHD grant no. 5 R01HD-05079 and NINCDS contract no. 4/USC252(2) (5).

REFERENCES

1. Lubs, H.A. and Ruddle, F.H. (1970) Science 169, 495.
2. Lubs, H.A. and Ruddle, F.H. (1970) Human Population Cytogenetics 119.
3. Patil, S.R., Lubs, H.A., Kimberling, W.J., Brown, J., Cohen, M., Gerald, P., Hecht, F., Moorhead, P., Myrianthopoulos, N. and Summitt, R. (1977) in _Population Cytogenetics_, E. Hook & I. Parker, eds., p. 103.
4. Lubs, H.A., Patil, S.R., Kimberling, W.J., Brown, J., Cohen, M., Gerald, P., Hecht, F., Moorhead, P., Myrianthopoulos, N. and Summitt, R. (1977) in _Population Cytogenetics_, E. Hook & I. Parker, eds., p. 133.
5. Lubs, H.A. and Lubs, M.L. (1972) in _Perspectives in Cytogenetics_ - The Next Decade, S. W. Wright, B. F. Crandall and L. Boyer, eds., p. 303.
6. Patil, S.R. and Lubs, H.A. (1977) Humangenetik, in press.
7. Lubs, H.A. and Ruddle, F.H. (1971) Nature 233, 134.
8. Boue, J., Taillemite, J.L., Hazael-Massieux, P., Leonard, C. and Boue, A. (1975) Humangenetik 30, 217.
9. Jacobs, P.A. (1977) _Progress in Medical Genetics_, in press.

NONRANDOM CHROMOSOMAL CHANGES IN HUMAN MALIGNANT CELLS

Janet D. Rowley

Department of Medicine, The University of Chicago,
and The Franklin McLean Memorial Research
Institute, Chicago, Illinois 60637

ABSTRACT. The role of chromosomal changes in human malignant cells has been the subject of much debate. The observation of nonrandom chromosomal changes has become well recognized in CML, and more recently in AML. In the present report, data are presented on the sites of duplication of chromosome No. 1 in hematologic disorders. Trisomy for region 1q25 to 1q32 was observed in every one of 34 patients whose cells showed duplication of some part of chromsome No. 1. Adjacent regions 1q21 to 1q25, and 1q32 to 1qter, also were trisomic in the majority of patients. Two patients had deletions, one of 1q32 to qter, and the other, of 1p32 to pter. The sites of chromosomal breaks leading to trisomy differ from those involved in balanced reciprocal translocations. Some of these sites are sometimes, but not always, vulnerable in constitutional chromosomal abnormalities. The nature of the proliferative advantage conferred on myeloid cells by these chromosomal changes is unknown.

INTRODUCTION

The role of chromosomal changes in human malignant cells has been debated for more than 60 years. The observation of a consistent chromosomal abnormality in chronic myelogenous leukemia (17), and more recently in meningioma (11), provides support for the concept, proposed by Boveri (5) in 1912, that chromosomal changes are among the fundamental alterations occurring in malignant cells. The majority of human tumors, however, appeared to have widely disparate karyotypes; therefore most investigators accepted the view that these abnormalities were merely epiphenomena (6). The presence of a normal karyotype in highly malignant cells further complicates our understanding of the significance of chromosomal changes (15, 29). Most of the evidence bearing on this question was obtained prior to the use of the new chromosomal banding techniques, which now permit the identification not only of individual chromosomes, but also of parts of chromosomes.

These techniques have been used in the analysis of the karyotypes of a wide variety of human malignant and potentially malignant cells. Analysis of cells from patients with various hematologic disorders, including leukemia, polycythemia vera, refractory anemia, and other chronic myeloproliferative states, provides the most abundant data on the frequency and types of chromosomal changes. These data have recently been reviewed (25, 26) and only a summary table (Table 1) is included here.

Some nonrandom rearrangements, such as the 9;22 translocation (22) in chronic myelogenous leukemia (CML) and the 8;21 translocation (23) in acute myelogenous leukemia (AML), are well recognized. Chromosomal banding techniques permit us to identify karyotypic aberrations with precision and to determine whether others, previously unrecognized, occur.

In this paper, I intend to focus on abnormalities affecting chromosome No. 1. The available information will be used in answers to the following questions: (1) Are there specific regions of chromosome No. 1 that frequently or invariably show the same abnormality? (2) Do breaks affect specific sites on the chromosome? (3) If such specific sites exist, can they be related to specific gene loci? (4) Can information regarding these chromosomal abnormalities be related to the problem of malignancy?

MATERIALS

Patients with hematologic diseases, whose cells have been studied with chromosome banding techniques in my laboratory between January, 1970 and January, 1976, provide the basis for this report. From this group, I selected all patients who showed any abnormality involving chromosome No. 1. There were three patients who had an additional No. 1, six patients who had structural rearrangements leading to a duplication of a part of No. 1, and six patients who had an apparently balanced translocation affecting chromosome No. 1, one of whom also had a deletion of part of No. 1.

To these patients I have added all those with hematologic disorders and abnormalities affecting No. 1 of whom I am aware. In a few instances, I have altered the sites of breakage listed in the published data. Finally, constitutional chromosomal abnormalities affecting chromosome No. 1, particularly with reference to break points, have been summarized from Borgaonkar's Catalog (2) and from his Repository of Chromosomal Variants and Anomalies in Man (3, 4).

The identification of chromosomes and band numbers used in this report is based on the nomenclature adopted at the Paris Conference (21), and the karyotypes are expressed as

TABLE I

CHROMOSOME ABNORMALITIES ASSOCIATED WITH HEMATOLOGIC DISORDERS*

Disorder	% of Patients Abnormal	Trisomy for all or long arm of No.1	5q-	-7	+8	8q[†] transl.	+9	9q+	15/17 transl.	17q+	20q-	Ph^1[§] (22q-)	Double Ph^1
Chronic myelogenous leukemia	85							+				+	
Acute phase	70[¶]				+			+		+		+	+
Acute non-lymphocytic leukemia	50	+		+	+	+	+			+			
Acute pro-myelocytic leukemia	100								+				
Polycythemia vera	25	+			+		+				+		
Refractory anemia	15?		+		+						+		
Myelosclerosis with or without myeloid metaplasia	15?	+		+			+						

*Adapted from table published in reference (25).

[†]Usually to No. 21; frequently associated with loss of an X or Y chromosome.

[§]Translocation to No. 9 in more than 90% of Ph^1 positive patients.

[¶]Other chromosome abnormalities superimposed on Ph^1 positive cell line.

recommended under this system. The total chromosome number is indicated first, followed by the sex chromosome constitution, and then by the gains, losses, or rearrangements of the autosomes. A + or - sign before a number indicates a gain or loss, respectively, of a whole chromosome; a + or - after a number indicates a gain or loss of part of a chromosome. The letters "p" and "q" refer to the short and long arms of the chromosome, respectively; "i" and "r" stand for "isochromosome" and "ring chromosome." "Mar" is marker, "del" is deletion, "ins" is insertion, and "inv" is inversion. Translocations are identified by "t", followed by the chromosomes involved in the first set of brackets; the chromosome bands in which the breaks occurred are indicated in the second brackets. Uncertainty of the chromosome or band involved is signified by "?."

RESULTS

The study included 36 patients who had abnormalities of No. 1 that altered the amount of genetic material of this chromosome in the cell; in 34 patients, the aberration led to duplication of part or all of No. 1. The data including karyotype, portion trisomic for No. 1, hematologic diagnosis, and reference are presented in Table 2 and Figure 1. The data from the two patients with deletions will be considered later. As can be seen from an analysis of the table, *every patient is trisomic for the region from 1q25 to 1q32*. Five patients are trisomic for all of No. 1, and 16 have part of No. 1 translocated to another chromosome, usually with two other normal No. 1's as well (exceptions are cases 1 and 4) (20). Five patients have an additional chromosome composed only of part of No. 1, as well as two normal No. 1's. Finally, 8 patients have complex structural rearrangements involving only chromosome No. 1 and leading to a duplication of part of the chromosome. These rearrangements usually produce a very large marker chromosome.

Except for the five patients who have an extra intact chromosome, 29 patients are trisomic for no more than the proximal one-third of the short arm and for the long arm. In addition to the trisomy for region 1q25 to 1q32 present in every patient, trisomy for 1q32 to qter or for 1q25 to 1q21 is present in 28 and 30 patients, respectively.

The 16 translocation trisomies involve No. 1 with chromosome Nos. 4, 6, 9, 11, 13, 15, 17, 22, and Y (Table 3). All breaks in No. 1 are in the centromere, or in q21 or q25. With three exceptions, all breaks in the centromere region involve translocations with acrocentric chromosomes (Nos. 13, 15, 22) which are also broken at the centromere. The

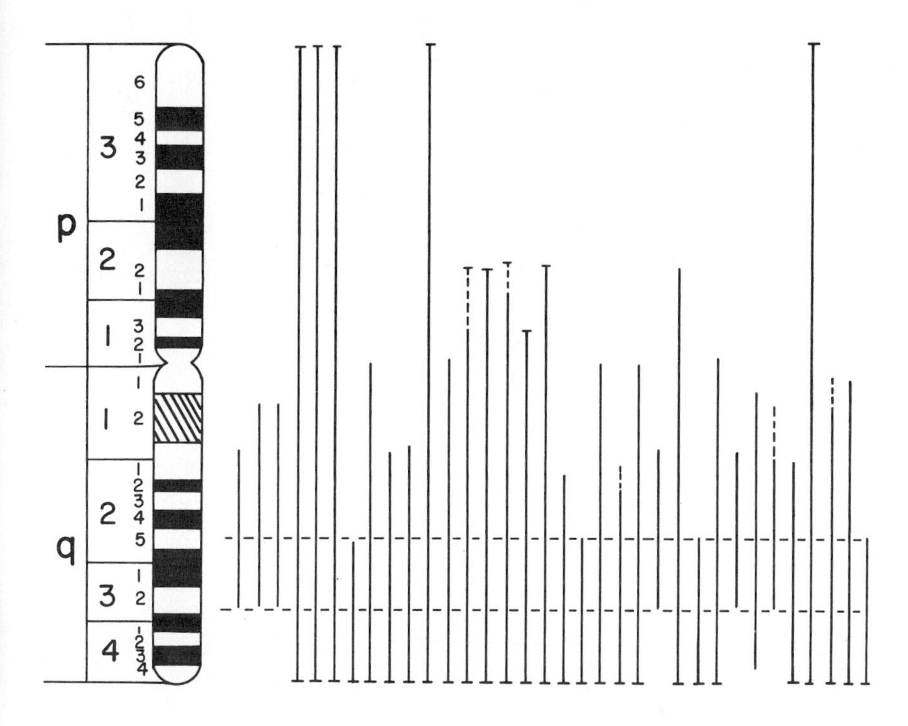

Fig. 1. The part of chromosome No. 1 that is trisomic in each patient listed in Table 2 is represented by a vertical line. The lines occur in the same order as the patients are listed in the table. Dashed vertical lines indicate uncertain break points. A short horizontal bar indicates the end(s) of the chromosome. The dashed horizontal lines enclose the segment that is trisomic in every one of the 34 patients.

exceptions involve (1) a translocation with 9p, a chromosome whose centromeric region contains repetitive DNA, in two patients, and (2) the end of the short arm of No. 1. Six of the eight tandem duplications of 1q have the distal break point in band 1q32; half of the proximal break points are in q21, and the other half are in bands q11 or q12.

Two patients had deletions of No. 1. In one, there was a 1;11 reciprocal translocation and a deletion of the same No. 1 distal to 1q32 (Table 4, case 42) (29); the other

TABLE 2

ABNORMALITIES OF CHROMOSOME NO. 1

Karyotype	Portion Trisomic for No. 1	Disease	Reference
46,XX,dup(1)(pter→q32::q21→qter	q21→q32	MMM	Case 1 (27)
50,XY,dup(1)(pter→q21::q32→q11::q21→qter t(4;?)(p16;?)+t(4;?)(p16;?),t(14;?)(q32;?), t(17;?)(p13;?),t(18;?)(q23;?),+20,+mar	q11→q32	HL	Case 2 (27)
47,XY,t(1*;10)(p22;p15),dup(1*)(p22→q32:: q11→qter),inv(6),del(11)(q23),14q+,21p+,+ring	q11→q32	PCL	Case 3 (27)
48,XX,+1,del(5)(q13-15),+11	whole chromosome	ANLL	Case 1 (24)
55,XX,+1,+1,+3,-4,-5,+6,+6,-8,-10,10q-,-11, +18,19q+,+4mar,+4min	whole chromosome	ANLL	Case 4 (27)
50,XX,+1,del(5)(q13-15),+6,-7,+8,+10,t(13q;14q), +21,+mar	whole chromosome	MM→AML	Case 5 (27)
46,XX,-6,+t(1;6)(q25;p25)	q25→qter	MMM	Case 6 (24)
46,XX,-15,+t(1;15)(cen;cen)	centromere→qter	PV→AML	Case 2 (24)
46,XY,t(12;17)(q13;p11)/ 47,X,t(1;Y)(q21;q12),+9	q21→qter	PV	Case 3 (24)

46,XX,-4,+t(1;4)(q21;q35)	q21→qter	ANLL	(8)
47,X,del(Yq),+1	whole	RA	(33)
46,XX,-15,+t(1;15)(cen;cen)	centromere→qter	PV	Case 32 (36)
47,XX,+del(1)(p13-21)	p13-21→qter	PV	Case 274 (36)
47,XY,+del(1)(p22)	p22→qter	MMM	Case E.I. (32)
47,XY,+del(1)(p22)	p22→qter	PV	Case J.W. (32)
47,XY,+del(1)(p13)	p13→qter	PV	Case E.K. (32)
47,XY,+del(1)(p21?)	p21?→qter	PV	Case E.D. (35)
47,XX,+t(1;9)(q21;q13)	q21→qter	PV	Case B.M. (35)
47,XX,-4,+t(1;4)(q25;p14-15),+13	q25→qter	ANLL	Case 3 (20)
43,X-Y,t(1;8)(p22;q24)-1,-22,+t(1q;22q), +t(1q;22q) and other complex abnormalities	centromere→qter	ANLL	Case 1 (20)
46,XX,del(7)(q22),-9,+t(1;9)(q21-22;p24) and other patterns	q21-22→qter	LS→ANLL	Case 2 (20)

TABLE 2 (Continued)

Karyotype	Portion Trisomic for No. 1	Disease	Reference
46,X?,t(9*;22)(q34;11),-1,+t(1;9*)(cen;p11-13)	centromere→qter	CML-blast	Case 4 (20)
46,XX,dup(1)(pter→q32?::q21?→qter)	q21?→q32?	MMM	Case 418 (18)
47,XX,t(1*;9)(cen?;cen?),+del(1q)(cen?), del(7)(q22),20q-	centromere?→qter	PV→MMM	Case 422 (18)
46,XX,-6,+t(1;6)(q25;p21)	q25→qter	MMM	(9)
46,XY,-13,+t(1;13)(cen;cen)	centromere→qter	PV	(9)
46,XX,dup(1)(pter→q32::21→qter)	q21→q32	MM	(30)
44,X,-21/45,XY,-21/46,XY,dup(1)(pter→q44:: q12→qter)	q12→q44	RA	Case 12 (37)
46,XY,dup(1)(pter→q32::q12 or 21→qter	q12 or 21→q32	ALL	Case 10 (37)
48,XXX,-15,-17,+t(1;i(17q))(q21;q25?),+2ring (17qter→17cen→17q25?::1q21→1qter)	q21→qter	LS	(7)
51,XX,+1,del(2)(q33),+6,+9,-19,+20q-,+mar1	whole chromosome	MMM	(16)

46,XY,-11,+t(1;11)(q11 or 12;q23)	q11 or 12→qter	RA	Case 2 (16)
47,XY,dup(1)(qter→q11::p36→qter),+ring	q11→qter	PV	Case 2 (16)
46,XX,-6,+t(1;6)(q25?;p23 or 25?)	q25→qter	PV	Case 5 (16)

*Same chromosome affected in both rearrangements.

ABBREVIATIONS:

MMM = Myelosclerosis with or without myeloid metaplasia
HL = Histiocytic lymphoma
PCL = Plasma cell leukemia
ANLL = Acute nonlymphocytic leukemia
MM = Multiple myeloma
PV = Polycythemia vera
RA = Refractory anemia
LS = Lymphosarcoma
CML = Chronic myelogenous leukemia
ALL = Acute lymphoblastic leukemia

TABLE 3

TRANSLOCATIONS OF 1q LEADING DIRECTLY TO TRISOMY*

Abnormality	Reference
t(1;Y)(q21;q12)	Case 3 (24)
t(1;4)(q25;p14-15)	Case 3 (20)
t(1;4)(q21;q35)	(8)
t(1;6)(q25;p25)	Case 6 (27)
t(1;6)(q25;p23-25)	Case 5 (16)
t(1;6)(q25;p21)	Case 1 (9)
t(1;9)(q21;q13)	Case B.M. (35)
t(1;9)(q21-22;p24)	Case 2 (20)
t(1;9)(cen-q11;p11-13)	Case 4 (20)
t(1;11)(q21;q23)	Case 1 (16)
t(1;13)(cen;cen)	Case 2 (9)
t(1;15)(cen;cen)	Case 2 (24)
t(1;15)(cen;cen)	Case 32 (36)
t(1;i(17q))(q21;q25)	(7)
t(1;22)(cen;cen)	Case 1 (20)

*Data taken from Table 2.

patient had a complex 3-way translocation that resulted in the loss of part of No. 1 from 1p32 to pter (12).

In addition to these 36 patients, nine patients with hematologic disorders have been reported who have reciprocal translocations involving chromosome 1 which do not lead to either gain or loss of No. 1. Four of the 36 patients considered earlier (cases 3 [27], 42 [29], 1 [20], 422 [18]) also have balanced rearrangements of No. 1 as well as trisomy

TABLE 4

OTHER REARRANGEMENTS OF NO. 1

Abnormality	Disease	Reference
Deletions		
45,XY,t(1;4;7)(p32;q31;q11 and q32)†	MMM	(12)
46,XX,t(8;17),t(1;*11); del(1)*(q32)§	ANLL	Case 42 (29)
Reciprocal Translocations		
t(1;5)(p36;q31)	PV¶	Case A.O. (34)
t(1;9?)(p36;q13?)	ANLL	Case 49 (29)
t(1*;11)(p36;q13),del(1*)(q32)	ANLL	Case 42 (29)
t(1;13)(p34-36;q14)	ANLL	Case M.S. (19)
t(1;17)(p36;q21)	ML→ANLL	Case 3 (28)
t(1;?)(p36?;?)	ML→ANLL	Case 8 (28)
t(1;?)(p36;?)	ALL	(1)
t(1;?)(p36;?)	PV	Case 8 (16)
t(1;9)(p34?;q34)	ANLL	Case 31 (29)
t(1;8)(p22;q24)	ANLL	Case 1# (20)
t(1;9)(cen;cen),+t(1q;9p)	PV→MMM	Case 422# (18)
t(1*;10)(p22;p15)	PCL	Case 3# (27)
t(1;6;11)(q12;q23;p15)	ANLL	Case 14 (37)

*Same chromosome involved in rearrangement.

†Deletion of No. 1 p32 pter.

§Deletion of No. 1 q32 qter.

¶Following treatment with chloramphenicol.

#Also listed in Table 2.

or deletion. In two cases, the reciprocal rearrangement involves the affected No. 1. The findings for these 13 cases are summarized in Table 4. Except in case 14 (37), the break points are located in the short arm of No. 1, specifically in bands p22, p34, and p36.

In marked contrast, the break points found in constitutional abnormalities (e.g., translocations, inversions, deletions) are distributed along the length of the chromosome, although there is some evidence of clustering. The data on 80 break points in Borgaonkar's registries (2-4) show 10 in band 1q32, 8 in band 1p32, 7 each in bands 1p36 and 1q21, 6 in 1q12, and 5 each in 1q11, 1q42, and 1q44. The rest of the 27 breaks are distributed in 13 bands.

DISCUSSION

The first two questions raised in the introduction can be answered affirmatively. There is one specific region of chromosome No. 1 that is *invariably* present in the trisomic state, and the two regions bounding this invariant region are trisomic in the majority (28 or 30/34) of patients. Except in patients who have an extra whole chromosome, trisomy for the short arm, bands p22 to pter, was never seen. Chromosomal studies of the malignant cells have been reported in only a few patients with cancer. Abnormalities of No. 1, often leading to partial trisomy, such as an isochromosome for the long arm of No. 1 or an extra 1q chromosome, have been observed (10, 31).

The sites of breakage show a very curious distribution. Breakage which leads directly to the trisomic state, through tandem duplication, translocation to another chromosome, or the production of an extra 1q chromosome, is confined to bands p22, cen, q12, q21, q25, and q32 (Fig. 2). All breaks in the short arm of No. 1 (p13 and p22) lead to an extra, separate chromosome that results in trisomy for the entire long arm. Breaks at the centromere usually involve a translocation with the long arm of an acrocentric chromosome. Of the other translocations which lead directly to trisomy for part of 1q, all have breaks in q21 or q25. All of the six breaks seen in band 1q32 were identified in patients with tandem duplications of No. 1.

A different pattern emerges if one scores for sites of breakage in reciprocal translocations that do not directly lead to trisomy. In this situation, the breaks are primarily in 1p36, with a smaller number occurring in 1p34 and 1p22; not a single break near the centromere or in the long arm was observed in these patients. These two sets of data, namely, breaks involved in trisomy and those involved in

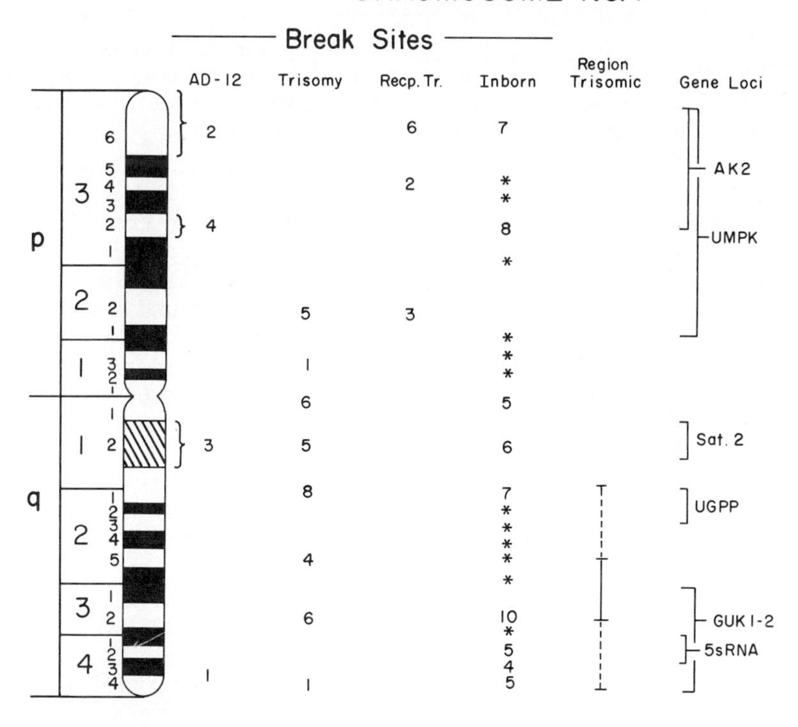

Fig. 2. A diagrammatic representation of chromosome No. 1, showing the number of breaks in a particular band related to (a) sites of breaks in adenovirus 12-infected cells, first vertical column of numbers. No. 1 is the site of the highest and No. 4 the lowest frequency of breaks. (b) break points leading to partial trisomy, second vertical column; data summarized from Table 2 and Fig. 1. (c) break points in balanced reciprocal translocations, third column; data summarized from Table 4. (d) break points for structural rearrangements noted in inborn chromosomal abnormalities, abstracted from refs. 2-4. (e) portion of No. 1 trisomic in hematologic disorders; solid line - trisomic in all patients, dashed line - trisomic in most patients. (f) location of selected gene loci on No. 1 (14).

reciprocal translocations, provide strong support for the no-otion of a specific association between the regions affected and the type and result of the structural rearrangement.

A comparison of the sites of breakage in the somatic mutations surveyed here and the break points seen in individuals with constitutional chromosomal abnormalities reveals a general similarity, as well as some major discrepancies. Thus, a high incidence of breaks is found in both types of patients in bands 1p36, the centromere, 1q12, 1q21, and 1q32. Bands 1p32 and 1q42-44 are frequently broken in constitutional rearrangements, whereas bands 1p22 and 1q25 are affected in somatic mutations and rarely in inborn abnormalities.

The mechanisms that produce these breaks and rearrangements are unclear. It may be significant that all of the four sites of breakage along chromosome No. 1 in adenovirus-type 12 (AD-12)-infected cells, namely, bands p36, p32, q12, and q42-44, are frequently involved in constitutional rearrangements (13, and McDougall, in preparation). Exceptions to this are the centromere and bands q21 and q32, which are often affected in constitutional rearrangements, but not in virus-induced breaks. In somatic mutations, 1p36 is affected in reciprocal translocations, and 1q12 in translocation trisomy; no breaks are seen in 1q42 or 43, or in 1p32, which are, respectively, the sites of the greatest (20%) and lowest (3%) frequency of AD-12-induced breaks.

Finally, I come to the last two questions concerning specific gene loci affected and their possible relationship to malignancy. The locations of selected gene loci are indicated in Figure 2. Guanylate kinase 1 and 2 (GUK 1 and 2) are located from 1q31 to qter, uridyl diphosphate glucose pyrophosphorylase (UGPP) and peptidase C (pep C) are located on bands 1q21 to 1q23, and the genes for 5 S RNA are on bands 1q41-42 (14). Fumerase hydratase 1 and 2 and possibly uridine-cytidine kinase (UCK) are on bands 1q42 to qter (14). It may be significant that some of these genes are involved in nucleic acid metabolism. At present we have no evidence to link abnormalities of any of these genes with malignancy.

In my view, the question no longer is, "Are there non-random chromosomal changes?" The signficant question for the future is, "What is the role of nonrandom changes in malignant transformation?"

ACKNOWLEDGEMENTS

This research was supported in part by the National Institutes of Health Grant No. CA 16910, the Leukemia Research Foundation, and the National Foundation of the March of Dimes. The Franklin McLean Memorial Research Institute is operated by The University of Chicago for the U.S. Energy Research and Development Administration under Contract No. EY-76-C-02-0069.

REFERENCES

1. Berger, R., Weisgerber, C. and Bernard, J. (1973) *Nouv. Rev. Franc. Hematol.* 13, 229.
2. Borgaonkar, D.S. (1975) *Chromosomal Variation in Man: A Catalog of Chromosomal Variants and Anomalies.* Johns Hopkins University Press, Baltimore.
3. Borgaonkar, D.S. and Bolling, D.R. (1975) *International Registry of Abnormal Karyotypes.* Johns Hopkins University Press, Baltimore.
4. Borgaonkar, D.S. and Bolling, D.R. (1976) *Repository of Chromosomal Variants and Anomalies in Man.* Johns Hopkins University Press, Baltimore.
5. Boveri, T. (1912) *Beitr. Pathol.* 14, 249.
6. DiPaolo, J.A. and Popescu, N.C. (1976) *Am. J. Pathol.* 85, 709.
7. Fleischman, E.W. and Prigogina, E.L. (1975) *Humangenetik* 26, 235.
8. Fukuhara, S. Personal communication.
9. Hsu, L.Y.F., Pinchiaroli, D., Gilbert, H., Weinreb, N. and Hirschhorn, K. (1976) *Society for Pediatric Research* (abstract).
10. Kakati, S., Oshimura, M. and Sandberg, A.A. (1976) *Cancer* 38, 770.
11. Mark, J. (1973) *Hereditas* 75, 213.
12. Marsh, W.L., Chaganti, R.S.K., Gardner, F.H., Mayer, K., Nowell, P.C. and German, J. (1974) *Science* 184, 966.
13. McDougall, J.K. (1971) *J. Gen. Virol.* 12, 43.
14. McKusick, V.A. and Ruddle, F.H. (1977) *Science*, in press.
15. Mitelman, F., Levan, G. and Brandt, L. (1975) *Hereditas* 80, 291.
16. Najfeld, V. Personnal communication.
17. Nowell, P.C. and Hungerford, D.A. (1960) *Science* 132, 1197.
18. Nowell, P.C., Jensen, J., Gardner, F., Murphy, S., Chaganti, R.S.K. and German, J. (1976) *Cancer* 38, 1873.
19. Oshimura, M., Hayata, I., Kakati, S. and Sandberg, A.A. (1976) *Cancer* 38, 748.
20. Oshimura, M., Sonta, S. and Sandberg, A.A. (1976) *J. Nat. Cancer Inst.* 56, 183.
21. Paris Conference. (1972) *Standardization in Human Cytogenetics.* The National Foundation--March of Dimes.

22. Rowley, J.D. (1973) *Nature (Lond.)* 243, 290.
23. Rowley, J.D. (1973) *Ann. Genet.* 16, 112.
24. Rowley, J.D. (1975) *Cancer* 36, 1748.
25. Rowley, J.D. (1976) *Blood* 48, 1.
26. Rowley, J.D. (1977) In: *Population Cytogenetics: Studies in Humans,* Ed. by Hook, E.B. and Porter, I.H. Academic Press, New York, p. 187.
27. Rowley, J.D. Unpublished.
28. Rowley, J.D., Golomb, H.M. and Vardiman, J. (1977) *Blood* (submitted for publication).
29. Rowley, J.D. and Potter, D. (1976) *Blood* 47, 702.
30. Spriggs, A.I., Holt, J.M. and Bedford, J. (1976) *Blood* 48, 595.
31. Tiepolo, L. and Zuffardi, O. (1973) *Cytogenet. Cell Genet.* 12, 8.
32. VanDyke, D.L. (1976) Personal Communication.
33. Warburton, D. and Bluming, A. (1973) *Blood* 42, 799.
34. Westin, J. (1976) *Scand. J. Haematol.* 17, 197.
35. Westin, J., Wahlstrom, J. and Swolin, B. (1976) *Scand. J. Haematol.* 17, 183.
36. Wurster-Hill, D. (1976) *Semin. Hematol.* 13, 13.
37. Yamada, K. and Furasawa, S. (1976) *Blood* 47, 679.

Author Index

(Article numbers are shown following the names of contributors. Affiliations are listed on the title page of each article)

L

M

N

O

P

R

S

V

W

Y

Subject Index

(Citations are to article number. The article numbers appear in the articles on top of the left hand pages in front of the names of the contributors.)

D

E

A
B 7
C 8
D 9
E 0
F 1
G 2
H 3
I 4
J 5